Analog Filters Using MATLAB

Lars Wanhammar

Analog Filters Using MATLAB

Springer

Lars Wanhammar
Department of Electrical Engineering
Division of Electronics Systems
Linköping University
SE-581 83 Linköping
Sweden
larsw@isy.liu.se

ISBN 978-1-4899-8309-1   ISBN 978-0-387-92767-1 (eBook)
DOI 10.1007/978-0-387-92767-1
Springer Dordrecht Heidelberg London New York

Printed on acid-free paper

Springer is part of Springer Science+Business Media (www.springer.com)

Preface

This book was written for use in a course at Linköping University and to aid the electrical engineer to understand and design analog filters. Most of the advanced mathematics required for the synthesis of analog filters has been avoided by providing a set of MATLAB functions that allows sophisticated filters to be designed. Most of these functions can easily be converted to run under Octave as well.

The first chapter gives an overview of filter technologies, terminology, and basic concepts. Approximation of common frequency selective filters and some more advanced approximations are discussed in Chapter 2. The reader is recommended to compare the standard approximation with respect to the group delay, e.g., Example 2.5, and learn to use the corresponding MATLAB functions. Geometrically symmetric frequency transformations are discussed as well as more general synthesis using MATLAB functions.

Chapter 3 deals with passive *LC* filters with lumped elements. The reader may believe that this is an outdated technology. However, it is still being used and more importantly the theory behind all advanced filter structures is based on passive *LC* filters. This is also the case for digital and switched-capacitor filters. The reader is strongly recommended to carefully study the principle of maximum power transfer, sensitivity to element errors, and the implications of Equation (3.26). MATLAB functions are used for the synthesis of ladder and lattice structures. Chapter 4 deals with passive filters with distributed elements. These are useful for very high-frequency applications, but also in the design of corresponding wave digital filters.

In Chapter 5, basic circuit elements and their description as one-, two-, and three-ports are discussed.

Chapter 6 discusses first- and second-order sections using single and multiple amplifiers. The reader is recommended to study the implication of the gain-sensitivity product and the two-integrator loop. Chapter 7 discusses coupled forms and signal scaling, and Chapter 8 discusses various methods for immitance simulation. Wave active filters are discussed in

Chapter 9 and leapfrog filters in Chapter 10. Finally, tuning techniques are discussed in Chapter 11.

Text with a smaller font is either solved examples or material that the reader may skip over without losing the main points.

Linköping
Sweden   Lars Wanhammar

Contents

Note to Instructors

The solutions manual for the book can be found on the author's webpage at http://www.es.isy.liu.se/publications/books/Analog Filters Using MATLAB/. Supplementary information can also be found on the author's webpage.

Chapter 1
Introduction to Analog Filters

1.1 Introduction

Signal processing techniques involve methods to extract information from various types of signal sources but also methods to protect, store, and retrieve the information at a later date. In, for example, a telecommunication system we are interested in transmitting information from one place to another, whereas in other applications, e.g., MP3 players, we are interested in efficient storing and retrieving of information. Note that storing information for later retrieval can be viewed as transmitting the information over a transmission channel with an arbitrary long time delay. In many cases, for example in the MP3 format, signal processing techniques have been used to remove nonaudible (redundant) information in order to reduce the amount of information that needs to be stored.

In, for example, a radio system, we need to generate different types of signals and modify the signals so that the information can be transmitted over a radio channel, e.g., by frequency modulation of a high-frequency carrier. Analog filters are key components in these applications.

Figure 1.1 illustrates a simple digital transmission system where analog filters are key components. Computer A acts as a digital signal source that generates a sequence of ASCII symbols. The symbols are represented by 8-bit words. In order to transmit a symbol over a telephone line, we must represent the bits in the symbol with a physical signal carrier that is suitable for the transmission channel at hand. Here we use a sinusoidal voltage with two different frequencies as signal carrier and use so-called frequency shift keying for representing the information.

In modem A (*mo*dulator/*dem*odulator), we let a "zero" bit correspond to 980 Hz and a "one" correspond to 1180 Hz. Hence, modem B has to determine if the received frequency is 980 or 1180 Hz in order to determine if a zero or one was transmitted. Two bandpass filters that let either of the sinusoidal signals pass can be used to resolve the frequency of the received signal by comparing the amplitudes of outputs of the two filters. In a similar way, modem B sends information to modem A, but instead uses the frequencies 1650 and 1850 Hz. Hence, filtering is an essential part of the modems.

The transmission system discussed above is now outdated. However, modern transmission systems with higher transmission capacity use similar techniques. For example, high-definition TV (HDTV), wireless local network (WLAN), and asymmetric digital subscriber line (ADSL) use several carriers and more advanced modulation methods. However, in these systems, different types of filters are also key components.

1.2 Signals and Signal Carriers

Examples of common signals and signal processing systems are speech, music, image, EEG, ECG, and seismic signals and radio, radar, sonar, TV, phone, and digital transmission systems. Characteristic for signal processing systems is that they store, transmit, or reduce the information. The concept "information" has a strict scientific definition, but we will

L. Wanhammar, *Analog Filters Using MATLAB*, DOI 10.1007/978-0-387-92767-1_1,
© Springer Science+Business Media, LLC 2009

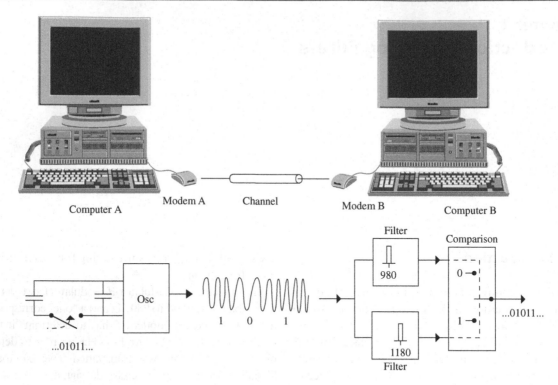

Fig. 1.1 Computer-to-computer communication over phone line

here interpret the concept "information" in its everyday sense, for example, representing what is said in a phone conversation. Moreover, the information is interpreted as what we consider to be of interest, e.g., what is said, but not who is speaking. In a different context, the relevant information may be the identity of the speaker.

1.2.1 Analog Signals

The information in a signal processing system is represented in the form of signals, which often are continuous in both time and amplitude. A signal carrier with continuous amplitude and time and that varies "in the same way as the information" is called an *analog signal*. For example, the signal from a microphone varies analogously with the sound pressure.

1.2.2 Continuous-Time Signals

In this case, the information and the signal do not vary analogously, i.e., one-to-one, but instead the information is embedded in the signal in a more complicated way. For example, the frequency of the output signal from an FM transmitter represents the information, i.e., the frequency varies in the same way as the information (speech, music, etc.).

Generally, a signal that is continuous in both amplitude and time but does not vary analogously with the information is referred to as a *continuous-time signal*. Hence, an analog signal belongs to a subset of continuous-time signals. Here we will only discuss analog signals and systems, although the analog filters that are discussed are often useful for continuous-time signals as well.

In this context, it is usually sufficient to assume that the signals can be considered as deterministic, i.e., they can be described with a function $x(t)$. However, in many cases, it is necessary to study signal processing systems using stochastic signals. Such signals, e.g., representing noise on a phone line, contain random variations, which cannot be described with ordinary mathematical functions, and statistical methods must be used instead.

Fig. 1.2 Discrete-time and digital signals

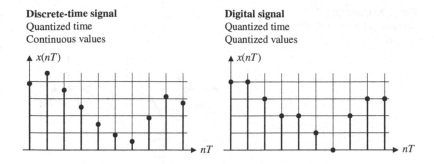

1.2.3 Signal Carriers

A signal is an abstract concept and is associated with a *signal carrier*. For continuous-time or discrete-time signals, which are discussed in the next section, the signal carrier is always a physical quantity. Typical signal carriers are currents, voltages, and charges in electrical circuits, but also mechanical vibrations and stress in crystals are common. Piezoelectric materials are used to convert between electrical and mechanical quantities.

In the literature, there are circuits referred to as voltage mode and current mode circuits. The difference is that the first uses negative feedback to reduce the effect of component errors, distortion, etc., whereas the latter only uses a low amount of feedback. This means that voltage mode circuits cannot be used for as high frequencies as current mode circuits, whereas the latter has higher sensitivity for errors in the components and larger signal distortion.

The terms signal and signal carrier are often misused. It is, however, often important to distinguish signals, which contain the information, from signal carrying quantities.

1.2.4 Discrete-Time and Digital Signals

Modern signal processing systems often use signals that are only defined at discrete time instances. Such discrete-time signals are often acquired through sampling of continuous-time signals, i.e., the discrete-time signal is a sequence of measurement values. Normally the samples are taken with the same time distance, T, i.e., the sampling is uniform. We distinguish discrete-time signals with continuous values from those that are quantized.

A signal, as shown to the left in Fig. 1.2, is only defined at discrete times and has continuous values is called a *discrete-time signal*. If the signal also has quantized values, as illustrated to the right in Fig. 1.2, the signal is called a *digital signal*. Note that we unfortunately do not distinguish between a discrete-time and a digital signal in English literature.

Of course, the signals may not necessarily originate from sampling of a continuous-time signal. In fact, it may not have to do with time at all. For example, a discrete-time or digital signal may be obtained by sampling the height of a mountain at various places. The corresponding signal is a real function of the coordinates, i.e., a two-dimensional signal.

Example 1.1 Consider the operation of the circuit shown in Fig. 1.3.

The switches, which can be implemented using MOS transistors, switch back and forth with the period $2T$. When the lower switch is in the left position, the capacitor C is charged to the voltage $v_{in}(t)$. When the switch at time $t = nT$ switches to the other position, the capacitor remains charged and the output voltage from the voltage follower changes to the new value $v_{out}(t) = v_{in}(nT)$ and remains thereafter constant during the remaining part of the clock phase. The upper switch, with its capacitor, works in the same manner, but in opposite phase. The output voltage will thus be a sequence of measured values, $v_{in}(nT)$, of the input signal. The output signal is apparently a discrete-time signal, but it is represented by a physical signal carrier; the stair-shaped voltage $v_{out}(t)$, which of course is continuous in time.

Fig. 1.3 Concept of sampling, signal, and signal carrier

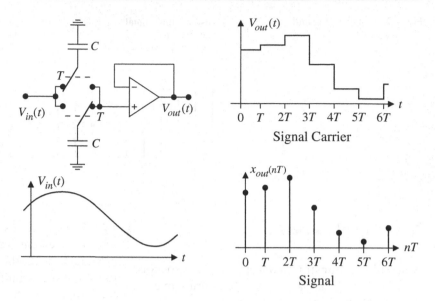

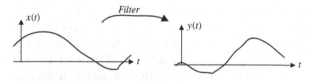

Fig. 1.4 Filter as a mapping

1.3 Filter Terminology

With the term *filter* we refer to a (mathematical) mapping of an input signal to an output signal. This mapping is normally linear and the superposition principle for signals is therefore valid. Unfortunately, the term filter is often given a much wider interpretation.

1.3.1 Filter Synthesis

We use the term filter *synthesis* for the process of determining this mapping. Here we limit ourselves to *time-invariant filters*, i.e., the filter properties do not vary over time.

The most common filter types are frequency selective, i.e., they let some frequencies pass and reject others. A historically important use of frequency selective filters was in radio receivers and in carrier frequency systems for transmission

of telephony; see Section 1.4.1. Frequency selective filters are used, among other things, as anti-aliasing filters; see Section 1.4.2, when sampling analog signals. Such filters are an essential part in interfaces between analog and digital systems, e.g., in GSM phones between the microphone and the A/D converter. Analog filters are also used to filter the output signal of D/A converters.

An example of time-variable filters are adaptive filters, which normally operate on time-discrete or digital signals, and are used to, e.g., equalize and correct for errors in the transmission channel. Adaptive filters are a major part of ADSL modems and cellular phones. Another type of filters is *matched filters*, which are used to detect if and when a given waveform occurs in a signal. Matched filters are used in radar and digital transmission systems to detect the arrival time for the echo and which of several symbols has been received, respectively.

1.3.2 Filter Realizations

A filter, as mentioned above, is a mathematical mapping of input signal to the output signal. We use the term *realization* of the filter to describe in detail how the output is computed from the input signal. There exists virtually an infinite number of possible ways to perform and organize these computations. Although they perform the same mapping and cannot be distinguished from each other by only observing the input and output signal, they may have very different properties.

In general, different realizations require different number of components and have different sensitivity to errors in the components. A realization with low sensitivity may meet the performance requirements with cheaper components with large tolerances. One of the main problems is therefore to find such low sensitive and thereby low cost realizations.

A filter realization can be described in several, but equivalent ways. Here we are concerned with analog filters, which use currents or voltages as signal carriers. The realization can therefore be described in terms of a set of coupled differential-integral equations as shown below. For example, an inductor with the inductance L is represented in the equation $v(t) = L\, di/dt$.

We may use the representation shown below, which uses signals in the time domain.

$$\begin{cases} v_{in}(t) & = Ri(t) + v_C + v_{out}(t) \\ v_C & = \frac{1}{C}\int\limits_0^t i(t)dt \\ v_{out}(t) & = L\frac{d}{dt}i(t) \end{cases}$$

A more common, however, is to use the equivalent representation in the Laplace domain shown below

$$\begin{cases} V_{in} & = RI + V_C + V_{out} \\ V_C & = \frac{I}{sC} \\ V_{out} & = sLI \end{cases}$$

Traditionally we do not use differential equations; instead, we use an equivalent graphical description with resistors, inductors, capacitors symbols, which corresponds to elementary equations, i.e., generic circuit theoretical elements. We will later introduce additional circuit elements for realization of analog filters. Figure 1.5 shows a filter in terms of these symbols that is equivalent to the two representations above.

There are several synonyms used: *realization, structure, algorithm,* and *signal-flow graph* for describing how the output is computed from the input signal.

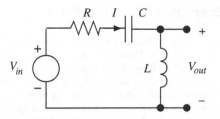

Fig. 1.5 Schematic representations of a filter realization

1.3.3 Implementation

The physical apparatus that performs the mapping (the filtering), i.e., executes the computations that are needed to compute the output signal according to the realization, is called an *implementation*. In an analog implementation, there is an input and an output signal carrier, which vary analogous with the input and the output signal.

A realization of *RLC* type consists of a network with inductors, capacitors, resistors, and a voltage or current source, which vary analogous with the input signal. The output signal carrier is either a current or a voltage. These circuit elements can (approximately) be implemented with coils, capacitors, and resistors. Unfortunately, we do not in the English literature always distinguish between a circuit element and its implementation. The meaning of the terms must therefore be inferred from the context. In other cases, there are a physical device and no corresponding circuit theoretical element, e.g., operational amplifier.

Table 1.1 shows a compilation and the recommended usage of different terms. VCVS and VCCS denote voltage-controlled voltage-source and voltage-controlled current-source, respectively. These and other circuit elements will be discussed further in Chapter 5.

Table 1.1 Components, circuit elements, and parameters

Physical component	Circuit element	Parameter
Resistor	Resistor	Resistance, R
Coil	Inductor	Inductance, L
Capacitor	Capacitor	Capacitance, C
Transformer	Transformer	$n:1$
–	Gyrator	r
Operational amplifier	VCVS	A
Transconductor	VCCS	Conductance, g_m

1.4 Examples of Applications

In this section, we will briefly describe some typical applications of analog filters. Here we will only discuss filtering of signals and not, e.g., filters for attenuation of harmonics in an AC/DC converter. Such filters for filtering large currents and voltages are also used in the electric power grid.

Historically, filters for use in telephone systems have had a large impact on the development of both filter theory and different types of filter technologies. Some of these filters must meet very strict requirements. Nowadays different types of analog filters in, e.g., cellular phones and hard drives are important applications that push the development forward as these analog filters are manufactured in great numbers annually.

1.4.1 Carrier Frequency Systems

In older parts of the telephone network, FDM (frequency division multiplex) is used for transmission over vast distances. To transmit many calls on the same transmission channel, the voice channels are placed next to each other in the frequency spectrum using modulation and filtering techniques.

When modulating a voice channel with a carrier frequency, two sidebands are created according to Fig. 1.6. By connecting a filter after the modulator, one of the sidebands can be filtered out, so that a signal spectrum, according to

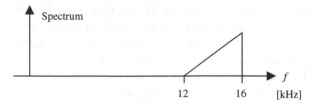

Fig. 1.7 Extracted side band

Fig. 1.7, is maintained and the frequency band that is occupied is minimized. The filter passes frequencies in the band 12–16 kHz and blocks frequencies in the band 0–12 kHz and above 16 kHz [60].

Figure 1.8 illustrates how three voice channels can be translated in frequency and then combined into a 3-group. Figure 1.9 shows the principle of combining four 3-groups into a 12-group. The filter, which is needed to filter out a 12-group, must comply with a specification that is among one of the toughest filter specifications that occur in practice.

In a similar way, higher-order channels are successively combined into groups of 3, 12, 60, 300, 900, 2700, and 10,800 channels. A carrier frequency system with 10,800 channels, corresponding to six analog TV channels, was first introduced in Sweden in 1972 and is referred to as a 60 MHz system.

The receiver side consists of corresponding demodulation and filtering stages to successively extract the different channels. A carrier frequency system thus contains a large number of frequency filters. For example, Ericsson manufactured a

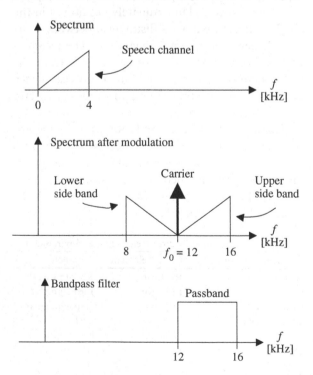

Fig. 1.6 Modulation of a voice channel

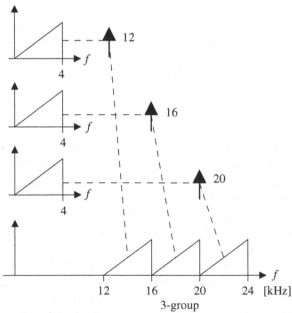

Fig. 1.8 Generation of a 3-group

Fig. 1.9 Generation of a 12-group

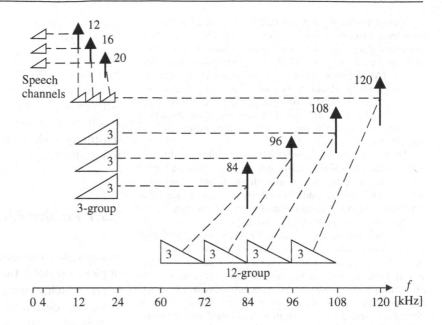

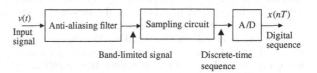

filter for formation of a 12-group containing a large number of inductors and capacitors, but also a crystal, which is a very stable resonance circuit. The volume was approximately $2\,l$ and it was contained inside a temperature-stabilized enclosure, which required a large space and an expensive cooling system.

These filters have also been implemented as crystal filters, whereas Siemens among others used metallic resonators instead of crystals. The requirements on these filters were very strict and the number of manufactured filters per year was large. During the late 1980s, approximately 5 million 12-group filters were manufactured annually.

Nowadays, instead of carrier frequency systems, more effective and cheaper digital transmission systems are used, using digital filter techniques, which can be implemented in integrated circuits at a much lower cost. With a digital transmission system, the available bandwidth can be used more effectively than for the corresponding analog systems. Analog systems have therefore successfully been replaced with digital transmission systems. Note that even these systems contain many analog filters, not as complex though.

1.4.2 Anti-aliasing Filters

When sampling an analog or continuous-time signal, it must be band limited in order to preserve the information intact in the discrete-time or digital signal. Otherwise so-called aliasing distortion occurs and the information is lost. Therefore, an anti-aliasing filter must be placed between the analog signal source and the sampling circuit according to Fig. 1.10.

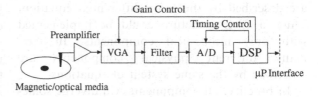

Fig. 1.10 Sampling of a continuous-time signal

1.4.3 Hard Disk Drives

An economically important application of analog filters is in the read channel of hard disk drives, as many hundred of millions of disk drives are manufactured annually. One of the major filtering tasks in the read channel is to equalize the frequency response so that subsequent pulses are not smeared out in time and overlap. This problem is referred to as intersymbol interference.

Figure 1.11 shows a block diagram of a typical mixed-mode[1] read channel. The signal obtained from the magnetic

Fig. 1.11 Read channel in a typical disk drive

---

[1] A mixed-mode system uses both continuous-time and discrete-time signals.

or optical media is first amplified by a preamplifier and then by a variable gain amplifier (VGA). The analog filter performs signal equalization, noise reduction, and band limiting before it is sampled. The analog-to-digital conversion (A/D) block includes a sample-and-hold stage and it has typically about 6 bits of resolution. The digital signal processor (DSP) core performs, if necessary, additional equalization. It also performs the data detection, controls gain and timing, as well as communicates with the µP interface.

The filter must also be programmable to allow for different bandwidths and gains requirements to accommodate for the change in data rate when reading from the inner and outer tracks of the disk. In addition, a tuning process is needed to determine the optimal cutoff and gain and compensate for temperature and power supply variations.

Partitioning the equalization between analog and digital filtering involves trade-off between the complexity and performance of the analog filter and the complexity and power consumption of the digital filter for a given chip area and power consumption. It is often favorable, whenever possible, to use digital over analog circuits, as cost, chip area, and power consumption as well as robustness of the design is better. Thus, the analog filter could be simplified to just perform anti-aliasing and the equalization could be performed entirely in the digital domain. However, in this approach the quantization noise generated by the A/D will be amplified by the digital equalization filter and result in an increased resolution requirement for the A/D in order to reduce the quantization noise contribution.

Current implementations of the equalization task therefore range from fully analog through mixed analog-digital to fully digital approaches.

1.5 Analog Filter Technologies

To implement an analog filter structure, many different technologies may be used. For an inductor, which corresponds to the differential equation $v(t) = L \, di/dt$, a coil can be used, but also mechanical springs, as their length and force are described by the same differential equation. Thus, a filter structure could be implemented with only mechanical components. In fact, many different physical components are described by the same system of equations.

In practice, all components will diverge somewhat from the ideal, i.e., they will not act as a simple circuit element. For example, a coil has losses due to resistance in the wires. In addition, unwanted parasitic (stray) capacitances are always present and affect the filters frequency

response. In integrated circuits, it is very hard to implement good inductors and resistors and we will therefore try to replace these with equivalent circuits.

The different technologies are impaired with different types of errors in the components. Hence, it is important to select a filter structure with low sensitivity to the errors in the intended implementation technology.

1.5.1 Passive Filters

Historically, the term *passive filter*[2] was used for implementations that only used passive components, which cannot generate signal energy, e.g., coils, capacitors, transformers, and resistors.

Nowadays, the term *passive filters* is used for filters that are realized using only passive, or lossless, circuit elements, i.e., inductors, capacitors, transformers, gyrators, and resistors, which cannot increase the signal energy. Most of these circuit elements have corresponding passive implementations. The circuit element gyrator, however, which is a lossless circuit element, can only be implemented using active components that amplify the signal energy. Gyrators and other more advanced circuit elements will be discussed further in Chapter 5.

Passive filters play an important role from a theoretical point of view, as they are used in the design of more advanced filters, but they are also widely used and implemented with passive components. Passive filters are often integrated into the printed circuit board (PCB) board in order to reduce the cost and size.

A new type of mechanical filters that are based on so-called MEMS technology (*microelectromechanical system*) has been developed in recent years. If a piezoelectric material is subjected to pressure, a voltage that is proportional to the pressure appears between the two pressure surfaces. If a voltage is applied, then the size of the material changes proportionally to the

---

[2]In the literature, the more restricted term *LC* is (wrongly) used to represent a filter that contains both *R*, *L* and *C* elements.

voltage. The piezoelectric effect can be used for converting between electrical and mechanical quantities (vibration that corresponds to pressure variations). The piezoelectric effect is also used in certain cigarette lighters to ignite the gas. In Chapters 3 and 4, we will discuss the design and implementation of passive filters in more detail.

At microwave frequencies, various types of transmission lines and components based on ferrite materials are used.

1.5.2 Active Filters

Historically, active filters were introduced to replace inductors impaired by a number of undesirable properties, i.e., non-linearity, losses, large physical size and weight, and they are only possible to integrate for very high frequencies. The term *active filter* comes from the active (amplifying) circuit elements that can generate signal energy in order to distinguish from filters that only consist of passive element. Active filters are therefore potentially unstable.

The first active filters used electron tubes as amplifying elements (1938) and later on, in the 1950s discrete transistors were used. Typically, the components were soldered on a circuit board made of thin film or thick film type. Those active filters had a significantly smaller physical volume than corresponding passive filters, especially for low (audio) frequencies, but suffered from high sensitivity for variations in the amplifying components compared with passive filters.

The modern theory for active filters is considered to have begun with a paper by J.G. Linvill (1954). This led to an increasing interest in research in element sensitivity and it was discovered that some of the *LC* filters that were used were optimal from an element sensitivity point of view. This issue will be discussed in detail in Chapter 3.

In the beginning of the 1970s, the operational amplifier had become so cheap that it could replace the transistor. Operational amplifier-based active filters were easier to design, especially for low

(audio) frequencies, and it therefore became the dominant technology. The usable frequency range was, however, limited to a few MHz. Nowadays, active filters can be implemented with bandwidths of several hundreds of MHz.

1.5.3 Integrated Analog Filters

The event of integrated analog filters makes integration of a complete system on a single chip possible. Normally a system on a single chip contains both digital and analog parts, e.g., anti-aliasing filters in front of A/D converters. Integrating a whole system on a single chip drastically reduces the cost.

Operational amplifiers and capacitors can relatively easily be implemented in CMOS processes, but the gain, the bandwidth of the amplifiers, and the capacitance vary strongly and have to be controlled by a controller circuit. Resistors with relatively low resistance values, but relatively high tolerances, can also be implemented. Different techniques, based on active elements, have therefore been developed to also remove the need for resistors.

The need for control of the filter frequency response is not only a problem, but also a necessity in some applications, e.g., in the read channel for hard drives and magnetooptic disks. The disk spins with constant speed and every bit occupies a fix space of the track, which means that the data rate will vary depending on which track is being read. In the read channel there is an analog filter that at the same time serves two purposes, one is band limiting the signal before the A/D converter (anti-aliasing filter) and the second is for equalizing the read channel (*equalizer*), i.e., shaping the frequency response of the read channel so that the reading of successive bits do not interfere. This phenomenon is called intersymbol interference. The bandwidth of the analog filter must also be able to vary with a factor of at least 3, which causes additional problems.

Controllable active integrated analog filters have during the 1990s, due to the large economic significance, been a driving force behind the development

of integrated active filter technology. Other important applications that are the driving force behind
technology development are anti-aliasing filters
that are used in front of the A/D converters and
filters to attenuate spurious elements after a D/A
converter. A/D and D/A converters are used in the
interfaces to digital signal processing systems, e.g.,
cellular phones, CD, DVD, MP3 players, and LAN
(*local area networks*). Because the filters in these
high-volume applications are often battery powered, the cost and the power consumption are of
major concern.

1.5.4 Technologies for Very High Frequencies

The time for propagation of electrical signals
becomes important in realization and implementation of analog filters for very high frequencies.
This time becomes significant when the component's physical size is $\lambda/4$ (a quarter of a wavelength of the highest frequency) or larger. For
example, an electrical signal in vacuum has a
wavelength of approximately 300 mm at
1 GHz. If instead the material is silicon oxide
with the relative dielectric constant 10.5, the
wavelength becomes $300/\sqrt{10.5} \approx 93$ mm. Thus,
a component of the size 23 mm or larger cannot
be considered to be small at 1 GHz and has to
be described with a more advanced circuit theoretical model, i.e., *distributed circuit element* [53].
In Chapter 4, we will discuss passive filters that
use transmission lines as the basic component.

If the components are small, we can, however,
use ordinary *lumped circuit elements*.

1.5.5 Frequency Ranges for Analog Filters

Filtering is a fundamental operation in most
electronic signal processing systems. It is therefore important to have a general knowledge of
limitation of different filter technologies. Some
of the most important analog filter technologies
and their typical usable frequency ranges are:

Passive Filters	Frequency range
Discrete LC components	100 Hz to 2 GHz
Distributed components	500 MHz to 50 GHz
Mechanical Filters	
Crystal filters	
Quartz – monolithic	1 MHz to 400 MHz
Quartz – non-monolithic	1 kHz to 100 MHz
Ceramic filters	200 kHz to 20 GHz
Metal resonator filters	10 kHz to 10 MHz
Surface acoustic wave filters	10 MHz to 4 GHz
Bulk acoustic wave filters	2 GHz to 20 GHz
Electrothermal filters	0.1 Hz to 1 kHz
Active filters	
Active RC filters	
Discrete components	0.1 Hz to 50 MHz
Integrated circuits	10 kHz to 500 MHz

Note that the frequency ranges given above are
not absolute limits; they just indicate typical frequency ranges. The usable frequency range is also
affected by the requirements of the filter. Crystal
filters, e.g., can only be used for bandpass filters
with very narrow passbands. In the microwave
domain there is a number of different filter technologies, but these will not be discussed in this book.

Power consumption is an important issue in many
applications. Generally, the power consumption is
proportional to the bandwidth, signal-to-noise ratio,
and inversely proportional to the distortion.

The choice of filter technology for a certain
application is, of course, dependent upon the filter
requirements and the acceptable manufacturing
cost. The cost of the filters depends to a high
degree on the number of manufactured filters. To
lower the cost, it is preferred to use technologies
that require little labor, i.e., can be manufactured
automatically and for this reason is suitable for
mass production. This is one of the most important reasons to develop filter technologies that
allow filters to be implemented in integrated circuits. Digital filters and integrated active RC and
SC filters are suitable for this. The development in
IC technology has made it possible to integrate
complete signal processing systems, e.g., a complete cellular phone on a single chip.

It is worth noting that there is no indication that older filter technologies, e.g., *LC* filters, are disappearing completely — there are certain cases where they are competitive, e.g., in the frequency range 1–2 GHz. Even "classic" components such as inductors and capacitors are still developed and improved.

In order to implement continuous-time filters for high frequencies, it is necessary to reduce the physical size of the components, i.e., whole filters must be implemented in an integrated circuit. This also makes it possible to implement circuits with both analog filters and digital circuits on the same silicon plate. Suitable technologies are CMOS and BiCMOS, which is a CMOS process with the possibility to implement bipolar transistors. Filters are also integrated in GaAs technology. The two later technologies are considerably more expensive than the standard CMOS technologies that are used for digital circuits.

1.6 Discrete-Time Filters

Implementation of discrete-time filters has mainly used charges as signal carriers. The earliest technologies, charged-coupled devices, use charges that were stored under plates on top of a silicon die and a digital clock to transfer the charges between different plates. Another technology, called bucket-brigade circuit, used MOS switches to transfer charges between different storage elements. Both these filter technologies have now disappeared, but charged-coupled devices are used in many image detectors, cameras, etc.

Yet another technology, *switched current circuits,* are circuits using currents as signal carriers (*current mode*) and has a potential greater frequency range compared to circuits based on ordinary operational amplifiers (voltage mode) because the latter uses less or no feedback.

Today, the main filter technology for discrete-time filters is the so-called switched capacitor techniques.

1.6.1 Switched Capacitor Filters

At the end of the 1970s a new type of discrete-time filter was developed, so-called *SC* filters (*switched capacitance filters*) [2], which could be integrated in a single IC circuit. This makes it possible to implement *SC* filters together with digital circuits, i.e., *SC* technology makes it possible to integrate complete systems on a chip (*system-on-chip*). In CMOS

technology, good capacitors and switches can easily be implemented. A MOS transistor is a good switch with a small resistance when it conducts (a few kΩ) and as an open-circuit when it does not conduct. Furthermore, good operational amplifiers can be implemented in CMOS.

Using switches, the capacitor network can be switched between several configurations. A control signal (clock) is used to switch between two different configurations. Signal carriers are the charges on the capacitors. Using this technique, a system of difference equations can be solved, i.e., a discrete-time filter can be implemented. The power consumption by *SC* filters is relatively low but increases with increasing clock frequency. The bandwidth of *SC* filters can be altered by changing the clock frequency. An enabling feature of *SC* filters is that the ratio of capacitances can be very accurate and therefore no trimming of the frequency response is needed.

SC filters is a mature technique used in a large number of applications, e.g., hearing aids, pacemakers, and A/D converters, but they are now often replaced by analog filters, especially for high frequency applications. The sampling circuit shown in Fig. 1.3 is an example of an *SC* circuit.

Integrated circuits with *SC* filters exist for different standard applications. For example, the integrated circuits MAX7490 and 7491 contain two second-order sections in a 16-pin package. The sections can realize transfer functions of lowpass, highpass, bandpass, and bandstop type. The circuits use power supply voltages of + 5 V and + 2.7 V, respectively, and consume only 3.5 mA. The center frequency, which is determined by the clock frequency, can be controlled from 1 Hz to 30 kHz.

1.6.2 Digital Filters

Digital filters developed quickly when cheap digital circuits were made available in the beginning of the 1970s. NMOS and TTL circuits had, however, too large power consumption and therefore only very simple circuits could be implemented. CMOS circuits were more suitable for integration of large and complex circuits, but the power consumption and the cost was large in comparison with the more mature technology based on operational amplifiers. In addition, an analog filter does not require A/D and D/A converters.

The development during the 1980s and 1990s of CMOS technology and digital signal processors, and the fact that many digital signal processing systems often include digital filters, made them more competitive [134]. Today, digital filters are usually preferred in applications that require high dynamic signal range, e.g., more than 50 dB, and sample frequencies of less than a few hundred MHz. Analog filters have their advantages in applications with less demands on the dynamic signal range and for higher frequencies. Of course, discrete-time and continuous-time filters are not direct competitors as they are more suitable in their own environments.

1.7 Analog Filters

In this section, we will discuss some of the characteristic properties of frequency selective analog filters.

1.7.1 Frequency Response

The properties of an analog filter can be described by the output signal for various input signals. In fact, the filters of interest here can be completely described by the output signal in response to a sinusoidal input signal with the angular frequency ω. The ratio of the Fourier transforms of the output signal, $Y(j\omega)$, and the input signal, $X(j\omega)$, is called frequency response.

Definition 1.1 The frequency response of a linear, time-invariant system is defined as

$$H(j\omega) \triangleq \frac{Y(j\omega)}{X(j\omega)}. \tag{1.1}$$

Henceforth we will assume that the analog filter's input and output voltages corresponds directly to the input and output signals, respectively. Hence, we do not strictly differentiate between signals and signal carriers, i.e., we assume that $X(j\omega) = V_1(j\omega)$ and $Y(j\omega) = V_2(j\omega)$. This distinction becomes more essential for discrete-time filters.

1.7.2 Magnitude Function

The frequency response $H(j\omega)$ is a complex function of ω. For this reason it is interesting to study

both the value of $H(j\omega)$ and the phase $\Phi(\omega)$. $H(j\omega)$ can be written as

$$H(j\omega) = |H(j\omega)|e^{j\Phi(\omega)} \tag{1.2}$$

or

$$H(j\omega) = H_R(\omega) + jH_I(\omega) \tag{1.3}$$

where $H_R(\omega)$ and $H_I(\omega)$ are real (even) and imaginary (odd) functions of ω, respectively.

Definition 1.2 The magnitude function is defined as

$$|H(j\omega)| \triangleq \sqrt{H_R^2(\omega) + H_I^2(\omega)}. \tag{1.4}$$

Normally we express the magnitude function using a logarithmic scale,

$$20\log(|H(j\omega)|) \text{ [dB]}.$$

Thus, $|H(j\omega)| = 1$ and 0.01 correspond to 0 dB and -40 dB, respectively. Some works in the literature use $20\ln(|H(j\omega)|)$ [Neper]. Figure 1.12 shows the magnitude function of a fifth-order lowpass Cauer filter.

A sinusoidal signal with an angular frequency of less than 1 rad/s will pass through the filter almost unaffected while frequencies are reduced to less than 1% if the angular frequency is larger than approximately 1.2 rad/s. Hence, this filter is a lowpass filter.

1.7.3 Attenuation Function

Instead of using the magnitude function, it is more common to use the attenuation function.

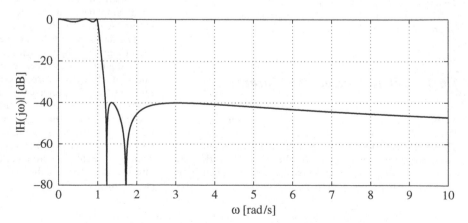

Fig. 1.12 Magnitude function for a fifth-order lowpass filter of Cauer type

Definition 1.3 The attenuation is defined as

$$A(\omega) \overset{\Delta}{=} - 20 \log(|H(j\omega)|) \text{ dB}. \qquad (1.5)$$

$|H(j\omega)| = 1$ corresponds to the attenuation 0 dB, i.e., no attenuation of the input signal. The attenuation for the same fifth-order lowpass filter of Cauer type is shown in Fig. 1.13. Note that the attenuation function and magnitude function (in dB) differ only in terms of the sign.

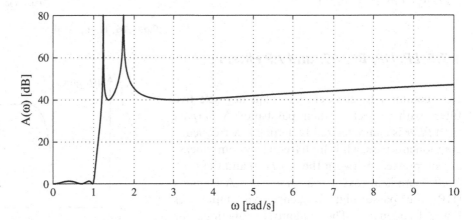

Fig. 1.13 Attenuation for a fifth-order lowpass filter of Cauer type

Typical attenuation in the stopband for analog filters are in the range 20–80 dB, which corresponds to values on the magnitude function in the interval 0.1–0.0001.

In order to simplify the design of the filter, the gain of the filter is normalized by dividing the mag-

required passband gain is adjusted to its desired value after the filter has been synthesized.

1.7.4 Phase Function

The frequency response is a complex function of ω and it is therefore necessary to also consider the phase of the frequency response.

Definition 1.4 The phase function[3] is defined as

$$\Phi(\omega) \overset{\Delta}{=} \arg\{H(j\omega)\} = \text{atan}\left(\frac{H_I(\omega)}{H_R(\omega)}\right). \qquad (1.6)$$

Figure 1.14 shows the phase function for the same fifth-order Cauer filter as before.

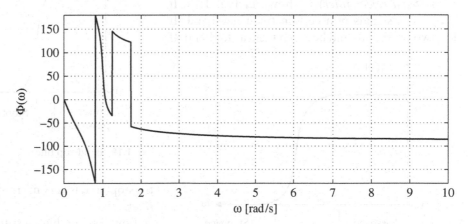

Fig. 1.14 Phase response for a fifth-order lowpass filter of Cauer type

nitude function with its largest value. Thus, the normalized gain in the passband is equal to 1, which corresponds to the attenuation 0 dB. The

---

[3]Note that in the literature, the phase is sometimes defined with a negative sign compared to Equation (1.6).

The phase is usually drawn between $-180°$ and $+180°$. This means that the apparent discontinuity (jump) in the phase function at $\omega \approx 0.8$ rad/s is not a discontinuity. In fact, it is an artifact of the plotting. However, the discontinuities at $\omega \approx 1.25$ rad/s and $\omega \approx 1.75$ rad/s are real discontinuities of $-180°$. The phase function decreases with $180°$ at a discontinuity. In many, but not all, applications these discontinuities may be neglected.

Note that the phase for high frequencies always approaches a multiple of $90°$. For $\omega = 0$, the phase is always a multiple of $90°$.

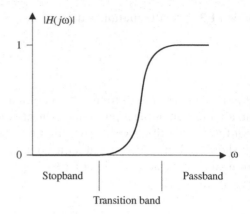

Fig. 1.16 Highpass filter

1.7.5 LP, HP, BP, BS, and AP Filters

It is common to characterize frequency selective filters with respect to their passbands A *lowpass* (LP) *filter* is characterized by letting low frequency components pass, while high frequency components are suppressed. Between the *passband* and the *stopband*, there is always a *transition band*. A *highpass* (HP) *filter* passes high frequencies and suppresses lower frequencies. The magnitude functions for a lowpass and a highpass filter are illustrated in Fig. 1.15 and Fig. 1.16, respectively.

The magnitude function for a *bandpass* (BP) *filter* is illustrated in Fig. 1.17. There are two stopbands and in between a passband. Bandpass filters are very common.

The magnitude function for a *bandstop (BS) filter* (*band reject filter*) is shown in Fig. 1.18. It suppresses signals in a certain frequency band. It has two passbands and between them a stopband. If

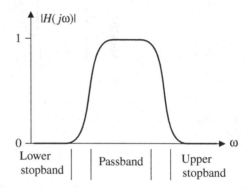

Fig. 1.17 Bandpass filter

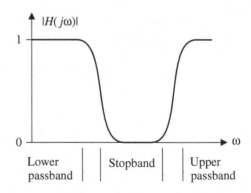

Fig. 1.18 Bandstop filter

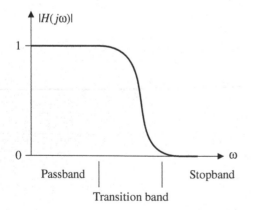

Fig. 1.15 Lowpass filter

the stopband is very narrow, it is often called a *notch filter*.

Lowpass and highpass filters with narrow transition bands together with bandpass and bandstop filters with narrow passbands and stopbands, respectively, are more difficult and more costly to implement. The cost for realizing the filters increases with decreasing transition band.

Figure 1.19 shows the magnitude function and the phase function for an allpass (AP) filter. Characteristic of allpass filters is that all frequencies pass through the filter with the same or no attenuation. However, different frequency components are delayed differently, which leads to distortion of the waveform. Allpass filters are therefore often used to equalize the delay of a system so the delay becomes equal for all frequencies.

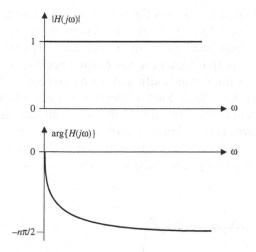

Fig. 1.19 Frequency response for an allpass filter

It is convenient, during the synthesis, to normalize the attenuation to 0 dB. After the synthesis has been completed, the gain of the filter is adjusted.

1.7.6 Phase Delay

Definition 1.5 The phase delay is defined as

$$\tau_f(\omega) \underset{=}{\overset{\Delta}{=}} - \frac{\Phi(\omega)}{\omega}. \tag{1.7}$$

The magnitude function and phase function describes how the filter affects the input signal in a steady state. The output signal $Y(j\omega)$ from a filter with the transfer function $H(j\omega)$ and the input signal $x(t) = A\, e^{j\omega t}$, i.e., a complex sinusoidal signal with amplitude A and angular frequency ω, is

$$y(t) = H(j\omega)x(t) = H(j\omega)Ae^{j\omega t} = |H(j\omega)|Ae^{j(\omega t + \Phi(\omega))}$$
$$= |H(j\omega)|Ae^{j\omega\left(t + \frac{\Phi(\omega)}{\omega}\right)} = |H(j\omega)|Ae^{j\omega(t - \tau_f(\omega))}.$$

How much a frequency component is delayed by the filter is given by the phase delay, which is a function of ω.

Figure 1.20 shows the phase delay for the same fifth-order Cauer filter as before. Note that the two discontinuities in the phase response cause a discontinuities in phase delay. The phase delay can be negative within a certain limited frequency band.

To investigate the filter's influence at fast variations in the input signal, we use the square wave shown in Fig. 1.21 as input signal. The period is 62.832 s, which corresponds to $\omega_0 = 0.1$ rad/s.

The square wave can be described by the Fourier series

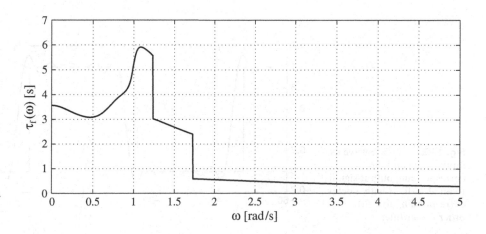

Fig. 1.20 Phase delay for a fifth-order lowpass filter of Cauer type

Fig. 1.21 Square wave

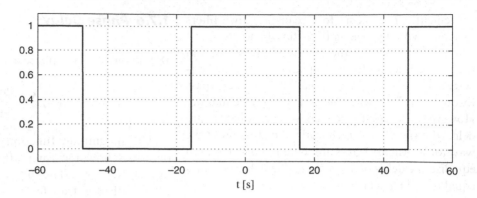

$$v(t) = v_0 \left(\frac{1}{2} + \frac{2}{\pi}\cos(\omega_0 t) - \frac{2}{3\pi}\cos(3\omega_0 t) + \frac{2}{5\pi}\cos(5\omega_0 t) - \dots \right)$$

$$v(t) = v_0 \left(\frac{1}{2} - \frac{2}{\pi}\sum_{n=1}^{\infty}\frac{(-1)^n}{2n-1}\cos((2n-1)\omega_0 t) \right).$$

(1.8)

Thus, the square wave only contains odd frequency components. A filter with a non-linear phase delay will delay the different frequency components differently.

Figure 1.22 shows the output signal for an ideal lowpass filter, which lets all frequencies up to $9\omega0$ pass unaffected and without any delay. The flanks of the output signal are less distinct because of the filter's finite bandwidth and a ringing occurs after every pulse flank. Such a filter is noncausal, which is evident from the ringing in the output signal, which occurs before (anticipates) the pulse flanks.

Figure 1.23 shows the output signal when all of the frequency components up to $9\omega0$ pass the filter

Fig. 1.22 A square wave as input signal and the corresponding output signal to an ideal filter without delay

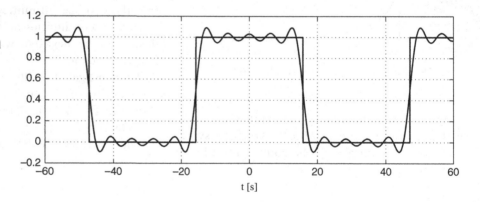

Fig. 1.23 A square wave as input signal and the corresponding output signal to an ideal filter with a delay corresponding to a fifth-order Cauer filter

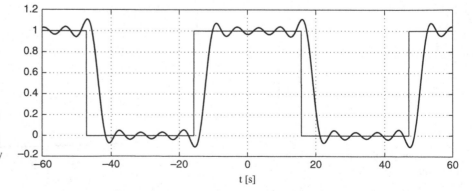

without any attenuation, but delayed corresponding to the delay of a fifth-order lowpass Cauer filter.

If the interesting information in the input signal is in the curve shape, the different frequency components must be delayed equally by the filter in order to leave the information, i.e., the waveform, undistorted. It is for this reason desirable that $\tau f(\omega)$ is constant so all frequency components are delayed with the same amount. An equivalent way of expressing this is saying that a filter has linear phase response. The magnitude function and phase function for a causal filter depend on each other.

1.7.7 Group Delay

A more useful measure of delay of a filter is the group delay.

Definition 1.6 The group delay is defined as

$$\tau_g(\omega) \triangleq -\frac{\partial \Phi(\omega)}{\partial \omega}. \qquad (1.9)$$

Figure 1.24 shows the group delay for a fifth-order Cauer filter. Note that the group delay varies strongly within the passband and has its peak at or slightly above the passband edge, $\omega = 1$ rad/s.

The group delay is an even, rational function of ω. Applications that require a small variation in the group delay are, e.g., video, EKG, EEG, FM (frequency modulated) signals, and digital transmission systems, where it is important that the waveform is retained.

To further study the delay properties of the filter, we consider two sinusoidal signals with the angular frequencies ω_1 and ω_2. Figure 1.25 shows the input signal and the corresponding output signal of the same fifth-order lowpass Cauer filter as discussed before. Both frequency components pass through the filter unaffected.

The input signal can be written as

$$x(t) = \sin(\omega_1 t) + \sin(\omega_2 t)$$
$$= 2\cos\left(\frac{\omega_1 - \omega_2}{2}t\right)\sin\left(\frac{\omega_1 + \omega_2}{2}t\right). \qquad (1.10)$$

Hence, the input signal will be perceived as an amplitude modulated carrier with the angular frequency $(\omega_1 + \omega_2)/2$ and with a slowly varying amplitude $2\cos[(\omega_1 - \omega_2)t/2]$. In Fig. 1.25 we have $\omega_1 = 0.9895$ rad/s, and $\omega_2 = 0.8995$ rad/s, which yields $(\omega_1 + \omega_2)/2 = 0.9445$ rad/s and $(\omega_1 - \omega_2)/2 = 0.045$ rad/s.

The components in the output signal, which has been phase shifted $\Phi_1(\omega_1)$ and $\Phi_2(\omega_2)$, respectively, can be written

$$y(t) = \sin(\omega_1 t + \Phi_1) + \sin(\omega_2 t + \Phi_2)$$
$$= 2\cos\left(\frac{\omega_1 - \omega_2}{2}t + \frac{\Phi_1 - \Phi_2}{2}\right)\sin\left(\frac{\omega_1 + \omega_2}{2}t + \frac{\Phi_1 + \Phi_2}{2}\right)$$
$$= 2\cos\left(\frac{\omega_1 - \omega_2}{2}\left(t + \frac{\Phi - \Phi_2}{\omega_1 - \omega_2}\right)\right)\sin\left(\frac{\omega_1 + \omega_2}{2}\left(t + \frac{\Phi_1 + \Phi_2}{\omega_1 + \omega_2}\right)\right)$$

When $\omega_1 \to \omega_2$, we get

$$y(t) = 2\cos\left(\frac{\omega_1 - \omega_2}{2}\left(t + \frac{d\Phi}{d\omega}\right)\right)\sin\left(\frac{\omega_1 + \omega_2}{2}\left(t + \frac{\Phi}{\omega_2}\right)\right) \quad (1.11)$$

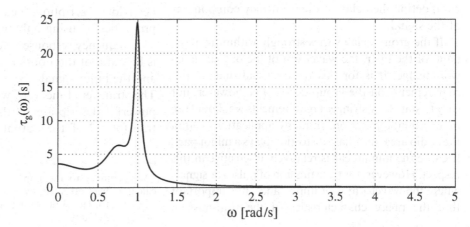

Fig. 1.24 Group delay for a fifth-order Cauer filter

Fig. 1.25 Input signal and corresponding output signal to a fifth-order Cauer filter

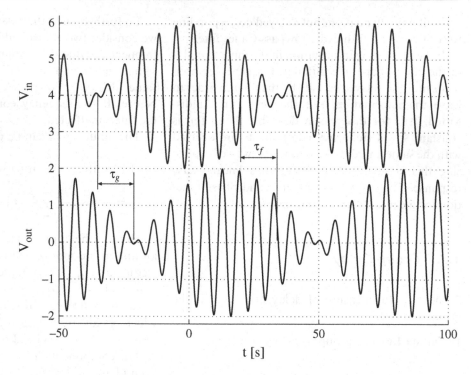

and

$$y(t) = 2\cos\left(\frac{\omega_1 - \omega_2}{2}(t - \tau_g(\omega))\right)\sin\left(\frac{\omega_1 + \omega_2}{2}(t - \tau_f(\omega))\right)\ (1.12)$$

where $\tau_f(\omega)$ is the phase delay and $\tau_g(\omega)$ is the group delay. For this filter, we have $\tau_f(\omega_0) = 4.536$ s and $\tau_g(\omega_0) = 12.898$ s.

The group delay describes the delay suffered by the modulating time function, i.e., the envelope (the LF signal), and the phase delay describes the delay of the carrier wave. For unmodulated (baseband, video) signals, the variations of the phase delay $\tau_f(\omega)$ define the delay of the frequency components of the signal.

If the group delay varies strongly within the passband of the filter, the waveform of the output signal will change. It is for this reason usual that we put requirements on the group delay. It is, however, not easy to state how stringent requirements we should use in a certain application. In many applications within the audio area, the phase distortion plays a minor part, because the human ear is relatively insensitive in this respect. However, for transmission of pulses or signals where the waveform is of importance, it is important that the phase characteristics of the transmission

system are linear, i.e., the group delay is constant, or else the waveform will be distorted.

The group delay is more commonly used than the phase delay, as it is a more sensitive indicator of deviations from the ideal linear-phase behavior than the phase delay. In addition, it has a simpler mathematical form and it is easily measured.

1.8 Transfer Function

A common method of describing a system is using a behavior description, i.e., describing the system properties by using only input and output signals. The frequency response is such a description, which is the ratio of the Fourier transforms of the output and the input signals for a sinusoidal input signal. The transfer function, which is another more powerful description, is the ratio of the Laplace transforms[4] of the output and the input signals.

---

[4] A forerunner to the Laplace transform, the operational calculus, was invented by Oliver Heaviside (1850–1925). The basis for Heaviside's calculus was later found in writings of Laplace (1780).

Here we consider transfer functions that can be realized with lumped elements. In Chapter 4 we will discuss more general transfer functions that require distributed elements for their realization.

Definition 1.7 The transfer function for an analog filter that can be realized with lumped elements is

$$H(s) = \frac{N(s)}{D(s)}. \tag{1.13}$$

$H(s)$ is a rational function in s where N(s) and D(s) are polynomials in s.

The degree[5] of the numerator polynomial for analog filters must be less than or equal to the degree of the denominator polynomial to make the filter realizable. The order of a transfer function of an analog filter is equal to the denominator order.

1.8.1 Poles and Zeros

It is useful to describe $H(s)$ using the numerator and denominator polynomial roots. The roots of the numerator are called *zeros* and the roots of the denominator are called *poles*. The transfer function can be written as

$$H(s) = G\frac{(s-s_{z1})(s-s_{z2})(s-s_{z3})\cdots(s-s_{zM})}{(s-s_{p1})(s-s_{p2})(s-s_{p3})\cdots(s-s_{pN})} M \le N. \tag{1.14}$$

The poles and zeros and the gain constant G is sufficient to fully describe the transfer function. The passband gain is, from a filtering point of view, uninteresting, as it does not vary with frequency and all frequency components are effected in the same way. We will later discuss how the gain constant G shall be determined in order to make the output signal of appropriate size.

A necessary condition for a filter to be stable is that the output signal is bounded for every limited input signal. Moreover, all poles must lie in the left half plane for a stable filter. Zeros, however, can lie anywhere in the s-plane, but for frequency selective filters, the zeros typically lie on the $j\omega$-axis.

---

[5] In the literature, the terms order and degree are used interchangeably, but the former refers to the order of the corresponding differential equation whereas the later refers to the degree of the polynomial.

Furthermore, there must for every complex pole s_p (zero s_z) exist a corresponding complex conjugate pole s_p^* (zero s_z^*).

The reason for this is that both the numerator and the denominator polynomials in the transfer function can only have real coefficients to make the filter realizable with real circuit elements. Thus, the poles and the zeros occur as complex conjugating pairs. However, simple poles and zeros can appear on the real axis in the s-plane. The magnitude function and phase function can easily be determined based on poles and zeros.

Definition 1.8 All roots of a Hurwitz[6] polynomial lie in the left half plane or on the $j\omega$-axis whereas an anti-Hurwitz polynomial has all roots in the right half plane. For a polynomial to be Hurwitz, it is necessary but not sufficient that all of its coefficients are positive.

If the denominator in Equation (1.13) has higher order than the numerator, i.e., $N > M$, then the transfer function has $(N-M)$ zeros at infinity because the transfer function asymptotically approaches zero in the same manner as the function

$$\frac{G}{s^{(N-M)}} \tag{1.15}$$

for large values of s.

. Figure 1.26 shows the poles and the zeros for a fifth-order Cauer filter, which has four finite zeros and one zero at $s = \infty$. A semi-circle with the radius ω_c = passband edge angular frequency has been marked in the figure.

Theorem 1.1 *For a stable analog filter, we have*
Number of poles = Number of finite zeros + Number of zeros at $s = \infty$.

Consider the transfer function in factorized form

$$H(s) = G\frac{(s-s_{z1})(s-s_{z2})(s-s_{z3})\cdots(s-s_{zM})}{(s-s_{p1})(s-s_{p2})(s-s_{p3})\cdots(s-s_{pN})} \tag{1.16}$$

where G is the gain factor. The frequency response is obtained by replacing s with $j\omega$,

---

[6] Adolf Hurwitz (1859–1919), Germany.

Fig. 1.26 Poles and zeros for a fifth-order LP Cauer filter

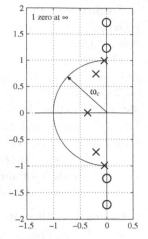

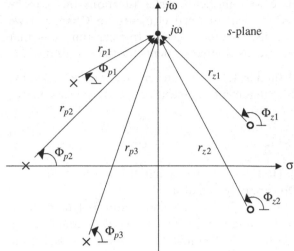

Fig. 1.27 Vector-based computation of the magnitude and phase functions

$$H(j\omega) = G \frac{(j\omega - s_{z1})(j\omega - s_{z2})(j\omega - s_{z3})\cdots(j\omega - s_{zM})}{(j\omega - s_{p1})(j\omega - s_{p2})(j\omega - s_{p3})\cdots(j\omega - s_{pN})}. \quad (1.17)$$

The factors can be written $j(\omega) - a_i - jb_i = -a_i + j(\omega - b_i) = r_i e^{j\Phi_i}$ where $a_i + jb_i$ correspond to either a pole or a zero where

$$r_i = \sqrt{a_i^2 + (\omega - b_i)^2}$$

$$\Phi_i = \arctan\left\{\frac{\omega - b_i}{-a_i}\right\} \quad (1.18)$$

Insertion in the expression for the frequency response yields

$$H(j\omega) = G \frac{(r_{z1}e^{j\Phi_{z1}})(r_{z2}e^{j\Phi_{z2}})(r_{z3}e^{j\Phi_{z3}})\cdots(r_{zM}e^{j\Phi_{zM}})}{(r_{p1}e^{j\Phi_{p1}})(r_{p2}e^{j\Phi_{p2}})(r_{p3}e^{j\Phi_{p3}})\cdots(r_{pN}e^{j\Phi_{pN}})}$$

$$= G \frac{r_{z1}r_{z2}r_{z3}\cdots r_{zM}e^{j(\Phi_{z1} + \Phi_{z2} + \Phi_{z3}\cdots\Phi_{zM})}}{r_{p1}r_{p2}r_{p3}\cdots r_{zN}e^{j(\Phi_{p1} + \Phi_{p2} + \Phi_{p3} + j\Phi_{pM})}}. $$

$$(1.19)$$

By considering vectors in the s-plane, we can determine the magnitude and the phase functions. Vectors are drawn from the poles and zeros to a common point on the $j\omega$-axis according to Fig. 1.27.

The pole-zero configuration corresponds to a lowpass filter with three poles and two finite zeros. We obtain, according to Equation (1.19), the magnitude response at the angular frequency ω, except for gain constant G, by multiplying the magnitude

of the vectors, which originate from the zeros, and dividing with the product of the magnitude of the vectors, which originate from the poles.

We get

$$|H(j\omega)| = |G|\frac{r_{z1}r_{z2}r_{z3}\cdots r_{zM}}{r_{p1}r_{p2}r_{p3}\cdots r_{pN}}. \quad (1.20)$$

We get the phase by adding the angles from the zeros and subtracting the angles from the poles according to

$$\Phi(\omega) = \arg\{G\} + \Phi_{z1} + \cdots + \Phi_{zM} - \Phi_{p1} - \cdots - \Phi_{pN} \quad (1.21)$$

The above method has been implemented in the MATLAB function PZ_2_FREQ_S(G, Z, P, W). The function is part of the accompanying toolbox, and it is significantly more accurate to perform all computations using the poles and zeros than by using the MATLAB function freqs(N, D, w), which uses the denominator and numerator polynomials N and D.

1.8.2 Minimum-Phase and Maximum-Phase Filters

Consider the four possible pole-zero configurations shown in Fig. 1.28. By considering the vectors

Fig. 1.28 Pole-zero configurations with the same magnitude function but with different phase responses

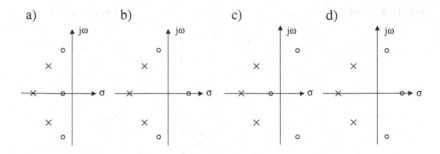

from the poles and zeros to an arbitrary point on the $j\omega$-axis, it is understood that their length is equal in the four cases, i.e., the magnitude functions are the same. The angles according to Equation (1.21) are however different. The phase characteristics and the group delays are different in the four cases.

Definition 1.9 A minimum-phase filter has all zeros in the left half plane or on the $j\omega$-axis.

This is applied in the case (a) shown in Fig. 1.28. This pole-zero configuration has *minimum-phase* and the smallest group delay of the four filters.

There exists a unique relationship between magnitude and phase response for a minimum-phase system. Hence, we cannot have conflicting requirements on the two responses.

Transfer functions with minimum phase are of special interest because good filter structures, e.g., *LC* ladder network, which are insensitive to errors in the component values can be used to realize transfer functions of minimum-phase type.

Any finite linear physical structure that is stable and where energy only travels through one path from the input to the output is normally a minimum-phase system.

Definition 1.10 A maximum-phase filter has all zeros in the right half plane.

The filter (d) in Fig. 1.28 is a maximum-phase filter. Allpass filters are an example of filters that are of the maximum-phase type.

1.9 Impulse Response

In previous sections, the filter properties were described by ratio of the Fourier or Laplace transforms of the output and input signals. It is also of interest to characterize the filter for other types of input signals such as steps and impulses [68]. It is often of theoretical importance to describe the filter using the Laplace transform and with an input signal that corresponds to $X(s) = 1$. This input signal corresponds to a Dirac function[7].

Definition 1.11 A filter's impulse response, h(t), is defined as

$$y(t) = h(t) \leftrightarrow Y(s) = H(s) \qquad (1.22)$$

$$x(t) = \delta(t) \leftrightarrow X(s) = 1. \qquad (1.23)$$

$h(t) = 0$ for $t < 0$ for a causal filter. Note that all definitions of filter properties that have been discussed in this chapter assume that the filter has no stored energy when the input signal is applied.

1.9.1 Impulse Response of an Ideal LP Filter

Consider the ideal LP filter shown in Fig. 1.29, which is not realizable in practice.

The frequency response is

$$H(j\omega) = |H(j\omega)|e^{j\omega\tau_0} \qquad (1.24)$$

and the magnitude function is

$$|H(j\omega)| = \begin{cases} 1 & |\omega| \leq \omega_c \\ 0 & |\omega| > \omega_c. \end{cases} \qquad (1.25)$$

The width of the transition band is zero, the phase function, $\Phi(\omega) = -\omega\tau_0$, is linear, i.e., the group delay

---

[7] Proposed by the nobel laureate Paul A. M. Dirac (U.K.) in 1927 (1902–1984).

Fig. 1.29 Ideal LP filter

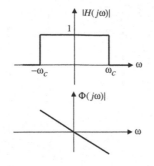

Theorem 1.2: Paley-Wieners Theorem *A necessary and sufficient condition for a magnitude function to be realizable with a causal analog filter is that*

$$\int_{-\infty}^{\infty} |H(j\omega)^2|d\omega < \infty$$

and

$$\int_{-\infty}^{\infty} \frac{|\ln|H(j\omega)||}{1+\omega^2}d\omega < \infty.$$

in the passband is constant and equal to τ_0. The filter is, thus, an ideal LP filter, but with a delay τ_0. In the literature, however, an *ideal LP filter* is often defined as an LP filter without delay, i.e., with $\tau_0 = 0$.

The impulse response for an ideal LP filter is

$$h(t) = \frac{1}{2\pi}\int_{-\infty}^{\infty} H(j\omega)e^{-j\omega t}d\omega = \frac{1}{2\pi}\int_{-\omega_c}^{\omega_c} e^{-j\omega\tau_0}e^{j\omega t}d\omega =$$

$$= \begin{cases} \frac{\omega_c}{2\pi}\frac{\sin(\omega_c(t-\tau_0))}{\omega_c(t-\tau_0)}, & t \neq \tau_0 \\ \frac{\omega_c}{2\pi}, & t = \tau_0. \end{cases} \qquad (1.26)$$

Figure 1.30 shows the impulse response for an ideal LP filter with $\omega_c = 1$ rad/s and $\tau_0 = 5$s. The filter is noncausal, as the impulse response is not 0 for $t < 0$. The maximum of the impulse response, which depends on the group delay, occurs at $t = \tau_0 = 5$s. Note that the period of the ringing is inversely proportional to bandwidth ω_c.

In order to make the filter realizable it is necessary, but not sufficient, that the impulse response is 0 for $t < 0$. A necessary and sufficient condition is stated in Theorem 1.2 [21].

Theorem 1.3 *A continuous-time filter with constant group delay, i.e., with linear phase characteristics, must have a symmetric impulse response.*

This implies that causal, analog filters, which are realized with lumped circuit elements, cannot have exact linear phase. Some filters, which are realized with distributed circuit elements, can, however, have linear phase.

Figure 1.31 shows the output signal from a realizable, causal LP filter with an impulse as input signal, i.e., the impulse response. Because the filter is casual, $h(t) = 0$ for $t < 0$. If the order of the numerator and denominator are equal, there will be an impulse at $t = 0$.

The length of the impulse response indicates for how long a time a disturbance at the input effects the output signal. It can be shown that a filter with rapid variations in the magnitude function or in the phase response results in an impulse response with long duration.

A signal carrier, i.e., a voltage, which corresponds to an impulse, can of course not be

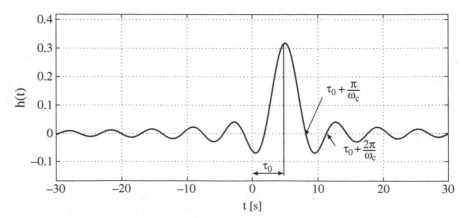

Fig. 1.30 Impulse response for an ideal LP filter

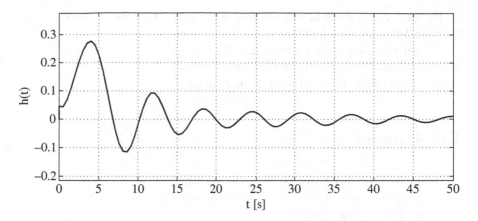

Fig. 1.31 Impulse response for a fifth-order LP filter of Cauer type

realized in practice. It is therefore not possible to directly measure the impulse response for an analog filter.

1.10 Step Response

Definition 1.12 A filter's step response, $s(t)$, is defined as

$$y(t) = s(t) \leftrightarrow Y(s) = H(s)\frac{1}{s} \qquad (1.27)$$

$$x(t) = u(t) \leftrightarrow X(s) = \frac{1}{s} \qquad (1.28)$$

where u(t) is the unit step function. The filter has no stored energy at the time when the step is applied.

u(t) in Equation (1.28) is called the Heaviside function. Figure 1.32 shows a typical output signal for an LP filter with a unit step as input signal, i.e., step response. We get an output signal, $s(t)$, that

increases from 0 to its final value, which corresponds to $|H(0)|$. The step response $s(t) = 0$ for $t < 0$ for a causal filter.

To describe how fast the output signal is growing, the term rise time is used. The *rise time* is defined as the time it takes for the output signal to grow from 10% to 90% of the final value with a unit step as input signal. For a filter of higher order than one, an overshoot is usually obtained.

The impulse response corresponds to the derivative of the step response.

$$h(t) = \frac{d}{dt}s(t). \qquad (1.29)$$

The step response is equal to the integral of the impulse response.

$$s(t) = \int\limits_{0}^{t} h(t)dt. \qquad (1.30)$$

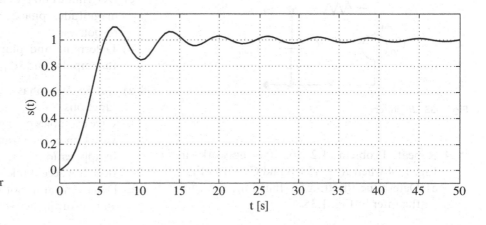

Fig. 1.32 Step response for a fifth-order LP filter of Cauer type

The step response will be delayed proportionately to group delay. The time for the step response to reach the value 0.5 is an approximate measure of the average group delay, and ringing in the step response indicates that the group delay varies strongly in the passband.

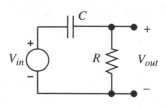

Fig. 1.34 *RC* filter

1.11 Problems

1.1 Describe the difference between the concepts signal and signal carrier as well as continuous-time and analog signals.

1.2 a) Determine the transfer function and frequency response for a first-order filter with a pole $s_p = -3$ rad/s and a zero $s_z = 0$.
   b) Sketch in the same diagram the magnitude and phase response and the group delay.
   c) Sketch in the same diagram the impulse and step responses.

1.3 a) Determine the transfer function for the *RC* filter shown in Fig. 1.33 when $R = 15$ kΩ and $C = 10$ nF.
   b) Determine and mark the position of the poles and zeros in the *s*-plane.
   c) Determine the frequency response.
   d) Determine and plot in the same diagram the magnitude and phase response and determine the type of filter.
   e) Determine and plot in the same diagram $\tau_f(\omega)$ and $\tau_g(\omega)$.
   f) Determine and plot in the same diagram $h(t)$ and $s(t)$.

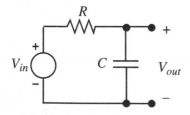

Fig. 1.33 *RC* filter

1.4 Repeat Problem 1.2 for the network in Fig. 1.34 when $R = 15$ kΩ and $C = 10$ nF.

1.5 a) Determine the transfer function, $H(s)$, for the filter in Fig. 1.35.

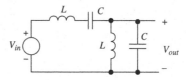

Fig. 1.35 *LC* filter

   b) Determine the magnitude function and phase angle at the angular frequency where the magnitude function has its maximal value.

1.6 a) Compute the gain of a filter at $\omega = \omega_0$ when the attenuation at the same angular frequency is 1.25 dB.
   b) Compute the gain of a filter when the attenuation at the same angular frequency is 40 dB.

1.7 The input to a filter is $v_{in}(t) = 0.5 \sin(\omega t + 0.4)$ V and the output signal is $v_{out}(t) = 0.75 \cos(\omega t + 5.2)$ V. Determine the magnitude and phase of the frequency response at that angular frequency.

1.8 a) Determine the transfer function for a second-order filter with a pole pair $s_p = -3 \pm 2j$ rad/s and a zero pair $s_z = \pm 3j$ rad/s.
   b) Determine the frequency response and the group delay for the filter.
   c) Determine and plot in the same diagram the magnitude, phase, and the group delay responses.
   d) Determine and plot in the same diagram the impulse and step responses.

1.9 a) Define the phase and the group delay functions.
   b) Give examples of applications where a small variation in the group delay is required and in applications where relatively large variations are acceptable.
   c) Determine and plot in the same diagram τ_f and τ_g for the network in Fig. 1.36.

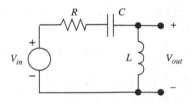

Fig. 1.36 *RLC* filter

1.10 Show that the phase response is

$$\Phi(\omega) = \text{atan}\left(j\left(\frac{H(-s)-H(s)}{H(-s)+H(s)}\right)\right) \text{for } s = j\omega$$

.

1.11 Show that the group delay is

$$\tau_g(\omega) = \frac{1}{2H(-s)}\frac{\partial H(-s)}{\partial s} - \frac{1}{2H(s)}\frac{\partial H(s)}{\partial s} \text{ for } s = j\omega.$$

1.12 a) Determine the area under the group delay expressed in the phase response at $\omega = 0$ and $\omega = \infty$.

  b) Determine all possible values for the phase response at $\omega = 0$ and $\omega = \infty$.

1.13 a) Determine the frequency response and group delay of a filter, with the transfer function

$$H(s) = \frac{s^2}{(s+0.4)(s+0.5)} = \frac{s^2}{s^2+0.9s+0.2}.$$

b) Plot the magnitude response and group delay in the same diagram using MATLAB.

c) Determine and plot in the same diagram the impulse and step responses using MATLAB.

1.14 Repeat Problem 1.13 for a filter with the transfer function

$$H(s) = \frac{s}{(s+0.5)(s+0.6)}.$$

1.15 Consider two filters with the following transfer functions

$$H(s) = \frac{s^2+16}{(s^2+2s+26)}$$

and

$$H(s) = \frac{s^2+16}{(s^2+2s+10)}$$

Fig. 1.38 Pole-zero configuration

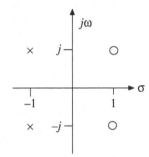

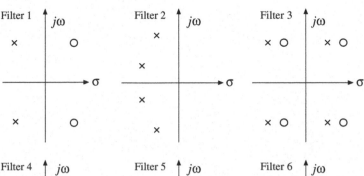

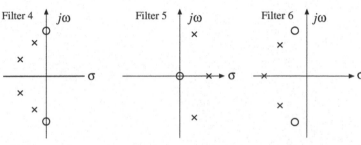

Fig. 1.37 Poles and zeros for six filters

a) Use the vector-based technique to sketch the pole-zero configuration in the s-plane.

b) Determine the order of the filter.

1.16 a) Determine the transfer function for the network in Fig. 1.36 when $R = 15$ kΩ, $C = 10$ nF, and $L = 10$ µH.

b) Plot the position of the poles and zeros in the s-plane.

c) Determine the frequency response.

d) Sketch using the vector-based technique the magnitude function, attenuation, and phase response of the network.

1.17 a) What is a minimum-phase filter?

b) What are the restrictions on the poles and zeros?

c) Give examples of two filters with identical attenuation, of which one of the filters is a minimum-phase filter whereas the other is not.

1.18 a) Which of the filters in Fig. 1.37 are minimum-phase filters?

b) Which filters are allpass filters?

c) Which filters are stable?

1.19 Consider a filter with the pole-zero configuration shown in Fig. 1.38.

a) Determine the transfer function, $H(s)$.

b) Sketch the magnitude function, $|H(j\omega)|$, for $\omega \geq 0$ rad/s.

c) Sketch the phase response, $\arg(H(j\omega))$, for $\omega \geq 0$ rad/s.

d) What type of filter is it?

e) Suggest a possible application for the filter.

1.20 Show that $L\{f(kt)\} = \dfrac{1}{k} F\left(\dfrac{s}{k}\right)$.

Chapter 2
Synthesis of Analog Filters

2.1 Introduction

In this chapter, we will discuss different methods to synthesize the transfer function of an analog filter, i.e., determine the required filter order and the coefficients in the transfer function, for different types of requirements. The coefficients, which determine the positions of the poles and zeros, will later be used to determine the component values in the filter realizations. Here, we will focus on frequency selective filters, which can be used to separate signals and noise that lie in different frequency bands. Moreover, in this chapter we will focus on transfer functions that can be realized with lumped circuit elements, i.e., the transfer functions are rational functions of s.

2.2 Filter Specification

Because an ideal lowpass filter cannot be realized, we must use an approximation of the ideal frequency response. Several standard approximations (frequency responses) have been proposed. The approximations, which usually are referred to as standard filters (approximations), have been optimized using different optimization criteria, i.e., to have the best possible performance from a certain point of view. Tables with standard filters, which are suitable for use in simple applications, are widely available [11, 100, 146]. Design of filters meeting more general requirements usually require numerical optimization techniques [29, 112, 124]. We will use programs, written in MATLAB^{TM} [75], both to synthesize standard filters and filters that meet more general requirements.

A filter specification contains all relevant performance requirements in terms of acceptable bounds within measurable quantities may vary. Typically, the specification contains information of acceptable bounds within measurable quantities may vary. Typically, the specification contains bounds on the acceptable passband ripple, stopband attenuation, cutoff frequency, temperature range, etc. Furthermore, the specification usually contains other types of constraints, e.g., date of delivery and an agreement between the buyer and the seller, as it is important that the responsibility for design and manufacturing errors, etc., is well defined.

2.2.1 Magnitude Function Specification

For an ideal LP filter, the magnitude function equals one in the passband and zero in the stopband. Because such filters cannot be realized, we have to use an approximation of the ideal filter. An acceptable approximation of the magnitude response of an ideal filter has a magnitude function with sufficiently small variation in the passband to be negligible and an attenuation that is sufficiently large in the stopband. In some approximations, there may also be requirements on the phase and group delay characteristics as well.

In order to give the designer full freedom to solve the *approximation problem*, it is common to specify the frequency response in terms of a specification of acceptable variations with the magnitude function and possibly the phase function and the group delay. We will discuss the specification of the group delay in Section 2.2.3.

L. Wanhammar, *Analog Filters Using MATLAB*, DOI 10.1007/978-0-387-92767-1_2,
© Springer Science+Business Media, LLC 2009

Typically, the tolerance ranges within the magnitude function may vary in the passband and the stopband, as shown in Fig. 2.1. In the transition band, the magnitude function is only required to be decaying.

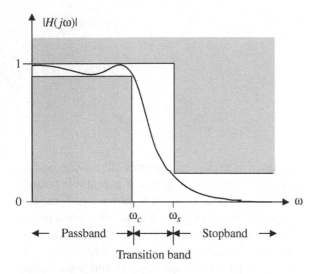

Fig. 2.1 Typical specification for an LP filter

Figure 2.1 shows a standard *specification* for the magnitude function for an LP filter with *cutoff angular frequency* ω_c and the *stopband edge angular frequency* ω_s. Sometimes the term *ripple edge* for ω_c is used. The *transition band* is the band between the stopband and passband edges, i.e., $\omega_s - \omega_c$.

A specification with a small transition band will require a filter of high order, and a filter that meets higher requirements than necessary will be more expensive to implement. Because the cost of a filter increases with the filter order, it is sensible to minimize the filter order. In many cases, however, it is advantageous to use a filter with slightly higher order than necessary.

2.2.2 Attenuation Specification

A typical specification of the attenuation requirements for a lowpass filter is shown in Fig. 2.2 where A_{max} is the allowed variation in the passband attenuation and A_{min} is the minimum required stopband attenuation.

From a filtering point of view, we are only interested in the relative attenuation in the passband and stopband. It is therefore convenient to normalize

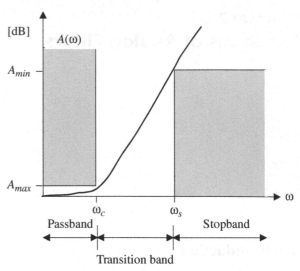

Fig. 2.2 Attenuation specifiction for an LP filter

the maximal passband gain in the passband to 1, which corresponds to an attenuation of 0 dB as was done in Figs. 2.1 and 2.2.

The real passband gain of the filter is then adjusted in the last step of the filter design.

2.2.3 Group Delay Specification

Sometimes there are also requirements on the phase function of a filter, but instead it is more common to use a specification for the group delay. Figure 2.3 shows a typical requirement on the group delay for a telephone channel.

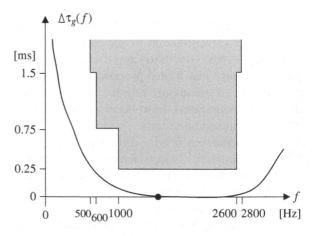

Fig. 2.3 Specification of variation in the group delay for a telephone channel

In this case, the variation of group delay in the passband may not be too large and the specification has therefore been normalized with respect to τ_{gmin}. Hence, we have

$$\tau_g(\omega) = \tau_{gmin} + \Delta\tau_g(\omega).$$

In some applications, the total group delay cannot be too large, e.g., as in a telephone connection via satellite. Typically, the communication channel is active in only one direction at a time, i.e., the direction from the speaker that starts to speak. When the speaker stops speaking, the other speaker may take command of the channel in the other direction. If the delay is too large, it will be difficult to have a conversation because the two speakers will tend to speak or listen at the same time.

The variation of the group delay in the passband cannot be too large either. Typically, the high frequency part of the speech signal is delayed more than the low frequency part. This is perceived as a high frequency time-compressed copy of the word at the end of the word. This is very disturbing to the speakers.

2.3 Composite Requirements

It is difficult to synthesize a transfer function, $H(s)$, which at the same time meets requirements on both the magnitude function and the group delay.

Traditionally, this problem has been solved by dividing the problem into two separate design problems, one involving the synthesis of a minimum-phase and the other of a maximum-phase filter, i.e., an allpass filter. This approach is illustrated in Fig. 2.4.

Definition 2.1 A minimum-phase filter has all zeros in the left half of the s-plane including the $j\omega$-axis.

First, we synthesize a minimum-phase transfer function, $H_m(s)$, which meets the requirement on the magnitude function. Then we synthesize a transfer function of allpass type, $H_{ap}(s)$, which corrects (adds to) the group delay of $H_m(s)$. This can be summarized

$$H(s) = H_m(s)H_{ap}(s) \tag{2.1}$$

$$|H(j\omega)| = |H_m(j\omega)H_{ap}(j\omega)| = |H_m(j\omega)| \tag{2.2}$$

$$\Phi(\omega) = \Phi_m(\omega) + \Phi_{ap}(\omega) \tag{2.3}$$

$$\tau_f(\omega) = \tau_{fm}(\omega) + \tau_{fap}(\omega) \tag{2.4}$$

$$\tau_g(\omega) = \tau_{gm}(\omega) + \tau_{gap}(\omega). \tag{2.5}$$

Figure 2.4 illustrates how the magnitude function and the group delay of the filter $H_m(s)$ and $H_{ap}(s)$ interact so that the combined filters meet the requirements on $H(s)$. However, this method is generally not optimal, i.e., the sum of the orders of the two filters is not always minimal, even though the minimum-phase filter and allpass filter both

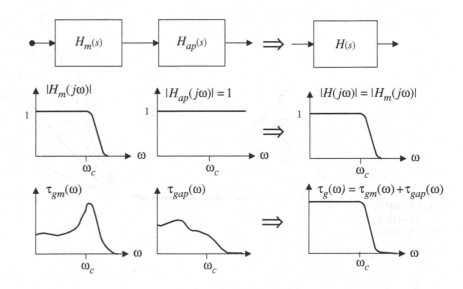

Fig. 2.4 Specification in terms of a minimum-phase and a group delay specification

have minimal order. Computer-based numerical optimization techniques may thus be used to synthesis optimal filters that simultaneously meet attenuation and group delay requirements.

2.4 Standard LP Approximations

Traditionally, we have derived different analytical solutions to the approximation problem, i.e., the problem to find a transfer function that meets the filter specification. Here we first consider the partial problem of finding a transfer function that meets the requirement on the magnitude function.

The analytical solutions, which will be discussed below, are optimal in a certain aspect. This means that one performance measure has been made as high as possible but often at the expense of other performance measures. In real applications, the filter requirements are often more complex. Often we require that the transfer function at the same time shall meet several different requirements, which even may be contradictory. This makes it difficult to synthesize a good filter. The availability of computer-based methods has changed the situation dramatically, and a transfer function that meets several different requirements can easily be found using an optimizing program. Hence, it is often possible to find more efficient and cheaper solutions than those found in filter tables [11, 100, 146].

In this chapter, the classic filter approximations for an LP filter and the corresponding computer-aided design methods are discussed first. In the following sections, we will discuss methods to transform an LP filter to a highpass, bandpass, or bandstop filter [68]. We also discuss computer-aided methods of synthesizing filters that meet non-standard requirements as well as correction of the group delay with the use of allpass filters.

The magnitude function squared for a filter can be written as

$$|H(j\omega)|^2 = \frac{1}{1 + \varepsilon^2 |C_N(j\omega)|^2} \qquad (2.6)$$

where $C_N(j\omega)$ is the *characteristic function*. $C_N(s)$ is an odd (even) function of s for an odd (even) order filter. The zeros of $C_N(s)$, which are called *reflection zeros*, typically lie in the passband. The poles of $C_N(s)$ lie in the stopband and are referred to as transmission zeros. ε is a constant that determines the variation in passband of the filter.

2.4.1 Butterworth Filters

The mathematically simplest and therefore most common approximation is Butterworth filters.[1] Butterworth filters are used mainly because they are easy to synthesize and not because they have particularly good properties [29, 135]. Figure 2.5 shows the attenuation for Butterworth filters of different

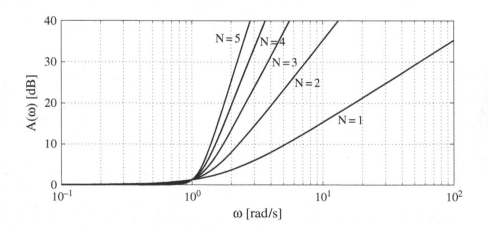

Fig. 2.5 Attinuation for Butterworth filters of different orders

---

[1]S. Butterworth, UK, 1930.

Fig. 2.6 Group delays for Butterworth filters of different orders

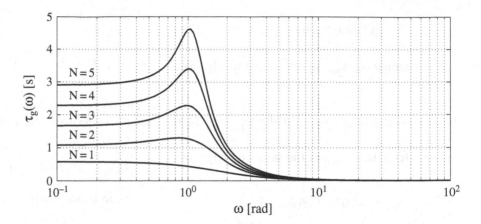

orders[2] and Fig. 2.6 shows the corresponding group delays. The allowed tolerance in the passband is in this case $A_{max} = 1.25$ dB, which is obtained at the passband edge $\omega_c = 1$ rad/s. Note that the overall group delay is larger and it has a sharper peak for filters of higher order. In an ideal filter, however, the group delay should be constant in the passband.

The area under the group delay is $N\pi/2$, where N is the filter order. Hence, the group delay in the passband will increase if the bandwidth is reduced and vice versa.

For a given attenuation requirement, the Butterworth approximation requires a relatively high filter order and hence the group delay becomes large in comparison to the standard approximations, which shall be discussed later. A high-order filter is expensive to implement and has large power consumption if the filter is implemented as an active *RC* filter.

Figure 2.7 shows the attenuation in the passband for Butterworth filters of different order. Note that a linear scale has been used for the ω-axis. For Butterworth filters, the attenuation and magnitude functions are *maximally flat* for $\omega = 0$ and $\omega = \infty$, and a maximum number of derivatives $(N-1)$ of the functions are zero at these angular frequencies. The magnitude function is monotonously decaying while the attenuation is monotonically increasing.

The magnitude function squared for a Butterworth filter can be written as

$$|H(j\omega)|^2 = \frac{1}{1 + \varepsilon^2 \left(\frac{\omega}{\omega_c}\right)^{2N}}. \qquad (2.7)$$

where ε is a constant that determines the variation in the passband, $0 - \omega c$. The characteristic function squared[3] for a Butterworth filter is

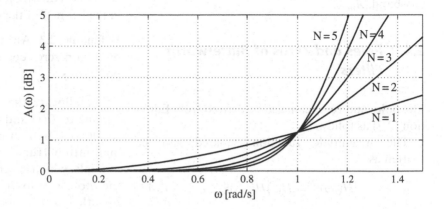

Fig. 2.7 Attenuation in the passband for Butterworth filters of different order

---

[2]The order of a rational function is the greater of the degree of the numerator and denominator polynomials. For a causal filter, the degree of the numerator is equal to or less than the degree of the denominator.

[3]In some literature, the factor ε^2 is included in the characteristic function.

$$|C_N(j\omega)|^2 = \left(\frac{\omega}{\omega_c}\right)^{2N}. \tag{2.8}$$

The attenuation A_{max} at the passband edge, i.e., $\omega = \omega_c$ is

$$A_{max} = -20\log(|H(j\omega_c)|) = -10\log\left(\frac{1}{1+\varepsilon^2}\right). \tag{2.9}$$

The attenuation A_{min} at the stopband edge, i.e., $\omega = \omega_s$ is

$$A_{min} = -20\log(|H(j\omega_s)|)$$

$$= 10\log\left(1 + \varepsilon^2\left(\frac{\omega_s}{\omega_c}\right)^{2N}\right). \tag{2.10}$$

The required *filter order*, which can be derived from Equations (2.9) and (2.10), is

$$N \geq \frac{\log\left(\frac{10^{0.1A_{min}}-1}{10^{0.1A_{max}}-1}\right)}{2\log\left(\frac{\omega_s}{\omega_c}\right)}. \tag{2.11}$$

The filter order must be an integer, and we therefore, but not always, select N to the nearest highest integer.

If the transition band $\omega_s - \omega_c$ is small, then the ratio ω_s/ω_c is small and the required filter order becomes large. Furthermore, the order is affected, but to a lesser degree, by the required stopband attenuation, A_{min}, and the allowed ripple in the passband, A_{max}.

2.4.2 Poles and Zeros of Butterworth Filters

The poles and zeros can also be derived from Equation (2.7) as follows.

The magnitude function squared can be written as

$$|H(j\omega)|^2 = H(s)H(-s)|_{s=j\omega}. \tag{2.12}$$

Equation (2.7) yields

$$H(s)H(-s) = \frac{1}{1 + \varepsilon^2\left(\frac{s}{j\omega_c}\right)^{2N}}. \tag{2.13}$$

The denominator is

$$D(s)D(-s) = 1 + \varepsilon^2\left(\frac{s}{j\omega_c}\right)^{2N}.$$

The roots to the denominator are evenly spaced along a circle with the radius, r_{p0}, with an angle spacing π/N.

$$s_{pk} = j\omega_c\varepsilon^{-1/N}e^{j\pi(2k-1)/2N} \quad \text{for } k = 1,2,\ldots,2N \tag{2.14}$$

which can be written as

$$s_{pk} = r_{p0}\left(-\sin\left(\frac{\pi(2k-1)}{2N}\right) + j\cos\left(\frac{\pi(2k-1)}{2N}\right)\right) \tag{2.15}$$

for $k = 1, 2,\ldots,2N$ where the *pole radius*, r_{p0}, is

$$r_{p0} = \omega_c\varepsilon^{-1/N} \tag{2.16}$$

and

$$\varepsilon = \sqrt{10^{0.1A_{max}} - 1}. \tag{2.17}$$

The denominator can be factorized into two polynomials $D(s)$ and $D(-s)$ belonging to $H(s)$ and $H(-s)$, respectively. $D(s)$ is a *Hurwitz*[4] *polynomial* and $D(-s)$ is an *anti-Hurwitz polynomial*.

Definition 2.2 A Hurwitz polynomial has all zeros in the left half of the s-plane or on the $j\omega$-axis whereas an anti-Hurwitz polynomial has its root in the right half of the s-plane or on the $j\omega$-axis. A strictly Hurwitz polynomial has all roots strictly in the left half of the s-plane.

Definition 2.3 An even (odd) polynomial has only non-zero coefficients for the even (odd) power of s.

We allocate the roots in the left half of the s-plane to $D(s)$ and those in the right half-plane to $D(-s)$. The factorization of an even polynomial into a Hurwitz and an anti-Hurwitz polynomials, which is very difficult to do accurately if the roots are closely clustered, can be computed by either of the programs HURWITZ_POLY and HURWITZ_ROOTS.

---

[4] Adolf Hurwitz, (1859–1919), Germany.

The roots belonging to $D(s)$, i.e., the poles of the Butterworth filter, are

$$s_{pk} = r_{p0}\left(\cos\left(\frac{\pi(N + 2K - 1)}{2N}\right)\right.$$
$$\left. + j\sin\left(\frac{\pi(N + 2k - 1)}{2N}\right)\right) \quad (2.18)$$

for $k = 1, 2,..., N$.

In tables, Butterworth filters are often denoted by PN, where P stands for the German word for exponent, i.e., "potenz," and N is the filter order (2 digits) [100]. Thus, $P08$, denotes an eight-order Butterworth filter. The poles of Butterworth filters, which are normalized with a passband edge = 1, are denormalized by multiplying with r_{p0}.

For each complex pole pair, we define its Q factor. In Chapter 3, we will discuss the Q factor in more detail. Here it is sufficient to note that it is favorable from an implementation point of view to have as low Q factors as possible.

Definition 2.4 The Q factor for a pole (pair) is defined as

$$Q = -\frac{|s_p|}{2Re\{s_p\}}. \quad (2.19)$$

The minimal Q factor is 0.5 and it occurs for a real pole and becomes infinite for a complex pole pair on the $j\omega$-axis.

For Butterworth filters, the Q factors are

$$Q = \frac{1}{2\cos\left(\frac{\pi(2k-1)}{2N}\right)} \quad (2.20)$$

and the largest Q factor for, e.g., 10th-order Butterworth filter, is $Q = 3.19623$, which is a relatively low value.

The finite zeros of a transfer function are the roots of the numerator. Hence, Butterworth filters have no finite zeros because the numerator is a constant. According to Theorem 1.1, the number of zeros is equal to the number of poles. Hence, Butterworth filters have N transmission zeros at $s = \infty$ and N reflection zeros at $s = 0$.

Figure 2.8 shows the poles for a fifth-order Butterworth filter with $A_{max} = 1.25$ dB and $\omega_c = 1$ rad/s. A semicircle with the radius ω_c has been marked in the s-plane.

An odd-order lowpass filter has one real pole and m complex conjugate pole pairs, i.e., $N = 2m + 1$ poles.

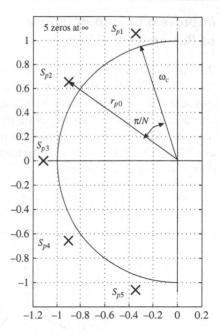

Fig. 2.8 Poles for a fifth-order Butterworth filter

Even-order filters have only m complex conjugate poles, i.e., $N = 2m$. The transfer function for a Butterworth filter has only poles and lacks finite zeros. According to Section 1.8.1, all five zeros lie therefore at $s = \infty$. Butterworth filters are therefore said to be of *all-pole type*.

The transfer function for a Butterworth filter can be written as

$$H(s) = \begin{cases} \dfrac{G}{(s-\sigma_0)(s^2 - 2\sigma_1 s + r_{p0}^2)...(s^2 - 2\sigma_m s + r_{p0}^2)} & N = \text{odd} \\[3mm] \dfrac{G}{(s^2 - 2\sigma_1 s + r_{p0}^2)...(s^2 - 2\sigma_m s + r_{p0}^2)} & N = \text{even} \end{cases}$$
$$(2.21)$$

where $\sigma_0 = -r_{p0}$.

Theorem 2.1 *A second-order polynomial with positive real coefficients has its roots in the left half-plane, and if any coefficient is negative or zero, there is at least one root in the right half-plane or on the $j\omega$-axis.*

Note that $\sigma_p < 0$ for all poles, and all of the coefficients in the denominator are positive in a stable analog filter.

The gain constant $G > 0$ is, in the last steps of the design, chosen so that the appropriate gain is obtained in the passband. During the synthesis of the transfer function, G is for the sake of simplicity normalized so that the largest gain within the passband, $|H(j\omega)|_{max}$, equals one. For lowpass filters of Butterworth type, we select G so that $|H(0)| = 1$, i.e., $A(0) = 0$ dB.

2.4.3 Impulse and Step Response of Butterworth Filters

Figure 2.9 shows corresponding impulse and step responses for a fifth-order Butterworth filter. The step response has a relatively small overshoot and the ringing decays relatively rapidly. Note the relation in Equation (1.30) between impulse and step responses. Thus, the step response is obtained by integration of the impulse response.

Example 2.1 Write a MATLAB program with the help of the included toolbox or Signal Processing Toolbox™ [75] that computes the required order, poles and zeros, and impulse and step response for a Butterworth filter that meets the specification shown in Fig. 2.10 where A_{max} = 0.28029 dB, A_{min} = 40 dB, ω_c = 40 krad/s, and ω_s = 56 krad/s.

Validate the result by plotting the magnitude function, group delay, and poles and zeros in the s-plane and compare the result with Equation (2.11) and Equation (2.18).

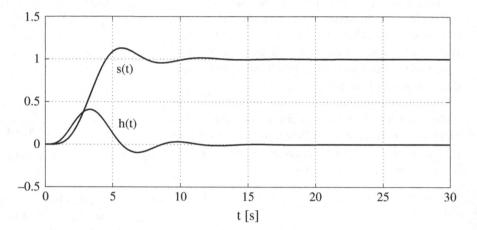

Fig. 2.9 Impulse and step response for a Butterworth filter with $N = 5$

Equation (2.11) yields

$$N \geq \frac{\log\left(\frac{10^{0.1A_{min}} - 1}{10^{0.1A_{max}} - 1}\right)}{2\log\left(\frac{\omega_s}{\omega_c}\right)} = \frac{\log\left(\frac{10^4 - 1}{10^{0.028029} - 1}\right)}{2\log\left(\frac{56}{40}\right)} = \frac{5.176}{0.29226} = 17.711.$$

The computation above gives $N = 17.711$, but the filter order must be selected to a larger integer. Often, but not always, the closest higher integer is chosen. Here we chose $N = 18$, which meets a slightly higher requirement. Thus, there is a so-called design margin. How the design margin may be exploited will be discussed later. Note that an analog filter of order $N = 18$ is very high.

The MATLAB function, which is a part of the accompanying toolbox,

```
N = BW_ORDER(Wc, Ws, Amax, Amin)
```

is used to determine required filter order. Because the order must be chosen to an integer, we get a design marginal. The MATLAB function

```
[G, Z, Zref, P, Wsnew] = BW_POLES(Wc, Ws, Amax, Amin, N)
```

is used to determine poles and zeros and the gain constant, G, so that $|H(j\omega)|_{max} = 1$. Z is an empty vector as Butterworth

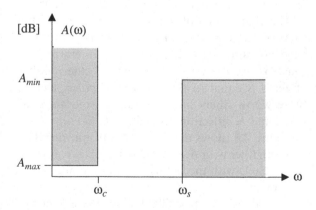

Fig. 2.10 Specification of an LP filter

filters lack finite zeros. The function BW_POLES uses the whole design marginal to reduce the stopband edge, i.e., so that the attenuation A_{min} is achieved at a lower frequency than ω_s. We get

```
% Synthesis of a lowpass Butterworth filter
Wc = 40000;  Ws = 56000;      % Requirement for the lowpass filter
Amax = 0.28029;   Amin = 40;
N = BW_ORDER(Wc, Ws, Amax, Amin);
N = 18;              % Re-run the program after selecting N = integer
[G, Z, Zref, P, Wsnew] = BW_POLES(Wc, Ws, Amax, Amin, N);
Q = -abs(P)./(2*real(P));
omega = linspace(0, 1e5, 1000);      % 1000 values between 0 and 1e5 rad/s
H = PZ_2_FREQ_S(G, Z, P, omega);     % Compute the frequency response
Att = MAG_2_ATT(H);                  % Compute the attenuation
Tg = PZ_2_TG_S(G, Z, P, omega);      % Compute the group delay
PLOT_A_TG_S(Att, Tg, omega, 60, 6*10^-4);
```

Figure 2.11 shows the resulting attenuation and the group delay.

The poles can, thus, be computed either according to Equation (2.18) or by using the above program and plotted with the following addition.

```
xmax = 10000;
xmin = -80000;
ymax = 80000;
PLOT_PZ_S(Z, P, Wc, Ws, xmin, xmax,
   ymax)
```

P = 1.0 e+04 *   Q =

-0.3758591 + 4.2960893i	0.5019
-0.3758591 - 4.2960893i	0.5019
-1.1161570 + 4.1655548i	0.5176
-1.1161570 - 4.1655548i	0.5176
-1.8225411 + 3.9084520i	0.5517
-1.8225411 - 3.9084520i	0.5517
-2.4735482 + 3.5325929i	0.6104
-2.4735482 - 3.5325929i	0.6104
-3.0493977 + 3.0493978i	0.7071
-3.0493977 - 3.0493978i	0.7071
-3.5325929 + 2.4735482i	0.8717
-3.5325929 - 2.4735482i	0.8717
-3.9084520 + 1.8225411i	1.1831
-3.9084520 - 1.8225411i	1.1831
-4.1655548 + 1.1161570i	1.9319
-4.1655548 - 1.1161570i	1.9319
-4.2960893 + 0.3758591i	5.7369
-4.2960893 - 0.3758591i	5.7369

G = 2.6614803 e+83

All zeros of a Butterworth filter lie at $s = \infty$ while the poles lie on a circle with the radius r_{p0}, which is shown in Fig. 2.12 where two semicircles with the radii ω_c and ω_s have been marked. The complex conjugate poles can be combined into a second-order equation according to

$$(s - s_p)(s - s_p^*) = s^2 - 2Re(s_p)s + |s_p|^2$$
$$= s^2 - 2\sigma_p s + r_p^2 .$$

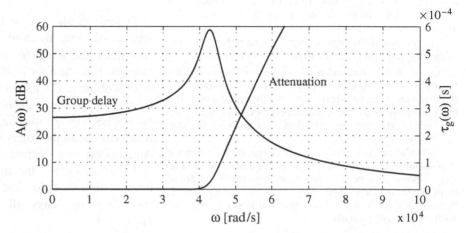

Fig. 2.11 Attenuation and group delay for an 18th-order Butterworth filter

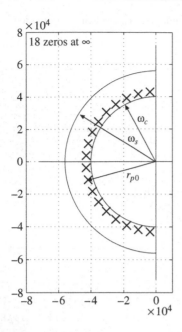

Fig. 2.12 Poles for an 18th-order Butterworth filter

$$H(s) = \left(\frac{1.8597653 \times 10^9}{s^2 + 85921.785s + 1.8597653 \times 10^9} \right)$$

$$\cdot \left(\frac{1.8597653 \times 10^9}{s^2 + 83311.095s + 1.8597653 \times 10^9} \right)$$

$$\cdot \left(\frac{1.8597653 \times 10^9}{s^2 + 78169.04s + 1.8597653 \times 10^9} \right)$$

$$\cdot \left(\frac{1.8597653 \times 10^9}{s^2 + 70651.858s + 1.8597653 \times 10^9} \right)$$

$$\cdot \left(\frac{1.8597653 \times 10^9}{s^2 + 60987.955s + 1.8597653 \times 10^9} \right)$$

$$\cdot \left(\frac{1.8597653 \times 10^9}{s^2 + 49470.963s + 1.8597653 \times 10^9} \right) \cdot$$

$$\cdot \left(\frac{1.8597653 \times 10^9}{s^2 + 36450.822s + 1.8597653 \times 10^9} \right)$$

$$\cdot \left(\frac{1.8597653 \times 10^9}{s^2 + 22323.141s + 1.8597653 \times 10^9} \right)$$

$$\cdot \left(\frac{1.8597653 \times 10^9}{s^2 + 7517.1822s + 1.8597653 \times 10^9} \right) \cdot$$

We use the following program for computing the impulse and step responses:

The transfer function is

```
Wc = 40000;    Ws = 56000;       % Requirement for the lowpass filter
Amax = 0.28029; Amin = 40;
tmax = 10^-3;
N = 18;
[G, Z, Zref, P, Wsnew] = BW_POLES(Wc, Ws, Amax, Amin, N);
t_axis = [0:0.01*tmax:tmax];
[h, dirac0, t_axis] = PZ_2_IMPULSE_RESPONSE_S(G, Z, P, t_axis);
[s_of_t, t_axis] = PZ_2_STEP_RESPONSE_S(G, Z, P, t_axis);
h_scale = 5*10^-5; ymin = -0.4; ymax = 1.3;
PLOT_h_s_S(h, h_scale, s_of_t, t_axis, tmax, ymin, ymax)
```

Figure 2.13 shows the corresponding impulse and step response, where the impulse response has been multiplied with a scaling factor in order to fit into the same diagram.

The area under the impulse response equals the final value of the step response, i.e., $\lim_{t \to \infty} s(t) = H(0) = 1$.

Hence, if the bandwidth of the filter is increased, the length of the impulse response becomes shorter and its amplitude becomes larger. Note that the impulse response has a relatively long ringing, which becomes larger and longer for higher-order Butterworth filers. The delay of the step response is almost 0.3 ms.

Note also the long ringing and the larger overshoot in the step response and compare this to the impulse response. The step response can be obtained from the impulse response through integration. The step response approaches 1, because we have normalized the Butterworth filter to $H(0) = 1$, i.e., $A(0) = 0$ dB.

2.4.4 Chebyshev I Filters

A Butterworth filter does not use the allowed passband tolerance efficiently. By allowing the magnitude function to vary within the acceptable passband bounds, a smaller transition band than for a Butterworth filter of the same order is obtained. For a Chebyshev I filter, the magnitude function varies between the two tolerance bounds, that is, the filter has *equiripple* variation, i.e., the error oscillates with equal peaks across the magnitude response in the passband. An equiripple error is optimal in the Chebyshev sense.

Figure 2.14 shows the attenuation for Chebyshev I filters of different orders and Fig. 2.15 shows the corresponding group delays. All filters have $A_{max} = 1.25$ dB and $\omega_c = 1$ rad/s.

Fig. 2.13 Impulse and step response for the Butterworth filter

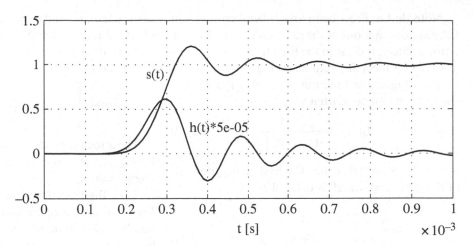

Fig. 2.14 Attenuation for Chebyshev I filters of different orders

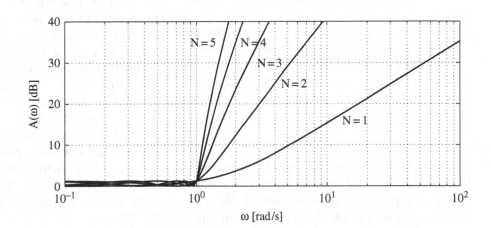

Fig. 2.15 Group delay for Chebyshev I filters of different orders

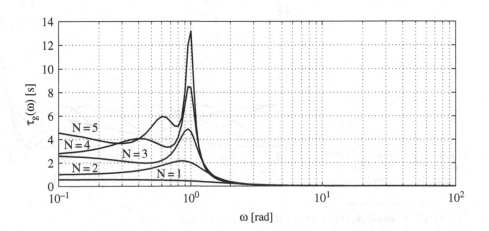

Note that a direct comparison between different filter approximations of the same order, which often, unfortunately, is done in the literature, is not correct as the filters meet different magnitude specifications.

The magnitude function squared for a Chebyshev I filter can be written as

$$|H(j\omega)|^2 = \frac{1}{1 + \varepsilon^2 T_N^2\left(\frac{\omega}{\omega_c}\right)} \quad (2.22)$$

where T_N is an Nth-order Chebyshev polynomial[5] of the first kind, which is defined as

$$T_N(x) = \begin{cases} \cos[N\mathrm{acos}(x)] & |x| \leq 1 \\ \cosh[N\mathrm{acosh}(x)] & |x| > 1. \end{cases} \quad (2.23)$$

Chebyshev polynomials are easily generated by using the recursion

$$T_{N+1}(x) = 2x T_N(x) - T_{N-1}(x) \quad (2.24)$$

with $T_0(x) = 1$ and $T_1(x) = x$. The Chebyshev polynomials oscillate between −1 and 1 for values of x between −1 and 1. Hence, the squared magnitude response oscillates in the passband between a maximum value of 1 and a minimum value of $1/(1 + \varepsilon^2)$. Hence,

$$\varepsilon = \sqrt{10^{0.1 A_{max}} - 1} \quad (2.25)$$

The use of Chebyshev polynomials was proposed by W. Cauer in 1931. Figure 2.16 shows the attenuation in the passband for Chebyshev I filters for different orders. A Chebyshev I filter has equiripple

in the passband and monotonically decreasing magnitude in the stopband.

The attenuation is obtained from Equations (2.22) and (2.23)

$$A(\omega) = 10\log\left[1 + \left[\varepsilon\cosh\left(N\mathrm{acosh}\left(\frac{\omega}{\omega_c}\right)\right)\right]^2\right]. \quad (2.26)$$

The attenuation at the passband edge, i.e., $\omega = \omega_c$, is A_{max}. Note that an even order filter has the attenuation A_{max} at $\omega = 0$.

The filter order can be determined from Fig. 2.16 as the order is equal to the sum of the number of minimum and maximum in the passband. Actually, the detailed variation in the passband is of little interest because we are only requiring the variation to be within the tolerance bounds. Furthermore, to obtain filters that are less sensitive to errors in the component values, we will require that the allowed passband variation is small, i.e., A_{max} is small. This issue will be further discussed in Chapter 3.

Note that the transition band is smaller and the group delay is larger and varies more within the passband for higher order filters. The group delay also varies more than for a Butterworth filter of the same order. However, we should avoid comparing filters of the same order because they do not perform the same amount of filtering.

The required *filter order* for a Chebyshev I filter, which can be determined with Equation (2.21) [29, 135], is

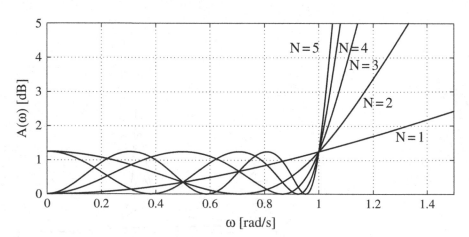

Fig. 2.16 Attenuation in the passband for Chebyshew I filter of different orders

[5] Pafnuty L. Chebyshev, (1821–1894), St. Petersburg, Russia.

$$N \geq \frac{\operatorname{acosh}\left(\sqrt{\dfrac{10^{0.1 A_{min}-1}}{10^{0.1 A_{max}}-1}}\right)}{\operatorname{acosh}\left(\dfrac{\omega_s}{\omega_c}\right)} \quad (2.27)$$

or alternatively we may use the function

```
N = CH_ORDER(Wc, Ws, Amax, Amin).
```

2.4.5 Poles and Zeros of Chebyshev I Filters

The transfer function is an all-pole function, i.e., all zeros are at infinity. The poles are obtained by solving the denominator of Equation(2.22) with $s = j\omega$

$$1 + \left[\varepsilon \cos\left(N \operatorname{acos}\left(\frac{s}{j\omega_c}\right)\right)\right]^2 = 0. \quad (2.28)$$

We get

$$\cos\left[N \operatorname{acos}\left(\frac{s}{j\omega_c}\right)\right] = \pm\frac{j}{\varepsilon}. \quad (2.29)$$

By letting

$$N \operatorname{acos}\left(\frac{s}{j\omega_c}\right) = x + jy \quad (2.30)$$

we get

$$\cos(x+jy) = \cos(x)\cos(jy) - \sin(x)\sin(jy) =$$
$$= \cos(x)\cosh(y) - j\sin(x)\sinh(y) = \pm\frac{j}{\varepsilon}$$

which yields the two equations

$$\cos(x)\cosh(y) = 0$$

and

$$-\sin(x)\sinh(y) = \pm\frac{1}{\varepsilon}.$$

Now, because $\cosh(y) > 0$, the first equation yields

$$\cos(x) = 0 \Rightarrow x = (2k-1)\frac{\pi}{2} \quad (2.31)$$
$$\text{for } k = 0, \pm1, \pm2, \dots$$

The second equation can now be solved.

$$y = \pm \operatorname{asinh}\left(\frac{1}{\varepsilon}\right). \quad (2.32)$$

Let s_{pk} denote the poles. Equations (2.29)–(2.32) yield

$$N \operatorname{acos}\left(\frac{s_{pk}}{j\omega_c}\right) = (2k-1)\frac{\pi}{2} \pm j \operatorname{asinh}\left(\frac{1}{\varepsilon}\right)$$

and

$$\frac{s_{pk}}{j\omega_c} = \frac{\sigma_{pk}+j\omega_{pk}}{j\omega_c} = \cos\left[\frac{\pi}{2N}(2k-1)\pm\frac{j}{N}\operatorname{asinh}\left(\frac{1}{\varepsilon}\right)\right].$$

The poles for an Nth-order Chebyshev I filter are [5, 135]

$$s_{pk} = -\omega_c\left(\operatorname{asin}\left(\frac{\pi(2k-1)}{2N}\right) + jb\cos\left(\frac{\pi(2k-1)}{2N}\right)\right) \quad (2.33)$$

for $k = 1, 2, \dots, N$ where[6]

$$a = \sinh\left(\frac{1}{N}\operatorname{asinh}\left(\frac{1}{\varepsilon}\right)\right) \quad b = \cosh\left(\frac{1}{N}\operatorname{asinh}\left(\frac{1}{\varepsilon}\right)\right) \quad (2.34)$$

$$\varepsilon = \sqrt{10^{0.1 A_{max}} - 1}. \quad (2.35)$$

All zeros lie at $s = \infty$. The poles to a Chebyshev I filter can also be computed with the function

```
[G, Z, P] = CH_I_POLES(Wc, Ws, Amax, Amin, N).
```

Z is an empty vector, as Chebyshev I filters lack finite zeros. P is a column vector with poles and G is the gain constant. Here and in corresponding programs, G is chosen for simplicity so that $|H(j\omega)|_{max} = 1$. After the filter has been synthesized, G can be multiplied with a suitable factor to obtain appropriate passband gain.

The poles positions are shown in Fig. 2.17 for a fifth-order Chebyshev I filter with $A_{max} = 1.25$ dB and $\omega_c = 1$ rad/s. A semicircle with the radius ω_c has been marked in the figure. Note that here we have chosen $A_{max} = 1.25$ dB, which is a relatively large value.

---

[6]$\operatorname{acosh}(x) = \ln(x+\sqrt{x^2-1})$ and $\operatorname{asinh}(x) = \ln(x+\sqrt{x^2+1})$.

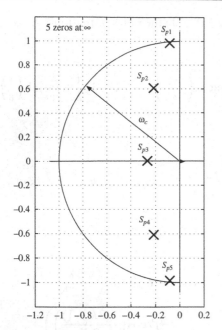

Fig. 2.17 Poles for the fifth-order Chebyshev I filter

We will show in Chapter 3 that an *LC* filter that has a large passband ripple will have high sensitivity for errors in the element values. Hence, we may often design a filter with lower passband ripple

than required by the application in order to be able to use less precise components. Moreover, both the Q factors for the poles and the group delay is reduced for Chebyshev I filters if the design margin is used to reduce A_{max}.

In tables, Chebyshev I filters are usually represented with $TN\rho$, where T stands for Chebyshev (Tshebycheff, according to German transcription), N is the filter order, and ρ is the reflection coefficient in the passband.

The poles of Chebyshev I filters, which are normalized with a passband edge $= 1$, are denormalized by multiplying with ω_c.

Definition 2.5 The reflection coefficient, ρ, which is given in %, is related to the ripple in the passband according to

$$A_{max} = -10\log(1 - \rho^2). \qquad (2.36)$$

Thus, the filter $T0815$ represents an eighth-order Chebyshev I filter with $A_{max} = 0.09883$ dB. Filter tables are often made with respect to the reflection coefficient, ρ, instead of with respect to A_{max}.

The transfer function for a Chebyshev I filter is

$$H(s) = \begin{cases} \dfrac{G}{(s - \sigma_0)(s^2 - 2\sigma_1 s + r_{p1}^2)\cdots(s^2 - 2\sigma_m s + r_{pm}^2)} & N = \text{odd} \\[4mm] \dfrac{G}{(s^2 - 2\sigma_1 s + r_{p1}^2)\cdots(s^2 - 2\sigma_m s + r_{pm}^2)} & N = \text{even} \end{cases} \qquad (2.37)$$

where $r_{pk} = \sqrt{\sigma_{pk}^2 + \omega_{pk}^2}$. The transfer function has only poles and lacks (finite) zeros, i.e., the filter is of all-pole type and all zeros are at $s = \infty$.

2.4.6 Reflection Zeros of Chebyshev I Filters

It is also of interest to determine points in the s-plane where the attenuation is minimum. These points are referred to as reflection zeros. For Butterworth filters, all N reflection zeros are at $s = 0$, and for Chebyshev I filters, we get the zeros by finding the minima of Equation (2.26). Hence,

$$\cos\left(N\mathrm{acos}\left(\frac{s}{j\omega_c}\right)\right) = 0$$

which yields the N reflection zeros for Chebyshev I filters

$$s_{rz} = i\omega_c \cos\left(\frac{\pi(2k-1)}{2N}\right) \qquad (2.38)$$

which all lie in the passband and on the $j\omega$-axis. As can be seen in Fig. 2.16, there is a reflection zero at $s = 0$ for filters of odd order.

2.4.7 Impulse and Step Response of Chebyshev I Filters

Figure 2.18 shows the impulse and step responses. Both the impulse response and step response have a larger and longer ringing than in the Butterworth filter

Fig. 2.18 Impulse and step responses for the fifth-order Chebyshev I filter

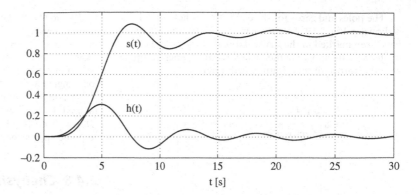

of the same order. However, we shall refrain from comparing different filter types with the same order, as they do not perform the same amount of filtering.

Note, however, that the Chebyshev I filter meets a stricter requirement. Thus, a comparison of different filter approximations of the same order is incorrect. Instead, different filter approximations should be compared when they meet the same specification of the attenuation.

The rise time of the step response is proportional to the width of the first half-period of the impulse response. The rise time and delay of the step response are also larger than for the Butterworth filter of the same order.

Example 2.2 Write a MATLAB program that determines necessary order, poles and zeros, and impulse and step responses for a Chebyshev I filter that meets the specification shown in Fig. 2.10 where $A_{max} = 0.28029$ dB, $A_{min} = 40$ dB, $\omega_c = 40$ krad/s, and $\omega_s = 56$ krad/s, i.e., the same specification as in Example 2.1. Verify the result by plotting the attenuation, the group delay, and poles and zeros in the s-plane and verify the result with Equations (2.21) and (2.22).

Equation (2.21) yields

$$N \geq \frac{\text{acosh}\left(\sqrt{\dfrac{10^4 - 1}{10^{0.028029} - 1}}\right)}{\text{acosh}\left(\dfrac{56}{40}\right)} = \frac{\text{acosh}(387.277)}{\text{acosh}(1.4)}$$

$$= 7.6726.$$

We modify the program in Example 2.1 according to the following:

```
% Synthesis of lowpass Chebyshev I filter
Wc   =40000;         Ws   =56000;
Amax=0.28029;        Amin=40;
N=CH_ORDER(Wc, Ws, Amax, Amin)
N=8;                 % Re-run the program after selecting an integer order
[G, Z, Zref, P, Wsnew]=CH_I_POLES(Wc, Ws, Amax, Amin, N);
Q    =-abs(P)./(2*real(P))
omega=linspace(0, 1e5, 1000);
H    =PZ_2_FREQ_S(G, Z, P, omega);
Att  =MAG_2_ATT(H);
Tg   =PZ_2_TG_S(G, Z, P, omega);
lw   =2; fs=16; fn='times';
PLOT_A_TG_S(Att, Tg, omega, 80, 8*10^-4);
set(gca, 'FontName', fn, 'FontSize', fs);
xtick([0:10000:100000])
```

which yields

```
N=8
P=1.0 e+04 *                    Q =
 - 0.2035170 + 4.0543643i       0.6366
 - 0.2035170 - 4.0543643i       0.6366
 - 0.5795674 + 3.4371241i       1.4151
```

```
 - 0.5795674 - 3.4371241i       1.4151
 - 0.8673839 + 2.2966129i       3.0071
 - 0.8673839 - 2.2966129i       3.0071
 - 1.0231491 + 0.8064632i       9.9733
 - 1.0231491 - 0.8064632i       9.9733

G=1.9829574 e+35.
```

The poles and zeros for the eight-order Chebyshev I filter is shown in Fig. 2.19. Two semicircles with the radii ω_c and ω_s have been marked in the figure. Note that the poles lie closer to the $j\omega$-axis and that the Q factors are higher compared with the corresponding Butterworth filter.

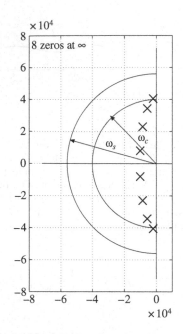

Fig. 2.19 Poles for an eigth-order Chebyshev I filter

The transfer function is

$$H(s) = \frac{1.6479289 \cdot 10^9}{(s^2 + 4070.3401s + 1.6479289 \times 10^9)}$$
$$\frac{1.2149721 \cdot 10^9}{(s^2 + 11591.348s + 1.2149721 \times 10^9)}$$
$$\frac{0.60267857 \cdot 10^9}{(s^2 + 17347.678s + 0.60267857 \times 10^9)}$$
$$\frac{0.16972169 \cdot 10^9}{(s^2 + 20462.981s + 0.16972169 \times 10^9)}$$

The constant G has been chosen so $|H(j\omega)|_{max} = 1$ in the passband and has been divided into four 0factors. These factors should be selected so that the internal signal levels in the corresponding realization are optimized. We will discuss this issue later. This optimization is referred to as scaling the signal levels in the filter realization.The attenuation and the group delay are shown in Fig. 2.20.

Note that the difference in the group delay of the filter shown in Fig. 2.20 and the filters in Fig. 2.14 is mainly caused by the difference in A_{max}. The group delay is largest just above the passband edge and it varies somewhat within the passband, whereas for Butterworth filters the group delay increases monotonically in the passband.

By modifying the program for computing the impulse and step responses for Butterworth filters, we can compute the impulse and step responses for Chebyshev I filters. Figure 2.21 shows the impulse and step responses. The step response can also be computed by integration of the impulse response. Note the long ringing in the impulse and step responses.

The step response approaches $|H(0)| = 10^{-0.05A_{max}}$ for a normalized Chebyshev I filter of even order. Here we have $|H(0)| = 0.968246$. However, for filters of odd order, the step response approaches $|H(0)| = 1$.

2.4.8 Chebyshev II Filters

The attenuation for Butterworth and Chebyshev I filters of LP type approaches infinity for high frequencies. Thus, the filters have a much larger attenuation in the stopband than necessary. In most practical applications, we only require that the attenuation in the stopband is sufficiently large; for example, at least 60 dB attenuation in the whole stopband. This extra attenuation comes at a higher cost.

A filter approximation that is similar to a Butterworth filter in the passband, but has equiripple stopband attenuation, is the *Chebyshev II filter*, which is also called *inverse Chebyshev filter*. Chebyshev II filters do not provide more attenuation than necessary in the stopband.

Figures 2.22 and 2.23 show the attenuation and the group delay for Chebyshev II filters of different orders.

The attenuation for Chebyshev II filters is monotonically increasing in the passband and has equiripple stopband attenuation. The Chebyshev II filter has finite zeros, which are situated on the $j\omega$-axis. In the passband, the filter resembles a Butterworth filter. Due to the finite zeros, the filter gets a smaller transition band than a Butterworth filter of the same order, but the same transition band as for a Chebyshev I filter of the same order. Figure 2.24 shows the corresponding attenuations in the passband for Chebyshev II filters of different order. All filters have $A_{max} = 1.25$ dB.

The magnitude function squared for a Chebyshev II filter is

$$|H(j\omega)|^2 = \frac{1}{1 + \varepsilon^2 \left(\dfrac{T_N^2 \left(\dfrac{\omega_s}{\omega_c} \right)}{T_N^2 \left(\dfrac{\omega_s}{\omega} \right)} \right)} \qquad (2.39)$$

Fig. 2.20 Attenuation and group delay for an eighth-order Chebyshev I filter

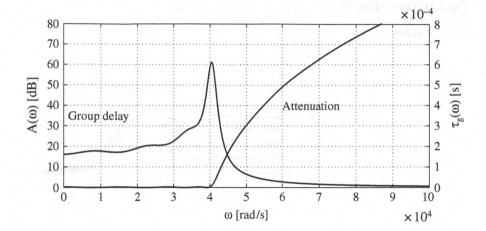

Fig. 2.21 Impulse and step response for the Chebyshev I filter

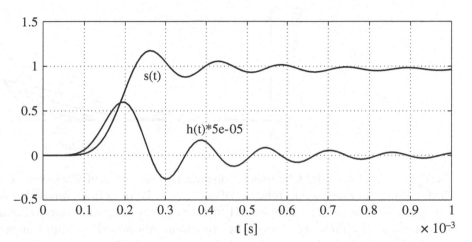

Fig. 2.22 Attenuation for Chebyshev II filters of different orders with $A_{max} = 1.25$ dB and $A_{min} = 40$ dB

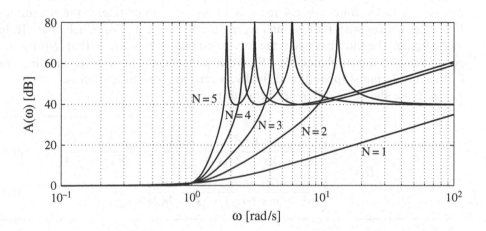

Fig. 2.23 Group delay for Chebyshev II filters to different orders

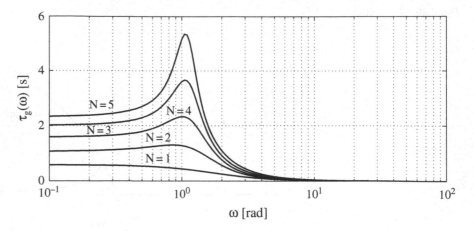

Fig. 2.24 Attenuation in the passband for Chebyshev II filters of different orders

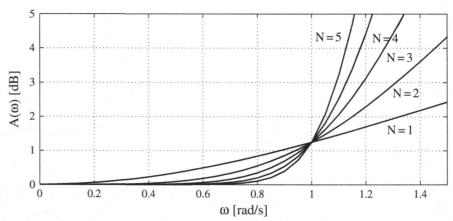

where T_N is an Nth-order Chebyshev polynomial, which is either an even or odd function of ω.

The required order for Chebyshev II filters can be determined either by using the function CH_ORDER, which is the same for Chebyshev I filters, or from Equation (2.21).

The pole-zero configuration for a fifth-order Chebyshev II filter with $A_{max} = 1.25$ dB, $A_{min} = 40$ dB, $\omega_c = 1$ rad/s, and $\omega_s = 1.4$ rad/s is shown in Fig. 2.25. Circles with radii ω_c and ω_s are marked in the figure. The filter has two finite zero pairs; one pair is close to the stopband edge, the other pair is further inside the stopband, and finally, a zero at $s = \infty$. Note the large difference between the pole

positions of Chebyshev I and Chebyshev II filters. Transfer functions with poles close to the $j\omega$-axis are more difficult to realize than if the poles lay far from the $j\omega$-axis. A simple measure of how difficult it is to realize a transfer function is Q factors for the poles.

The transfer function has finite zeros. For odd-order filters, there is a zero at $s = \infty$, but for even-order filters, the magnitude function approaches A_{min}. For Chebyshev II lowpass filters, we can choose G so that $|H(0)| = 1$.

The transfer function for a Chebyshev II filter can be written as

$$H(s) = \begin{cases} \dfrac{G(s^2 + r_{z1}^2) \cdots (s^2 + r_{zm}^2)}{(s - \sigma_0)(s^2 - 2\sigma_1 s + r_{p1}^2) \cdots (s^2 - 2\sigma_m s + r_{pm}^2)} & N = \text{odd} \\[4mm] \dfrac{G(s^2 + r_{z1}^2) \cdots (s^2 + r_{zm}^2)}{(s^2 - 2\sigma_1 s + r_{p1}^2) \cdots (s^2 - 2\sigma_m s + r_{pm}^2)} & N = \text{even.} \end{cases} \quad (2.40)$$

Fig. 2.25 Poles and zeros for a fifth-order Chebyshev II filter

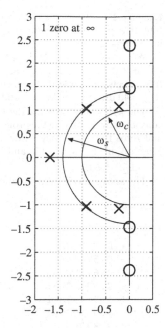

2.4.9 Poles and Zeros of Chebyshev II Filters

The squared magnitude response as given by Equation (2.39) can be rewritten as

$$|H(j\omega)|^2 = H(s)H(-s)$$

$$= \frac{T_N^2\left(j\dfrac{\omega_s}{s}\right)}{T_N^2\left(j\dfrac{\omega_s}{s}\right) + \varepsilon^2 T_N^2\left(\dfrac{\omega_s}{\omega_c}\right)} \quad \text{for } s = j\omega.$$

$$(2.41)$$

The transmission zeros are obtained when the numerator is zero, i.e.,

$$T_N^2\left(j\frac{\omega_s}{s_z}\right) = \cos^2\left[N a\cos\left(j\frac{\omega_s}{s_z}\right)\right] = 0.$$

All of the transmission zeros lie on the imaginary axis, and at these frequencies the attenuation is

infinite. Associating half of them to $H(s)$ and the other half to $H(-s)$ yields

$$s_{zk} = j\frac{\omega_s}{\cos\left[\frac{\pi}{2N}(2k-1)\right]} \quad \text{for } k = 1, 2, \cdots, N. (2.42)$$

For $N =$ odd, one zero is obtained at infinity.

The poles are obtained by solving for the roots of the denominator of Equation (2.41)

$$T_N^2\left(j\frac{\omega_s}{s_{pk}}\right) + \varepsilon^2 T_N^2\left(\frac{\omega_s}{\omega_c}\right) = 0$$

or

$$\cos^2\left[N a\cos\left(j\frac{\omega_s}{s_{pk}}\right)\right] + \varepsilon^2 \cosh^2\left[N a\cosh\left(\frac{\omega_s}{\omega_c}\right)\right] = 0.$$

$$(2.43)$$

Letting $K = \varepsilon \cosh\left[N a\cosh\left(\dfrac{\omega_s}{\omega_c}\right)\right]$ yields

$$\cos^2\left[N a\cos\left(j\frac{\omega_s}{s_{pk}}\right)\right] = -K^2.$$

In the same way as was done for Chebyshev I filters, we get

$$\frac{j\omega_s}{s_{pk}} = \cos\left[\frac{\pi}{2N}(2k-1) \pm \frac{j}{N} a\sinh(K)\right]$$

and by using a polar representation, i.e., $s_{pk} = r_{pk}e^{j\phi_k}$, we get

$$\begin{cases} \dfrac{\omega_s}{r_{pk}}\sin(\phi_k) = \cos\left[\dfrac{\pi}{2N}(2k-1)\right]\cosh[N a\sinh(K)] \\[2mm] \dfrac{\omega_s}{r_{pk}}\cos(\phi_k) = \pm\sin\left[\dfrac{\pi}{2N}(2k-1)\right]\sinh[N a\sinh(K)]. \end{cases}$$

Selecting the poles in the left-hand s-plane yields

$$\begin{cases} r_{pk} = \dfrac{\omega_s}{\sqrt{\cos^2\left[\dfrac{\pi}{2N}(2k-1)\right] + \sinh^2\left[\dfrac{1}{N} a\sinh(k)\right]}} \\[4mm] \phi_k = a\tan\left\{-\cot\left[\dfrac{\pi}{2N}(2k-1)\right]\coth\left[\dfrac{1}{N} a\sinh(K)\right]\right\} + \pi \\[2mm] \text{for } k = 1 \cdots N \end{cases}$$

$$(2.44)$$

The transfer function can now be written in the form

$$H(s) = \frac{G}{D_0(s)} \prod_{k=1}^{n} \frac{s^2 + \omega_{zk}^2}{(s^2 - 2\sigma_k s + r_{pk}^2)} \quad (2.45)$$

where $\omega_{zk} = |s_{zk}|$, $\sigma_k = r_{pk} \cos(\phi_k)$, $n = 2$ floor $(N/2)$, and

$$D_0(s) = \begin{cases} s - \sigma_{n+1} & \text{for } N \text{ odd} \\ 1 & \text{for } N \text{ even.} \end{cases} \quad (2.46)$$

Poles and zeros can be determined with the function CH_II_POLES. The reflection zeros for Chebyshev II filters are at $s = 0$.

2.4.10 Impulse and Step Response of Chebyshev II Filters

Figure 2.26 shows the impulse and step responses. Note that the impulse and step responses for Butterworth and Chebyshev II filters are similar and that the size of the ringing and its duration in the step response is small.

For Chebyshev II filters of even order, the impulse response has an impulse, $\delta(t)$ at $t = 0$, and the corresponding step response has a small initial step. The step response approaches 1 for high frequencies for Chebyshev II filters because $H(0) = 1$ for both odd and even filter orders.

Example 2.3 Determine poles and zeros for a Chebyshev II filter that meets the same attenuation requirement as the filter in Example 2.1.

We know that an eighth-order Chebyshev I filter meets the requirement and, hence, an eighth-order Chebyshev II filter meets the same requirement, i.e., $A_{max} = 0.28029$ dB, $A_{min} = 40$ dB, $\omega_c = 40$ krad/s, and $\omega_s = 56$ krad/s.

The poles and zeros for the Chebyshev II filters can be computed by modifying the program for computing the poles and zeros for Chebyshev I filters. We therefore modify the program in Example 2.2 and instead call the function CH_II_POLES.

The above, modified program yields

```
N = 8

Z = 1.0 e+05 *
  0 + 0.5709710i
  0 - 0.5709710i
  0 + 0.6735063i
  0 - 0.6735063i
  0 + 1.0079734i
  0 - 1.0079734i
  0 + 2.8704653i
  0 - 2.8704653i

P = 1.0 e+04 *              Q =
  -0.5295096 + 4.5906227i   0.5265
  -0.5295096 - 4.5906227i   0.5265
  -1.8485257 + 4.7708137i   0.7463
  -1.8485257 - 4.7708137i   0.7463
  -4.0650817 + 4.6840490i   1.3380
  -4.0650817 - 4.6840490i   1.3380
  -7.1772824 + 2.4619622i   4.1989
  -7.1772824 - 2.4619622i   4.1989

G = 0.0100000.
```

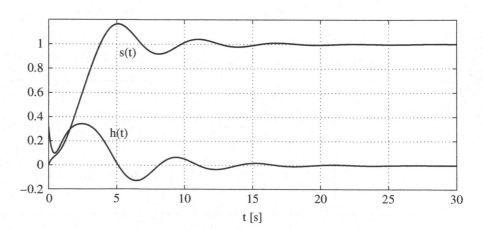

Fig. 2.26 Impulse and step response for a fifth-order Chebyshev II filter

The poles and zeros of the Chebyshev II filter are shown in Fig. 2.27. Note that the Q factors are lower than for the corresponding Chebyshev I filter. For $N =$ even, all zeros are finite, and for $N =$ odd, there are $(N-1)$ finite zeros and one zero at $s = \infty$. Two semicircles with the radii ω_c and ω_s have been marked in the figure.

The poles of Chebyshev II filters, which in tables often are normalized with a stopband edge $= 1$, are denormalized by multiplying with ω_{c3}, i.e., the 3-dB passband frequency.

The attenuation and the group delay are shown in Fig. 2.28. The constant G has been chosen so $|H(0)| = 1$.

The transfer function for the Chebyshev II filter is

$$H(s) = \frac{0.06987403(s^2 + 82.395710 \times 10^9)}{(s^2 + 143545.65s + 5.7574641 \times 10^9)} \frac{0.37858268(s^2 + 10.160103 \times 10^9)}{(s^2 + 81301.635s + 3.8465204 \times 10^9)}$$
$$\cdot \frac{0.57708826(s^2 + 4.5361070 \times 10^9)}{(s^2 + 36970.513s + 2.6177710 \times 10^9)} \frac{0.65506019(s^2 + 3.2600794 \times 10^9)}{(s^2 + 10590.192s + 2.1354197 \times 10^9)}.$$

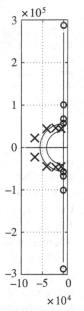

Fig. 2.27 Poles and zeros for an eighth-order Chebyshev II filter

The impulse response and step response for the Chebyshev II filter can be computed by modifying the previously discussed program for computing the impulse and step responses for Chebyshev I filters. Figure 2.29 shows the corresponding impulse and step responses.

The impulse response, in this case, has an impulse, $\delta(t)$ at $t = 0$, because the filter order is even. The impulse is thus so small that it does not show up in the figure. Note that the delay of the step response is smaller than for the two earlier filter approximations, which agrees with the fact that the group delay is smaller.

In order to realize a Chebyshev II filter of even order with a passive LC filter, the poles and zeros positions must be modified. A necessary requirement for an LC realizable lowpass filter is that there is at least one zero at $s = \infty$. This modification, which also is necessary for Cauer filters of even order, will be discussed later.

2.4.11 Cauer Filters

Cauer filters, also known as *elliptic filters or Zolotarev filters*, have equiripple in both the passband and stopband and therefore exploit the acceptable

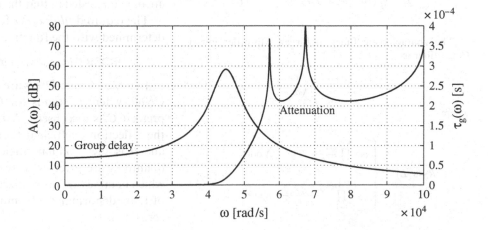

Fig. 2.28 Attenuation and group delay for the eighth-order Chebyshev II filter

Fig. 2.29 Impulse and step
response for the eighth-order
Chebyshev II filter

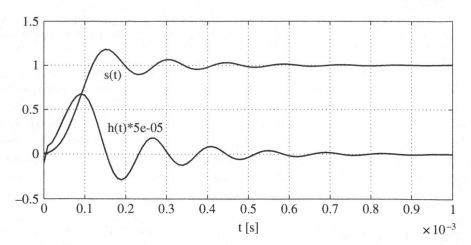

tolerances in the filter specification optimally. This means that a Cauer filter meets a standard magnitude specification with lower filter order than any other filter approximation. The main contribution to the development of Cauer, Chebyshev I, and Chebyshev II filters was made by the German scientist Wilhelm Cauer (1933).

The characteristic function for an Nth-order Cauer filter involves the Chebyshev rational function $R_N(x, L)$. R_N is a complicated function and its details are beyond the scope of this book, and the interested reader is referred to [5, 29, 96]. The Chebyshev rational function R_N has the following properties

- R_N is odd (even) for N odd (even).
- The N zeros lie in the interval $-1 < x < 1$ and the N poles lie outside this interval.
- R_N oscillates between ± 1 in the interval $-1 < x < 1$ and $R_N(1, L) = 1$.
- $1/R_N$ oscillates between $\pm 1/L$ in the interval $|x| > x_L = \omega_s/\omega_c$ where

$$L = \sqrt{(10^{0.1 A_{min}} - 1)/(10^{0.1 A_{max}} - 1)}. \quad (2.47)$$

The characteristic function for an Nth-order lowpass Cauer filter is

$$|C_N(j\omega)|^2 = R_N^2\left(\frac{\omega}{\omega_c}, L\right) \quad (2.48)$$

where

$$R_N(s, L) = \begin{cases} c_1 s \prod\limits_{i=1}^{N/2} \dfrac{s^2 + (x_L/x_i)^2}{s^2 + x_i^2} & N \text{ odd} \\[3ex] c_2 \prod\limits_{i=1}^{(N-1)/2} \dfrac{s^2 + (x_L/x_i)^2}{s^2 + x_i^2} & N \text{ even.} \end{cases} \quad (2.49)$$

The normalizing constants c_1 and c_2 are determined from $R_N(1, L) = 1$. Obviously, for $x = x_L/x_i$ we have a reflection zero, ω_{ri}, because $R_N(x_L/x_i, L) = 0$, i.e., $|H(j\omega)| = 1$, and a transmission zero for $x = x_i = \omega_{zi}$. Note that the reflection and transmission zeros are related according to

$$\omega_{ri} = \frac{\omega_s}{\omega_c} \frac{1}{\omega_{zi}}. \quad (2.50)$$

The expression for the transmission zeros, ω_{si}, involves the elliptic sine function, which is the reason why the name elliptic filter sometimes is used.

Figures 2.30 and 2.31 show the attenuation and the group delay for Cauer filters of different order. Figure 2.32 shows the passband for corresponding attenuations for Cauer filters of different order.

The order of a Cauer filter can be determined from the passband response as the sum of the number of maxima and minima in the passband. This is also true for Chebyshev I filters. Filters of even order have the attenuation A_{max} at $\omega = 0$ while odd-order filters have $A_{max} = 0$. Note, however, we are not really interested in the detailed behavior in the passband. In fact, we are only interested in that the requirement is met.

The required *filter order* for a Cauer filter can be determined with the function

```
CA_ORDER(Wc, Ws, Amax, Amin).
```

In tables, Cauer filters are usually represented by $CN\rho\theta$, where C stands for Cauer-Chebyshev (the prefix CC is also used), N is the filter order, ρ is the reflection coefficient (%), see Equation (2.36), and θ is the modular angle (degrees). The three quantities are given with two digits. Cauer filters, which in tables are normalized with a passband edge of 1, are denormalized by multiplying the poles and zeros with ω_c.

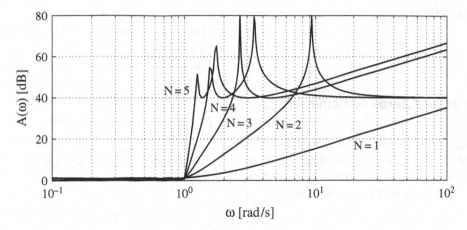

Fig. 2.30 Attenuation for Cauer filters with different orders with $A_{max} = 1.25\,\text{dB}$, $A_{min} = 40\,\text{dB}$, and $\omega_c = 1\,\text{rad/s}$

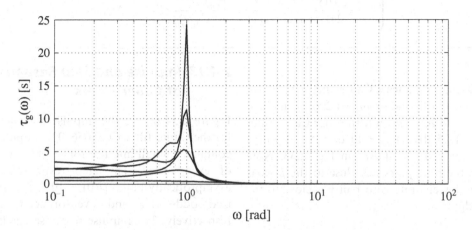

Fig. 2.31 Group delay for Cauer filters with different orders with $A_{max} = 1.25\,\text{dB}$, $A_{min} = 40\,\text{dB}$, and $\omega_c = 1\,\text{rad/s}$

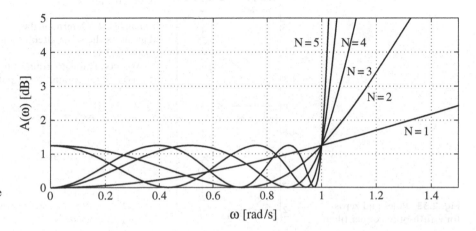

Fig. 2.32 Attenuation in the passband for Cauer filters with different orders

Definition 2.6 The modular angle is defined as

$$\theta = \arcsin\left(\frac{\omega_c}{\omega_s}\right). \qquad (2.51)$$

2.4.12 Poles and Zeros of Cauer Filters

The poles and zeros are complicated to derive. An algorithm for computing the poles and zeros is given in [5]. The transfer function has finite zeros. Filters of odd order have a zero at $s = \infty$, but for filters of

even order, the magnitude function approaches the stopband attenuation, A_{min}. The gain constant G is chosen in the programs so that $|H(j\omega)|_{max} = 1$.

The poles and zeros of Cauer filters, which are normalized with respect to ω_c, can be determined with the function

```
[G, Z, R_ZEROS, P, Wsnew] =
CA_POLES(Wc, Ws, Amax, Amin, N).
```

The transfer function for a Cauer filter can be written

$$H(s) = \begin{cases} \dfrac{G(s^2 + r_{z1}^2)\dots(s^2 + r_{zm}^2)}{(s - \sigma_0)(s^2 - 2\sigma_1 s + r_{p1}^2)\dots(s^2 - 2\sigma_m s + r_{pm}^2)} & N = \text{odd} \\[4mm] \dfrac{G(s^2 + r_{z1}^2)\dots(s^2 + r_{zm}^2)}{(s^2 - 2\sigma_1 s + r_{p1}^2)\dots(s^2 - 2\sigma_m s + r_{pm}^2)} & N = \text{even}. \end{cases} \qquad (2.52)$$

Thus, $C042056$, denotes a Cauer filter with $N = 4$, $A_{max} = 0.1772877$ dB, and $\omega_s/\omega_c = 1.2062 = \sin(56°)$.

The poles and zeros for a fifth-order Cauer filter with $A_{max} = 1.25$ dB, $A_{min} = 40$ dB, $\omega_c = 1$ rad/s, and $\omega_s = 1.205$ rad/s is shown in Fig. 2.33. Note that one of the pole pairs lies close to the $j\omega$-axis and that the lower finite zero pair lies close to the stopband edge.

2.4.13 Impulse and Step Response of Cauer Filters

Figure 2.34 shows the impulse and step response for the Cauer filter $C055056$. The impulse response contains a small impulse for $t = 0$ for Cauer filters of even order. The step response approaches asymptotically 1 and $|H(0)| = 10^{-0.05 A_{max}}$ for normalized odd-order and even-order Cauer filters, respectively. The impulse response has larger ringing than any of the previous filters, but note that they do not meet the same requirements on the magnitude function. Hence, we should not compare these filters. A proper comparison will be done in Section 2.4.14.

Example 2.4 Determine the poles and zeros for a Cauer filter that meets the same attenuation requirement as the filter in Example 2.1.

We modify the program by instead calling CA_ORDER and CA_POLES for computing of poles and zeros. We get

```
N = 5
Z = 1.0 e+04*
    0 - 7.9217042i
    0 + 7.9217042i
    0 - 5.4610294i
    0 + 5.4610294i

P = 1.0 e+04*                    Q =
   -1.2952788 - 3.0512045i     1.2795
   -1.2952788 + 3.0512045i     1.2795
```

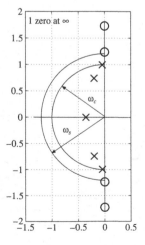

Fig. 2.33 Poles and zeros for a fifth-order Cauer filter

Fig. 2.34 Impulse and step response for the Cauer filter C055056

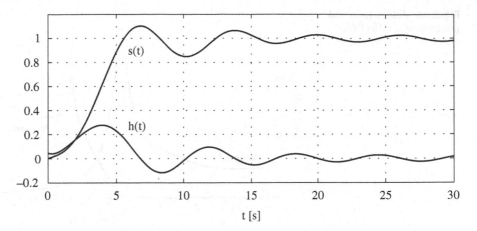

-0.3208531 - 4.1105968i   6.4249
-0.3208531 + 4.1105968i   6.4249
-2.1649281   0.5000

G = 2.1607653 e+03.

The locations of the poles and zeros for the Cauer filter are shown in Fig. 2.35.

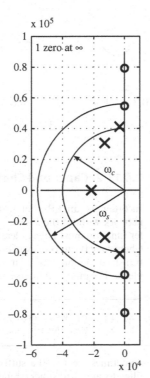

Fig. 2.35 Poles and zeros for a fifth-order Cauer filter

For $N =$ even, all zeros are finite, and for $N =$ odd, there is a zero at $s = \infty$. Two semicircles with radii ω_c and ω_s are also shown in the figure.

Note that is a zero pair, $s_{z3,4} = \pm j\,54610.294$, in the transition band, i.e., inside the utter semicircle because $54,610.294 < 56,000$. This is because the design margin has been used to reduce the stopband edge.

The transfer function can be written as

$$H(s) = \frac{21607653}{(s + 2164.9281)}$$
$$\cdot \frac{(s^2 + 62753397)}{(s^2 + 2590.5577s + 1.0987596 \times 10^9)} \qquad (2.53)$$
$$\cdot \frac{(s^2 + 29822842)}{(s^2 + 6417.0618s + 1.6999953 \times 10^9)}.$$

The two finite zero pairs can be combined with either of the two second-order sections to optimize the signal levels inside the filter realization, see Section 7.3.

The attenuation and the group delay are shown in Fig. 2.36. The constant G has been chosen so $|H(0)| = 1$. Note that the magnitude function approaches zero when $\omega \to \infty$ as there is a transmission zero at $s = \infty$.

By modifying the above program in the same way as done earlier, we get the impulse and step responses. Figure 2.37 shows the impulse and step responses. Note that the ringing has a somewhat longer duration than for corresponding Butterworth and Chebyshev I, and Chebyshev II filters that meet the same specification. The impulse and step responses for filters with large group delays are delayed proportionately to the group delay.

2.4.13.1 Cauer Filters with Minimum Q Factors

A less expensive circuit, with smaller element spread, is required to implement a pole pair with a low Q factor. The importance of Q factors will be further discussed in Chapter 6. Cauer filters with the following relationship between A_{max} and A_{min} have minimal Q factors [71]

$$A_{max} = 10\log\left(\frac{10^{0.1A_{min}}}{10^{0.1A_{min}} - 1}\right). \qquad (2.54)$$

Hence, for an arbitrary specification it may be favorable to modify the specification so that Equation (2.54) holds. For example, $A_{min} = 40$ dB yields $A_{max} = 0.0004343$ dB, which is a very small

Fig. 2.36 Attenuation and group delay for the fifth-order Cauer filter

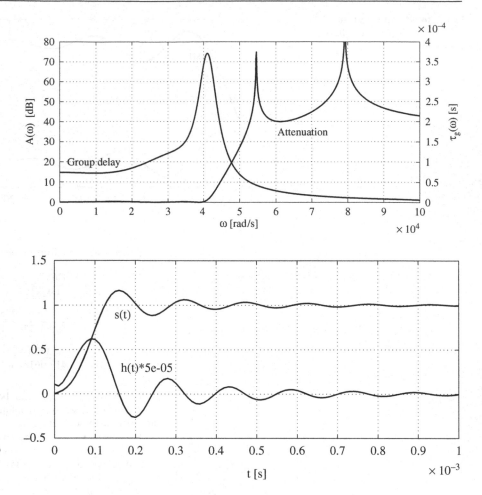

Fig. 2.37 Impulse and step response for the fifth-order Cauer filter

passband ripple. It may appear that this is an unreasonable small ripple, but in fact it is advantageous to design the filter for a smaller ripple than required, as it results in a less sensitive LC filter. This issue will be further discussed in Sections 3.3.8 and 3.3.9.

This special case is related to digital half-band filters where the poles lie on the imaginary axis in the z-plane.

2.4.13.2 Cauer and Chebyshev II Filters of Type B

In order for an LP filter to be realized with an LC ladder network without the couples coils (see Chapter 3), the transfer function must have at least one zero at $s = \infty$. Cauer and Chebyshev II filters of even order lack this zero. However, the transfer functions can be modified to circumvent these problems. The modified filters are denoted type b and type c [29].

```
[G, Z, R_ZEROS, P, Wsnew] = CH_II_B_POLES(G, Z, P, Wc, Ws, Amax, Amin)
[G, Z, R_ZEROS, P, Wsnew] = CA_B_POLES(G, Z, P, Wc, Ws, Amax, Amin)
```

Figure 2.38 shows the attenuation for a fourth-order Cauer filter and corresponding modified filter.

The highest zero pair has been moved to $s = \infty$ so the modified filter obtains a double zero at $s = \infty$. The remaining finite zeros are also moved slightly, which has the effect that the stopband edge is increased. Such filters are provided with a suffix b, e.g., C045042b. The passband edge, ω_c, and A_{max} and A_{min} do not change, but the transition band increases with this modification.

Sometimes we use the suffix a for the unmodified Cauer filter to avoid misunderstandings.

2.4.13.3 Cauer and Chebyshev I Filters of Type C

Cauer and Chebyshev I filters of even order must be modified so that $|H(0)| = 1$ in order to be able to use the same source and load resistances in a ladder network (see Section 3.3.3). In type b filters, the highest finite zero pair is moved to infinity as was

Fig. 2.38 Cauer filter, $C045042a$, and corresponding modified Cauer filter, $C045042b$

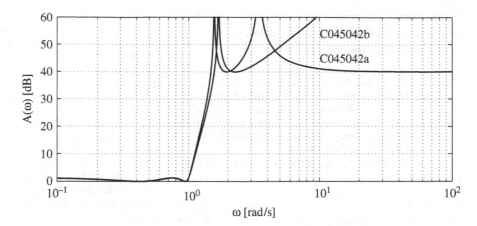

done in type b filters. In addition, the lowest reflection zero is moved to the origin. The passband edge ω_c and attenuations A_{max} and A_{min} are not affected by this modification, but the transition band becomes larger compared with $C045042b$. This modification is indicated with the suffix c, e.g., $C045042c$. Figure 2.39 shows how the attenuation is changed.

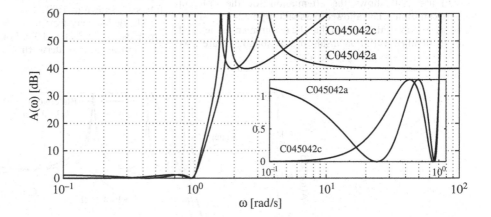

Fig. 2.39 Cauer filter, $C045042a$, and corresponding modified Cauer filter, $C045042c$

The programs

```
[G, Z, P, Wsnew] = CH_I_C_POLES(G, Z, P, Wc, Ws)
[G, Z, R_ZEROS, P, Wsnew] = CA_C_POLES(G, Z, P, Wc, Ws, Amax, Amin)
```

generate the modified pole and zeros for a type c filter. Note that for these two filter approximations, it is not valid that the sum of the number of maxima and minima in the passband is equal to the filter order.

2.4.14 Comparison of Standard Filters

In comparing the standard approximations Butterworth, Chebyshev I, Chebyshev II, and Cauer filters, we find that the two latter have less variation in the group delay. In the literature it is often stated that Cauer filters have larger variation in the group delay than, i.e., Butterworth filters and that this is a valid reason for using the Butterworth approximation. The mistake is that the two filter approximations are compared using the same filter order. This is obviously not correct, which is evident of the following example, as Cauer filters can handle a considerably stricter requirement on the magnitude function. Even the step response for a Cauer filter is better. The difference between Chebyshev II and

Cauer filters is however relatively small, the latter has a somewhat smaller group delay however the order is on the other hand larger.

Example 2.5 Compare Butterworth, Chebyshev I, Chebyshev II, and Cauer filters that meet the same standard LP specification: $A_{max} = 0.01\,\text{dB}$, $A_{min} = 40\,\text{dB}$, $\omega_c = 2\,\text{rad/s}$, and $\omega_s = 3\,\text{rad/s}$ [135]. Note that A_{max} has been chosen very small because this will make the element sensitivity in a corresponding LC realization small. This will be discussed in detail in Section 3.3.

We get the following filter orders with the four standard approximations.

Butterworth: $N_B = 18.846 \Rightarrow N_B = 19$
Chebyshev I and Chebyshev II: $N_C = 8.660 \Rightarrow N_C = 9$
Cauer: $N_{Ca} = 5.4618 \Rightarrow N_{Ca} = 6$

Note the large difference between the required orders for standard approximations that meet the same requirement.

Figure 2.40 shows the attenuation for the different approximations. The allowed passband ripple is very small, and we are only interested in that the requirement is met, and not in detailed variation inside the passband. Figure 2.41 shows the corresponding group delays.

The attenuation in the transition band and the stopband varies between the different filters. The Chebyshev II filter has a more gradual transition between the passband and stopband compared with the Chebyshev I filter in spite of the fact that they have the same order. Note that the Cauer filter has a smaller transition band than required. The order has been rounded up from 5.46 to $N = 6$.

The attenuation approaches infinity for all of the filters, except for the Cauer filter, as its order is even.

The differences in the group delays are large between different approximations. The peaks in the group delays lie above the passband edge $\omega_c = 2\,\text{rad/s}$. The Butterworth and Chebyshev I filters have larger group delay in the passband whereas the Chebyshev II and Cauer filters have considerably smaller group delay and the difference between the latter two is small.

In the literature, it is commonly stated that the Butterworth filter has the best group delay properties. This is obviously not correct and is based on an unfair comparison between the standard approximations of the same order. According to Fig. 2.41, Chebyshev II and Cauer filters have considerably better properties. The Q factors for the four filters are shown in Table 2.1.

The element sensitivity for an LC filter is proportional to the group delay, see Section 3.3.9. For example, by using a Cauer filter instead of a Butterworth filter, the group delay can be

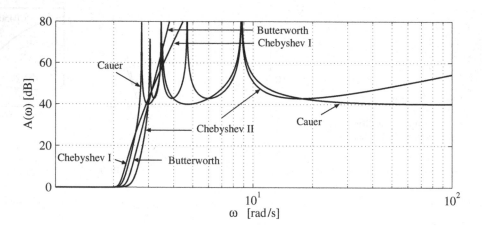

Fig. 2.40 Attenuation for Butterworth, Chebyshev I, Chebyshev II, and Cauer filters

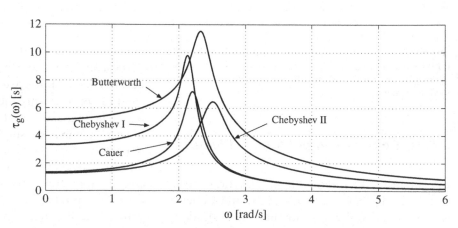

Fig. 2.41 The group delay for Butterworth, Chebyshev I, Chebyshev II, and Cauer filters

Table 2.1 Comparison of Q factors

Butterworth	Chebyshev I	Chebyshev II	Cauer
$N=19$	$N=9$	$N=9$	$N=6$
0.5000	0.5000	0.5000	No real pole
0.5069	0.6821	0.5980	0.5832
0.5286	1.1807	0.9067	1.4315
0.5685	2.2638	1.6395	6.0126
0.6336	7.2466	5.1370	
0.7382			
0.9142			
1.2447			
2.0368			
6.0548			

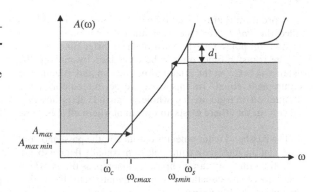

Fig. 2.42 Example of distribution of the design margin

reduced with about a factor 3 and the component tolerance is increased with the same factor. Components with large tolerances are considerably cheaper than those with small tolerances. Besides, the number of components is fewer. It is therefore important to use an approximation with small group delay. Cauer is often the preferred approximation because the require order is significantly lower than for Chebyshev II and the group delay and Q factors are similar.

The conclusion is that Cauer is the best approximation in most cases, i.e., when we have requirements on both the magnitude function, the group delay, and step response. We will later see that the Cauer filter has almost as low sensitivity to errors in the element values as Chebyshev II filters when it is realized with an LC filter.

2.4.15 Design Margin

The filter order must be an integer. Often, but not always, the order is chosen to the closest larger integer. The difference between theoretical and chosen filter order allows that the slightly more stringent requirement can be met. The design margin can then be used to reduce the passband and the stopband ripples and reduce the width of the transition band in excess of what is required in the specification.

For example, we first compute the case where the whole design margin is used to reduce the stopband edge, ω_{smin}, and afterwards choose a suitable reduction of this as shown in Figure 2.42.

In the same manner, we can then use the remaining part of the design margin to determine the largest possible attenuation in the stopband and afterwards choose a suitable increase of the attenuation requirement with d_1. Finally, in the same manner the remaining part of the design margin is used to increase the passband edge and reduce the passband ripple.

The intention is to obtain a safety margin for the errors, which always are present in the element

values in the circuit that realizes filter. By exploiting the design margin appropriately, we can minimize the number of filters that violates the specification with a given statistical error distribution of the component values.

It can be shown that the Q factors of the poles become smaller if we use the design margin to reduce the passband ripple, and, hence, the filter will be simpler to implement.

2.4.16 Lowpass Filters with Piecewise-Constant Stopband Specification

Cauer filters have equiripple in both the passband and stopband, i.e., they meet a filter specification with a maximum attenuation of A_{max} in the passband and has at least the attenuation A_{min} in the stopband. However, we often have different requirements in different parts of the stopband as illustrated in Fig. 2.43. In such cases, the Cauer filter is not the best filter approximation. The zeros must in these cases be computed using numerical optimizing methods.

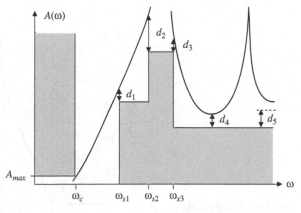

Fig. 2.43 Lowpass filter that meets a piecewise-constant stopband requirement

A program that is based on a well-known method [29] is called the *PolePlacer* where the ancient term pole refers to attenuation poles, i.e., the zeros of the transfer function.

The transfer function can be optimized by making the distance d_i between the attenuation function and attenuation requirement equal. In this way, we get an equally large attenuation margin in the whole stopband. It is however not certain that there exists an optimum where all d_i become equal.

The passband requirement is constant in the passband, and we can let the filter have equiripple or maximally flat passband.

In the same way can highpass and bandpass filters, which meet a piecewise-constant stopband requirement, be synthesized using the functions

```
POLE_PLACER_HP_MF_S
POLE_PLACER_HP_EQ_S
POLE_PLACER_BP_MF_S
POLE_PLACER_BP_EQ_S
```

where MF and EQ indicate maximally flat and equiripple passband, respectively. Note that we must provide reasonable initial values for the number of finite and non-finite zeros and their placement in order for the programs to find an optimal solution.

Example 2.6 Below is shown how the POLE_PLA CER_L-P_EQ_S program may be used.

```
Amax = 0.5;          % POLE_PLACER_LP_EQ_S
Wc = 1;
Amin = [75 55];      % Amin = 75 dB from 2 and 55 dB above 2.5 rad/s
Wstep = [2 2.5];     % Break frequencies
Wfi = [2.5 4];       % Initial guess of finite zeros ±j2.5 and ±j4 rad/s
NIN = 1;             % Number of zeros at infinity, N = 2Nfi + NIN
                     % Hence, a 5th-order LP filter with equiripple passband

[G, Z, Zref, P, dopt] = POLE_PLACER_LP_EQ_S(Amax, Wc, Amin, Wstep, Wfi, NIN)
W = (0:1000)*4*pi/1000;
H = PZ_2_FREQ_S(G, Z, P, W);
axis([0 8 0 100]), subplot('position', [0.1 0.4 0.88 0.5]);
PLOT_ATTENUATION_S(W, H)
```

If the program does not converge, a new initial placement of the finite zeros or an increase of the number of zeros may be tried. If *dopt* < 0, the number of zeros might be too small. If instead the function POLE_PLACER_LP_MF_S is used, a filter with maximally flat passband is obtained. However, this requires an increase of the filter order. Figure 2.44 shows the resulting attenuation function.

All design margins in the stopband are equal with *dopt*. The program yields the following poles and zeros. The design margin in the stopband is at least 2.27 dB. Zref is a vector with the reflection zeros. For the maximally flat case, all reflection zeros are at the origin.

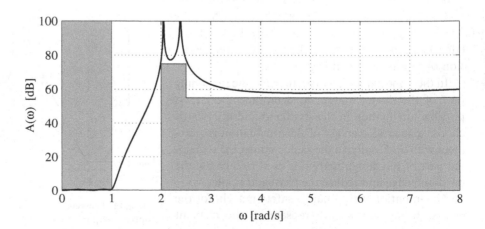

Fig. 2.44 Lowpass filter with piecewise-constant stopband requirement

```
N = 5

Z =                              P =
0 - 2.0472076i                   -0.0921637 - 1.0123270i
0 - 2.3868661i                   -0.0921637 + 1.0123270i
0 + 2.0472076i                   -0.2886563 + 0.6753131i
0 + 2.3868661i                   -0.2886563 - 0.6753131i
                                 -0.4023776

G = 0.009392200                  dopt = 2.2734442
```

2.5 Miscellaneous Filters

2.5.1 Filters with Diminishing Ripple

The sensitivity to errors in the element values in the double terminated ladder that implements an LP filter is low for $\omega = 0$ and increases for increasing frequency and becomes largest where the group delay is maximal, i.e., just above the passband edge. The sensitivity can be reduced by reducing the ripple, i.e., use an approximation where the ripple decays toward the passband edge. We will discuss this case in more detail in Chapter 3.

2.5.2 Multiple Critical Poles

Another way of reducing the sensitivity emanates from the fact that the most sensitive pole pair is the one that has the highest Q factor, see Section 6.4. By increasing the filter order, i.e., higher than necessary, we can make the pole pair with the highest Q factor a multiple pole pair and reduce their Q factors. This technique is called multiple critical root and can be used for both maximum flat passband ($MUCRMAF$) and equiripple passband ($MUCROER$) [94].

2.5.3 Papoulis Monotonic L Filters

In the literature, there exist a number of different types of all-pole approximations, e.g., Papoulis, parabolic, and Halpern approximations [68, 88]. These filters can have, in some cases, either better group delay, step response, or lower Q factors.

However, the interest for different standard approximation has diminished with the event of effective optimizing programs that are able to optimize several different parameters at the same time.

Papoulis developed this class of filters whose attenuation increases monotonically in the passband and has maximal attenuation rate at the cutoff edge. The approximation maximizes the rate of change of the magnitude function at $\omega = 1$ under the constraint of monotonic attenuation response.

The filters combine the desirable features of the Butterworth and Chebyshev responses. The step response of these filters is good because the magnitude response monotonically decreases. The characteristic function for these filters is an Legendre polynomial of the first kind, and they are therefore referred to as Monotonic L filters.

2.5.4 Halpern Filters

Halpern filters are closely related to Papoulis filters but optimize the shaping factor under the conditions of a monotonically decreasing response. Halpern filters have a monotonic step response and maximum asymptotic cutoff rate. The characteristic function is a polynomial related to the Jacobi polynomials.

From the point of view of the stopband attenuation, little can be gained from maximizing the asymptotic attenuation cutoff rate instead of maximizing the attenuation rate at the edge as done for the Papoulis filters. In addition, the Papoulis filters yield much smaller passband magnitude distortion than the Halpern filters, and the latter are therefore only of academic interest.

2.5.5 Parabolic Filters

This class of filters has all the poles located in a parabolic contour in the left-hand side of the s-plane. Parabolic filters have a monotonic passband magnitude response, similar to the Halpern filters, and the variation between the maximum and minimum values of their group delay is better, and the step response has the fastest response without overshoot [68].

2.5.6 Linkwitz-Riley Crossover Filters

Loudspeakers are not capable of covering the entire audio spectrum with acceptable loudness and distortion. However, high-quality loadspeakers can be manufactured for smaller frequency ranges. Audio crossover filters are therefore used to split the audio signal into separate frequency bands, which are sent to individual loudspeakers that have been optimized for those bands.

Linkwitz-Riley crossover filters, named after S. Linkwitz and R. Riley, consist of a parallel combination of a lowpass and a highpass filter. Each filter consist of two cascaded Butterworth filters. Because each Butterworth filter has 3 dB attenuation at the cutoff frequency, the resulting Linkwitz-Riley filter will have 6 dB attenuation at the cutoff frequency. The overall attenuation at the cutoff frequency of the lowpass and highpass filter will be 0 dB, and the crossover filter behaves like an allpass filter with a smoothly changing phase response.

2.5.7 Hilbert Filters

Hilbert filters are used in single-sideband modulation schemes. The Hilbert filter is an allpass filter that approximates the phase function $-90°$ for $\omega > 0$ and $90°$ for $\omega < 0$. Hilbert filters have the ideal frequency response

$$H(j\omega) = \begin{cases} -j & \omega > 0 \\ 0 & \omega = 0 \\ j & \omega < 0. \end{cases} \qquad (2.55)$$

Thus, a Hilbert filter generates an output signal that is $\pm 90°$ different compared to the input signal. It is, however, often simpler to realize the Hilbert filter as a filter with two outputs with $90°$ difference in their phase responses.

2.6 Delay Approximations

In previous sections, transfer functions that meet a given magnitude specification have been discussed. In this section, we will discuss filter approximations, which approximate linear phase or constant group delay. A group delay requirement is more stringent than a linear-phase requirement. There exist standard approximations with maximally flat or equiripple group delay.

2.6.1 Gauss Filters

Characteristic for a Gauss filter [11, 29, 68, 88, 146] is that the step response does not have any overshoot or ringing and that the impulse response is approximately symmetric around the time t_0, i.e., the phase function is almost linear.

Both Gauss and Bessel filters are nowadays of limited interest because filters that meet demands on both the magnitude function and the phase function can be determined with the help of optimization programs.

2.6.2 Lerner Filters

Lerner filters have approximately linear phase in most of the passband and a relatively small transition band [88].

2.6.3 Bessel Filters

A Bessel filter, which also is called a Thomson filter, approximates a lowpass filter with linear phase [88]. Bessel filters were first described by Z. Kiyasu 1943 and later by W.E. Thomson 1952. Bessel filters have a maximally flat group delay for $\omega = 0$. Figure 2.45 shows the magnitude function for Bessel filters with orders 1–5. The magnitude function decays monotonically and the filter has relatively poor attenuation in the stopband and a wide transition band. The passband edge depends on the filter order. There exists no simple expression for computing the required filter order of the poles given an attenuation and group delay requirement.

Figure 2.45 shows the attenuation and Fig. 2.46 shows the corresponding group delays for Bessel filters with orders 1–5. The filters have been normalized to have $\tau_g(0) = 1$ s. Note that the frequency range over which the group delay is approximately constant increases with increasing filter order. In fact, the bandwidth-group delay product[7] is fixed and increases with the filter order. Hence, we may obtain a larger group delay over a small bandwidth or vice versa.

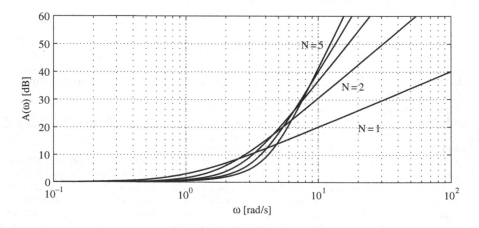

Fig. 2.45 Attenuation for Bessel filters of different orders

___
[7]This product is an alternative formulation of the Heisenberg uncertainty principle.

Fig. 2.46 Group delay for Bessel filters

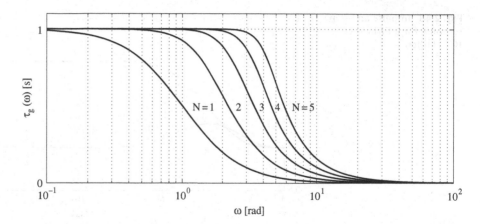

The transfer function for a Bessel filter is

$$H(s) = \frac{B_N(0)}{B_N(s)} = \frac{b_0}{\displaystyle\sum_{k=0}^{N} b_k s^k} \qquad (2.56)$$

where $b_k = \dfrac{(2N-k)!}{2^{N-k}k!(N-k)!}$ and $B_N(s)$ is a Bessel polynomial [140].

The Bessel polynomial can easily be computed by the recursion

$$B_n = (2n-1)B_{n-1} + s^2 B_{n-2} \qquad (2.57)$$

with $B_0 = 1$ and $B_1 = s + 1$. Bessel filters do not have finite zeros.

The poles for a fifth-order Bessel filter are shown in Fig. 2.47 with $\tau_g(0) = 3.93628$ s, and Fig. 2.48 shows the impulse and step responses of the corresponding filter.

Fig. 2.47 Poles for a fifth-order Bessel filter

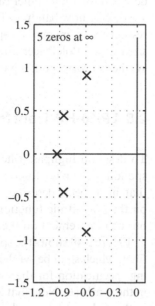

Note that a filter with linear phase has an impulse response that is symmetric around the time $t_0 \approx \tau_g$. The impulse response for the Bessel filter is almost symmetric, which results in an almost linear phase response. Note that because the impulse response only has a small undershoot, the ripple in the step response will also be small. The overshoot and the ringing in the step response are much smaller compared with the previously discussed filters.

The required order for a Bessel filter with a given group delay at $\omega = 0$, a maximal deviation in the group delay from $\tau_g(0)$, and a maximum ripple in the passband at ω_0, can be computed with the function BESSEL_ORDER.

Filters that are similar to Bessel filters are used in the read channel for hard drives to equalize both magnitude function and the group delay and to reduce the noise. The resulting read channel should have linear phase in order to reliably detect the ones and zeros. In this application, the filter's bandwidth varies depending on which track is being read. A typical filter has order seven and a bandwidth that can be varied between 10 and 100 MHz. The allowed variation in phase function is less than $\pm 0.05°$.

Because hard drives are manufactured in several hundred million units annually, these filters are economically important, and large development efforts have been made to integrate the filters on the same chip as other digital logic that is a part of the read channel.

It is common to use different compromising solutions between, e.g., Chebyshev II filter and Bessel filter. Such a filter can therefore obtain a relatively good group delay and at the same time a good attenuation and narrow transition band because the finite zeros give large stopband attenuation and do not affect the group delay [11, 29]. Note that zeros on the $j\omega$-axis do not affect the group delay. Moreover, the area under the group delay for an Nth-order all pole filter depends only on the filter order. We have

$$\int_0^\infty \tau_g(\omega)\,d\omega = \int_0^\infty -\frac{\partial\Phi}{\partial\omega}\,d\omega = -[\Phi(\infty) - \Phi(0)] = -\left[-N\frac{\pi}{2} - 0\right] = N\frac{\pi}{2}. \qquad (2.58)$$

Fig. 2.48 Impulse and step response for a fifth-order Bessel filter

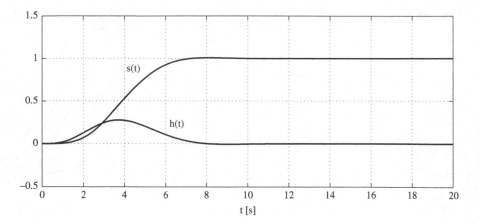

Hence, because the area under the group delay is constant, an increase in the bandwidth will result in a decrease in the group delay and vice versa.

2.6.4 Lowpass Filters with Equiripple Group Delay

Lowpass filters with equiripple group delay response can be designed by using numerical optimization routines. The equiripple group delay filters have a wider equiripple–group delay band and similar stopband attenuation as the corresponding Bessel filter of the same order and are therefore often preferred.

2.6.5 Equiripple Group Delay Allpass Filters

An allpass filter can be derived from any of the previously discussed allpole, lowpass filters by forming the transfer function using their denominator, i.e.,

$$H(s) = \frac{D(-s)}{D(s)} \qquad (2.59)$$

The group delay of this filter is twice that of the corresponding lowpass filter, e.g.,

$$\tau_{gAP} = 2\tau_{gBessel}.$$

The ripple will also be twice as large as in the lowpass filter.

2.7 Frequency Transformations

In the following sections, we will discuss a method to compute the poles and zeros of highpass, bandpass, or bandstop filters from the poles and zeros of a lowpass filter [68]. This technique is based on a frequency transformation. These transformations, which are also called *reactance transformations*, result in a filter that meets a magnitude specification. However, the resulting filters are in general not optimal, except for highpass filters, which are optimal if the corresponding lowpass filter is optimal. Hence, bandpass and bandstop filters designed using reactance transformations result in suboptimal solutions, but the technique is often used due to its simplicity. The group delay with the lowpass filter is not retained by the transformation. Thus, it is not meaningful to frequency transform Bessel filters because the group delay is distorted.

Optimal frequency selective filters can, however, be designed by using numerical optimization techniques. We discuss how bandpass filters that are optimal with respect to the magnitude specification can be synthesized with a PolePlacer program. It is also possible to synthesize corresponding bandstop filters.

2.8 LP-to-HP Transformation

To design a highpass filter, we first design a corresponding lowpass filter according to the methods that have been described earlier. The requirement on the magnitude function of the LP filter depends on the requirement on the highpass filter.

Figure 2.49 shows the specification for the highpass filter, which shall be synthesized, and the corresponding specification for the lowpass filter, which is used in the synthesis. Note that A_{max} and A_{min} are the same in the highpass filter and the lowpass filter.

To separate the frequency variable for the lowpass filter from the corresponding frequency variable for the desired frequency transformed filter, capital

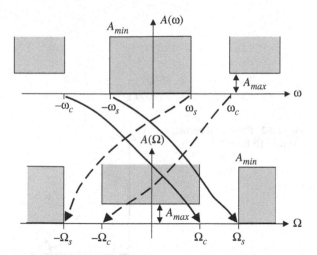

Fig. 2.49 Specificaiton for a highpass and corresponding lowpass specification

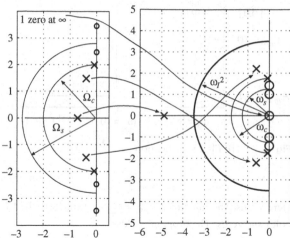

Fig. 2.50 LP-HP transformation of poles and zeros

letters and lowercase letters are used, respectively. The complex frequency for the lowpass filter and highpass filter is $S = \Sigma + j\Omega$ and $s = \sigma + j\omega$, respectively.

The relation between s and S is

$$S = \frac{\omega_I^2}{s} \qquad (2.60)$$

where ω_I is the *transformation angular frequency*. It is often favorable, as it simplifies the computations, to choose $\omega_I = \omega_c$.

Figure 2.50 shows how the poles and zeros are transformed when $\omega_I^2 = 3.5$. Poles on the real axis are mapped onto the real axis and zeros on the $j\Omega$-axis are mapped onto the $j\omega$-axis. A complex pole pair is mapped to a complex pole pair. Zeros in the LP filter at $S = \infty$ are mapped to $s = 0$.

The design process for highpass filter is as follows:

1. Determine the LP specification from the HP specification using Equation (2.60). That is, $\Omega_c = -\omega_I^2/\omega_c$, $\Omega_s = -\omega_I^2/\omega_s$ and with the same A_{max} and A_{min}.
2. Determine required filter order, poles and zeros for the lowpass filter.
3. Transform poles and zeros of the lowpass filter as well as any zeros at $S = \infty$ using Equation (2.60).

Example 2.7 Write a MATLAB program that determines necessary order and poles and zeros for a Butterworth filter that meets the specification shown in Fig. 2.51. Verify the result by plotting the attenuation and poles and zeros in the s-plane.

We get

```
Wc = 30000;    Ws = 8000;
Amax = 0.1;    Amin = 40;    % Requirements for the highpass filter
WI = Wc;                     % We select WI = Wc
Omegac = Wc;                 % Omegac = WI^2/Wc = Wc
Omegas = WI^2/Ws;
NLP = 5;       % Synthesis of 5th-order lowpass Butterworth filter
[GLP, ZLP, ZrefLP, PLP, Wsnew] = BW_POLES(Omeagc, Omegas, Amax, Amin, NLP);
QLP = -abs(PLP)./(2*real(PLP))
% Transform the LP to a HP filter
[GHP, ZHP, PHP] = PZ_2_HP_S(GLP, ZLP, PLP, WI^2)
N = NLP;                     % LP and HP filter has the same order
QHP = -abs(PHP)./(2*real(PHP))
```

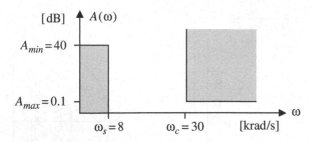

Fig. 2.51 HP filter specification

```
PHP = 1.0e+04 *                      QHP =
  -2.05989                            0.50000
  -1.666485  -  1.210772i             0.61804
  -1.666485  +  1.210772i             0.61804
  -0.6365407 -  1.959071i             1.618034
  -0.6365407 +  1.959071i             1.618034
```

Fig. 2.52 Poles and zeros for the HP filter

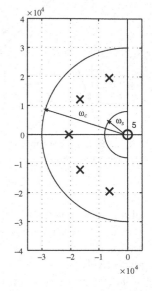

Here we use the function `PZ_2_HP_S` to transform the LP poles and zeros. The program yields

```
NLP = 5

PLP = 1.0e+04 *                      QLP =
 -4.369168                            0.50000
 -3.534731  -  2.568132i              0.61804
 -3.534731  +  2.568132i              0.61804
 -1.350147  -  4.155326i              1.618034
 -1.350147  +  4.155326i              1.618034
GLP = 1
GHP = 1
ZHP =

0
0
0
0
0
```

The poles and zeros for the highpass filter are shown in Fig. 2.52. The attenuation and the group delay for the highpass filter are shown in Fig. 2.53. The transfer function of the highpass filter is

$$H(s) = \frac{s^5}{(s + 20598.888)(s^2 + 12730.813s + 424314200)}$$
$$\cdot \frac{1}{s^2 + 33329.701s + 424314200}.$$

(2.61)

Note that the orders of the highpass and lowpass filters are the same and that a Butterworth filter of highpass type has N zeros at $s = 0$ while the corresponding lowpass filter has N zeros at $S = \infty$. The Q factors are the same in the lowpass and highpass filters.

Note that all poles and zeros are mirrored in a circle with radius ω_I^2. This means that if the poles with the LP filter lie on a circle with radius R_{p0}, as is the case for Butterworth filters, the poles with corresponding highpass filters will also lie on a circle, but with the radius r_{p0}

$$r_{p0} = \frac{\omega_I^2}{R_{p0}}.$$

An HP filter with piecewise-constant stopband requirements can be synthesized with the programs `POLE_PLACER_HP_MF_S` and `POLE_PLACER_HP_EQ_S` where MF and EQ denote maximally flat and equiripple passband, respectively.

2.8.1 LP-to-HP Transformation of the Group Delay

The group delay of the highpass filter can be expressed in terms of the group delay for the corresponding lowpass filter. We have for the highpass and lowpass filters

Fig. 2.53 Attenuation and group delay for the highpass filter

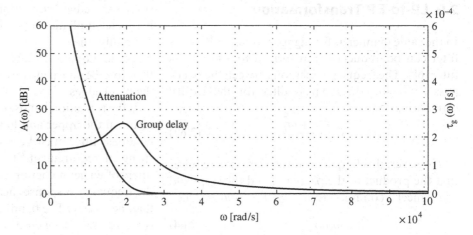

$$\tau_{gHP}(\omega) = -\frac{d}{d\omega}\Phi_{HP}(\omega) = -\frac{d}{d\omega}\Phi_{LP}(\Omega)$$

$$= -\frac{d\Omega d}{d\omega d\Omega}\Phi_{LP}(\Omega) = \frac{d\Omega}{d\omega}\tau_{gLP}(\Omega)$$

and the relation between the lowpass and highpass angular frequencies is, according to Equation (2.60), $\Omega = -\omega_I^2/\omega$. We get

$$\frac{d\Omega}{d\omega} = -\frac{d}{d\omega}\frac{\omega_I^2}{\omega} = \frac{\omega_I^2}{\omega^2}.$$

Finally, we get

$$\tau_{gHP}(\omega) = \frac{\omega_I^2}{\omega^2}\tau_{gLP}(\Omega). \tag{2.62}$$

The transfer function can be expanded into a partial faction expansion. Each term in this expression consists of a first- or second-order transfer function. The group delay for the lowpass filter is the sum of contributions for each individual pole pair

$$\tau_{gLP}(\Omega) = \sum_{n=1}^{N}\tau_{gn}(\Omega)$$

where $\tau_{g1}(\Omega) = \frac{-\sigma_p}{\Omega^2 + \sigma_p^2}$ and $\tau_{g2}(\Omega) = \frac{-2\sigma_p(\Omega^2 + r_p^2)}{\Omega^4 + 2(2\sigma_p^2 - r_p^2)\Omega^2 + r_p^4}$

for a first-order pole and second-order pole pair, respectively.

Replacing Ω with ω according with $\Omega = -\omega_I^2/\omega$ yields for a real pole

$$\tau_{gHP}(\omega) = \left(-\frac{\omega_I^2}{\sigma_p}\right)\frac{1}{\omega^2 + \frac{\omega_I^4}{\sigma_p^2}}$$

and for a second-order pole pair

$$\tau_{gHP}(\omega) = -2\frac{\sigma_p\omega_I^2}{r_p^2}\cdot\frac{\omega^2 + \frac{\omega_I^4}{r_p^2}}{\omega^4 + \frac{2\omega_I^4}{r_p^4}(2\sigma_p^2 - r_p^2)\omega^2 + \frac{\omega_I^8}{r_p^4}}.$$

The group delay for the highpass filter has a similar shape as the group delay for the lowpass filter, but distorted according to Equation (2.62). The group delay for the highpass filter approaches zero for $\omega \to \infty$ and toward a constant for $\omega \to 0$. Figure 2.54 shows the group delays for a fifth-order lowpass Butterworth filter with $A_{max} = 0.1$ dB and $\Omega_c = 1$ rad/s and the corresponding highpass filter, which is obtained with $\omega_I = \Omega_c$. Hence, they have the same passband edges. Note that the group delay for the highpass filter is small in the passband.

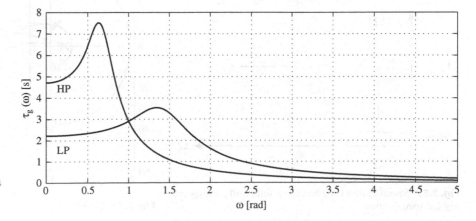

Fig. 2.54 Group delay for a lowpass and corresponding highpass filter

2.9 LP-to-BP Transformation

In the same manner as for highpass filters, a lowpass filter can be frequency transformed to a bandpass filter [68]. The frequency relation between the bandpass filter and the lowpass filter for the LP-BP transformation is

$$S = \frac{s^2 + \omega_I^2}{s}. \qquad (2.63)$$

However, this frequency transformation requires that the product of the passband and the stopband edges meet a condition for *geometric symmetry*, i.e.,

$$\omega_I^2 = \omega_{s1}\omega_{s2} = \omega_{c1}\omega_{c2} \qquad (2.64)$$

where ω_I is the *transformation angular frequency*.

Figure 2.55 shows the specifications for a bandpass filter and the specification for a corresponding lowpass filter.

The relations between the edges in the LP and BP filters are

$$\begin{aligned} \Omega_c &= \omega_{c2} - \omega_{c1} \\ \Omega_s &= \omega_{s2} - \omega_{s1}. \end{aligned} \qquad (2.65)$$

The bandwidth of the lowpass filter is equal to the bandwidth of the bandpass filter, and the stopband edge corresponds to the difference between stopband edges for the bandpass filter. Normally, the specification does not meet the condition in Equation (2.64). The specification must therefore be sharpened by changing at least one of the band edges.

If the transition bands or the attenuation requirements in the upper and lower stopbands differ, the bandpass filter must be designed to meet the most stringent requirement. This means that bandpass filters often get higher order than necessary because the frequency response meets a higher requirement than necessary. The bandpass filter that is obtained is a *geometric symmetric bandpass filter*.

Figure 2.56 illustrates how the poles and zeros are mapped by the LP-BP transformation. A complex conjugate pole pair is mapped to two complex conjugate pole pairs, both with the same Q factors.

From Fig. 2.56 it is obvious that this way of designing bandpass filters results in a filter with the same number of zeros in the upper and lower stopbands. An optimizing program must be used to design a more complicated bandpass filter with, for example, different number of zeros in the two stopbands.

The design process for geometric symmetrical bandpass filters is

1. Compute the LP specification from the BP specification. If needed, change the band edges so that the symmetry constraint is satisfied.

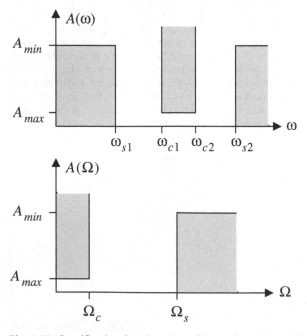

Fig. 2.55 Specification for a bandpass filter and corresponding LP specification

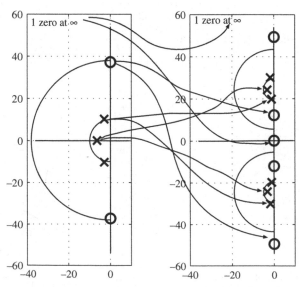

Fig. 2.56 LP-BP transformation of poles and zeros

2. Determine necessary order and the poles and zeros of the LP filter.
3. Transform the poles and zeros of the LP filter. Transform also the zeros at $S = \infty$.

Example 2.8 Write a MATLAB program that determines necessary order and poles and zeros for a Cauer filter that meets the specification shown in Fig. 2.57. Validate the result by plotting the attenuation and poles and zeros in the *s*-plane.

The requirement of geometric symmetry, Equation (2.64), is not met because $\omega_{c1}\omega_{c2} = 800$ [krad/s]^2 and $\omega_{s1}\omega_{s2} = 720$ [krad/s]^2. The symmetry requirement can be met, i.e., if the lower transition band is made smaller, i.e., ω_{s1} is changed so $\omega_{s1} = 13.333333$ krad/s.

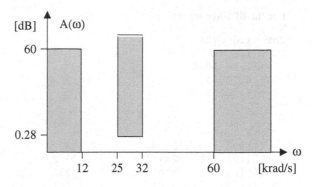

Fig. 2.57 BP filter specification

```
Wc1 = 25000;     Wc2 = 32000;              % Requirement for the bandpass filter
Ws1 = 12000;     Ws2 = 60000;
Amax = 0.28;     Amin = 60;
if Wc1*Wc2    <= Ws1*Ws2;                  % Modify band edges if needed
        Ws2 = Wc1*Wc2/Ws1;
else
        Ws1 = Wc1*Wc2/Ws2;
end
WI2 = Wc1*Wc2;
Omegac = Wc2 - Wc1;                        % Requirements for lowpass filter
Omegas = Ws2 - Ws1;
NLP = CA_ORDER(Omegac, Omegas, Amax, Amin)% Synthesis of LP filter
NLP = 3;                                   % Select the next higher integer
[GLP, ZLP, R_ZEROSLP, PLP, Wsnew] = CA_POLES(Omegac, Omegas, Amax, Amin, N);
QLP = -abs(PLP)./(2*real(PLP))
                                           % Transform the lowpass to a bandpass filter
[GBP, ZBP, PBP] = PZ_2_G_SYM_BP_S(GLP, ZLP, PLP, WI2);

figure(1)
PLOT_PZ_S(ZBP, PBP, 0,0, -40000, 10000, 80000)
ZBP
PBP
GBP
QBP = -abs(PBP)./(2*real(PBP))
W = [0:100:100000];
H = PZ_2_FREQ_S(GBP, ZBP, PBP, W);

figure(2)
axis_Amax = 80; axis_Tg_max = 1.6*10^-3;
Att = MAG_2_ATT(H);                        % Compute the attenuation
Tg = PZ_2_TG_S(GBP, ZBP, PBP, W);   % Compute the group delay
PLOT_A_TG_S(Att, Tg, W, axis_Amax, axis_Tg_max);
set(gca, 'FontName', 'times', 'FontSize', 16);
```

We use the function PZ_2_G_SYM_BP_S to synthesize a geometric symmetrical bandpass filter. The program yields

```
Ws1 = 13333.333
NLP = 3
ZLP = 1.0 e+04 *
  0 - 5.064487i
  0 + 5.064487i
```

```
PLP = 1.0 e+03 *            QLP =
  -2.563182 - 7.566248i       1.558365
  -2.563182 + 7.566248i       1.558365
  -5.257168                   0.500000

GLP = 1.3080498 e+02
```

For the BP filter we get

```
N = 6
ZBP =1.0 e+04 *
  0 - 1.264104i
  0 + 1.264104i
  0 - 6.328590i
  0 + 6.328591i
  0

PBP=1.0 e+04 *            QBP =
-0.111152 - 2.472474i     11.13352
-0.111152 + 2.472474i     11.13352
-0.145166 + 3.22910i      11.13352
-0.145166 - 3.22910i      11.13352
-0.262858 - 2.81619i      5.380143
-0.262858 + 2.81619i      5.380143

GBP=1.3080498 e+02
```

Poles and zeros and two semicircles, which indicated the passband and the stopband edges for the highpass filter, are shown in Fig. 2.58. Note that every pole (or zero) in the LP filter is mapped to a complex conjugate pole (or zero) pair in the BP filter while the zero at $S = \infty$ in the LP filter is mapped to $s = 0$ and $s = \infty$. There are equally many zeros in the upper and lower stopband. The order for the bandpass filter is twice as high as for the corresponding lowpass filter, and the Q factors are much larger. The transfer function is

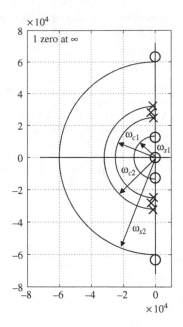

Fig. 2.58 Poles and zeros for the bandpass filter

$$H(s) = \frac{130.80050s(s^2 + 15979600)}{(s^2 + 2223.036s + 61254838)(s^2 + 2903.327s + 1044815429)}$$
$$\cdot \frac{(s^2 + 4005106465)}{(s^2 + 5257.168s + 8000000000)}.$$

(2.66)

Note that the design margin for Cauer filters is used so the stopband edge is reduced and the lower stopband edge has been increased so an infinite zero pair lies in the specification's lower transition band. The attenuation and the group delay for bandpass filter are shown in Fig. 2.59.

The group delay of the LP filter is mapped to two peaks for the BP filter, see Problem 2.34. Note that the group delay is largest at the lower end of the passband. The impulse response is shown in Fig. 2.60. The frequency of the ringing is related to the center frequency of the bandpass filter, and the envelope is related to the impulse response of the corresponding lowpass filter.

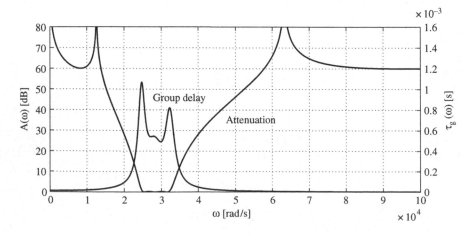

Fig. 2.59 Attenuation and group delay for the bandpass filter in Examples 2.8

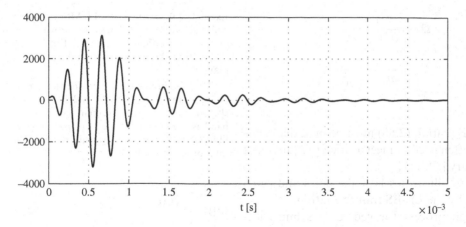

Fig. 2.60   Impulse response for the bandpass filter

2.10 LP-to-BS Transformation

Bandstop filters can be designed in a similar way as the bandpass filter [68]. Figure 2.61 shows the specifications for a bandstop filter and corresponding lowpass filter.

The frequency relation between the bandstop filter and lowpass filter using the LP-BS Ltransformation is

$$S = \frac{\omega_I^2 s}{s^2 + \omega_I^2}.\qquad(2.67)$$

The frequency transformation of BS filters yields a geometric symmetrical BS filter, i.e., the frequency transformation results in that the product of passbands and respective stopband edges will meet the *geometric symmetry* constraint

$$\omega_I^2 = \omega_{s1}\omega_{s2} = \omega_{c1}\omega_{c2}\qquad(2.68)$$

where ω_I is the *transformation angular frequency*. The relations between the LP and BS filters band edges are

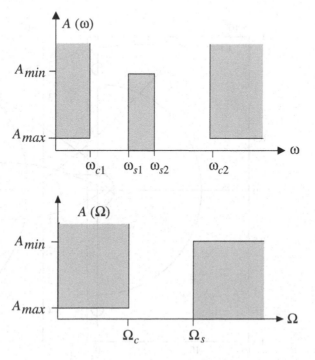

Fig. 2.61   Specification for a bandstop filter and corresponding LP specification

$$\Omega_c = \frac{\omega_I^2}{\omega_{c2} - \omega_{c1}}$$

$$\Omega_s = \frac{\omega_I^2}{\omega_{s2} - \omega_{s1}} \qquad (2.69)$$

Because Equation (2.68) must be met, the band-stop filters often need to meet a stricter requirement than necessary [29, 112].

Figure 2.62 illustrates how the poles and zeros are mapped by the LP-BS transformation.

The design process for geometric symmetrical bandstop filters is

1. Compute the LP specification from the BS specification. If needed, adjust the band edges to meet the symmetry constraint.
2. Determine necessary order and poles and zeros for the LP filter.
3. Transform the poles and zeros of the LP filter. Transform also any zeros at $S = \infty$.

Example 2.9 Write a MATLAB program that determines necessary order and poles and zeros for a Cauer filter that meets the specification shown in Fig. 2.63. Verify the result by plotting the attenuation and poles and zeros in the s-plane.

The requirement on geometric symmetry, Equation (2.64), yields $\omega_{c1}\omega_{c2} = 540$ [krad/s]^2 and $\omega_{s1}\omega_{s2} = 650$ [krad/s]^2. The symmetry requirement can be met if the transition band is made smaller, i.e., ω_{s1} is changed to $\omega_{s1} = 20.769$ krad/s, as it is the lower transition band that determines the filter order. We get

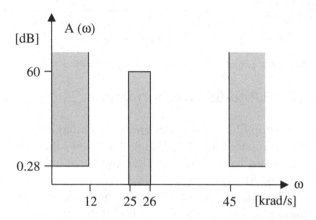

Fig. 2.63 BS filter specification

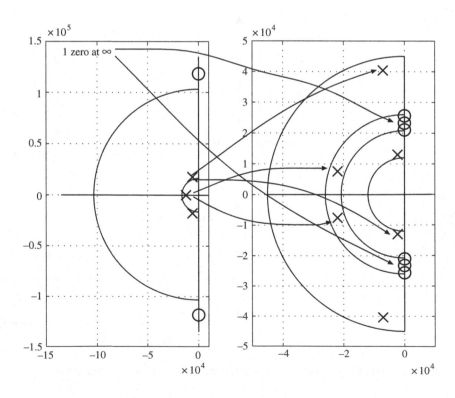

Fig. 2.62 LP-BP transformation of poles and zeros

```
Wc1 = 12000;    Wc2 = 45000;
% Requirement for the bandstop filter
Ws1 = 25000;    Ws2 = 26000;
Amax = 0.28;    Amin = 60;
% Modify band edges if needed
if Wc1*Wc2 >= Ws1*Ws2;
        Ws2 = Wc1*Wc2/Ws1;
else
        Ws1 = Wc1*Wc2/Ws2;
end
WI2 = Wc1*Wc2;
% Requirements for lowpass filter
Omegac = WI2/(Wc2 - Wc1)
Omegas = WI2/(Ws2 - Ws1)
% Synthesis of lowpass filter (Cauer)
NLP = CA_ORDER(Omegac, Omegas, Amax, Amin)
NLP = 3;                        % Select next higher integer
[GLP, ZLP, R_ZEROSLP, PLP, Wsnew] = CA_POLES(Omegac, Omegas, Amax, Amin, N);
QLP = -abs(PLP)./(2*real(PLP))
                                % Transform the lowpass to a bandstop filter
[GBS, ZBS, PBS] = PZ_2_G_SYM_BS_S(GLP, ZLP, PLP, WI2);
QBS = -abs(PBS)./(2*real(PBS))

figure(1)
PLOT_PZ_S(ZBS, PBS, 0,0, -50000, 10000, 50000);
alfa = linspace(pi/2, 3*pi/2, 200);
plot(Wc1*cos(alfa), Wc1*sin(alfa), '_', 'linewidth', 1);
plot(Wc2*cos(alfa), Wc2*sin(alfa), '_', 'linewidth', 1);
plot(Ws1*cos(alfa), Ws1*sin(alfa), '_', 'linewidth', 1);
plot(Ws2*cos(alfa), Ws2*sin(alfa), '_', 'linewidth', 1);
N = 2*NLP              % The BS filter has the order 2*NLP
ZBS
PBS
GBS
W = [0:10:100000];
H = PZ_2_FREQ_S(GBS, ZBS, PBS, W);

figure(2)
axis_Amax = 80;      axis_Tg_max = 0.8*10^-3;
Att = MAG_2_ATT(H);                        % Compute the attenuation
Tg = PZ_2_TG_S(GBS, ZBS, PBS, W);          % Compute the group delay
PLOT_A_TG_S(Att, Tg, W, axis_Amax, axis_Tg_max);
set(gca, 'FontName', 'times', 'FontSize', 16);
xtick([0:10000:100000])
```

We get

N = 6
ZBS = 1.0e+04 *
  0 - 2.563104i
  0 + 2.563104i
  0 - 2.106821i
  0 + 2.106821i
  0 - 2.323790i
  0 + 2.323790i

PBS = 1.0e+04 *
  -0.2257678 - 1.298334i
  -0.2257678 + 1.298334i

  -0.7020074 - 4.037086i
  -0.7020074 + 4.037086i
  -2.197004 - 0.7570833i
  -2.197004 + 0.7570833i

GBS = -1.000000e+00

QLP =	QBS =
1.558365	2.918536
	2.918536
1.558365	2.918536
	2.918536
0.500000	0.5288544
	0.5288544

The poles, zeros, and semicircles, which indicate the passband and stopband edges for the bandstop filter, are shown in Fig. 2.64. The Q factors are not affected by the transformation. Because the MATLAB routine for the Cauer filter uses the design margin to lower the stopband edge for the lowpass filter, the width for the corresponding stopband will increase, which is evident as two finite zeros lie in the transition band. The stopband thus becomes larger than necessary.

The attenuation and the group delay for the bandstop filter are shown in Fig. 2.65. Note that the group delay becomes largest in the lower transition band.

The transfer function is

$$H(s) = \frac{-(s^2 + 6.569500 \cdot 10^8)(s^2 + 5.400000 \cdot 10^8)}{(s^2 + 4.394007 \cdot 10^4 s + 5.400000 \cdot 10^8)(s^2 + 1.404015 \cdot 10^4 s + 16.79088 \cdot 10^8)}$$

$$\cdot \frac{(s^2 + 4.438669 \cdot 10^8)}{s^2 + 0.4515357 \cdot 10^4 s + 1.736657 \cdot 10^8}.$$

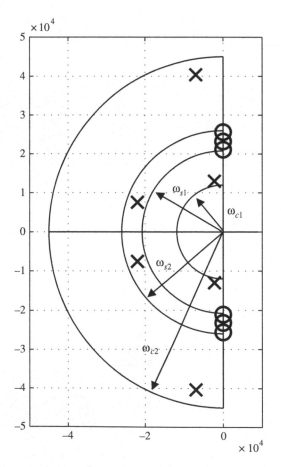

Fig. 2.64 Poles and zeros for the bandstop filter

2.11 Piecewise-Constant Stopband Requirement

Frequency transformations of lowpass filter to bandpass filter result in both stopbands meeting the same attenuation requirement A_{min}. Furthermore, one of the transition bands normally becomes smaller than necessary due to the geometric symmetry constraint. The filter, thus, gets unnecessary high order. In many applications, the requirements are different in the two stopbands and the requirements usually vary, which is shown in Fig. 2.66.

With the help of an optimizing program of the type *Pole-Placer*, the poles and zeros can be determined to a filter that meets a partially constant stopband requirement. The passband can be chosen with equiripple or as a maximum flat. By choosing a suitable number of zeros in both stopbands, the filter order can be minimized.

A way of optimizing the filter, so the design margin is used well, is by making the distances d_i between the attenuation function and the specification equal. It is however not certain that the program always will find such a solution. In such cases, we have to try with another set of the start values for the zeros or increasing the number of zeros if the filter order is too small. There are, however, in both cases certain limits on how the zeros can be placed in order for the transfer function to be realized with an LC network.

Example 2.10 Below is shown the use of POLE_PLA-CER_BP_MF_S for a maximum flat passband. Using the program POLE_PLACER_BP_EQ_S, a **BP** filter with equiripple passband can be synthesized. Note that the program is sensitive to the initial positions of the zeros.

Fig. 2.65 Attenuation and group delay for the bandstop filter in Example 2.9

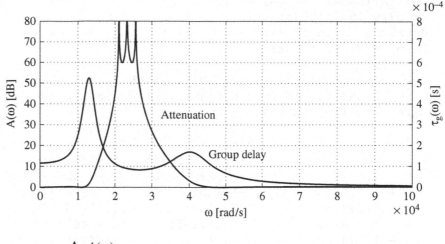

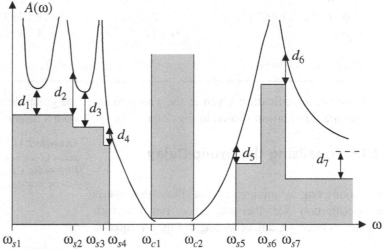

Fig. 2.66 Bandpass filter with piecewise-constant stopband requirement

```
% POLE_PLACER_BP_MF_S with maximally flat passband
Amax = 0.5;
Wc1 = 3;         Wc2 = 4;         % Passband edges 3 and 4 rad/s
Amin_low = [40];                  % 40 dB between 0 and 2 rad/s
Wstep_low = [2];
Amin_high = [35 60];              % 35 dB between 5 and 6 rad/s
                                  % 60 dB from 6 rad/s and higher
Wstep_high = [5 6];
Wi_low = [0.4];                   % Initial finite zeros in the lower stopband
Wi_high = [7 8];                  % and in the upper stopband
NIN = 1;                          % One zero at infinity
NZ = 1;                           % One zero at the origin
[G, Z, P, dopt] = POLE_PLACER_BP_MF_S(Amax, Wc1, Wc2, ...
Amin_low, Amin_high, Wstep_low, Wstep_high, Wi_low, Wi_high, NZ, NIN)
W = (0:1000)*15/1000;
H = PZ_2_FREQ_S(G, Z, P, W);
Att = MAG_2_ATT(H);
subplot('position', [0.08 0.4 0.90 0.5]);
```

```
PLOT_ATTENUATION_S(W, Att)
axis([0 15 0 100]); color=[0.7 0.7 0.7]; % Gray
patch([0 Wstep_low(1) Wstep_low(1) 0], [0 0 Amin_low(1) ...
Amin_low(1)], color);
V=axis;
patch([Wstep_high(1)V(2)V(2)Wstep_high(2)Wstep_high(2)...
Wstep_high(1)], [0 0 Amin_high(2) Amin_high(2) Amin_high(1)...
Amin_high(1)], color);
```

```
Z=                             P =
    0                             -2.328055e-01 - 4.103456e+00i
    0 - 1.777425i                 -2.328055e-01 + 4.103456e+00i
    0 + 1.777425i                 -6.190255e-01 + 3.779765e+00i
    0 - 5.678486i                 -6.190255e-01 - 3.779765e+00i
    0 + 5.678486i                 -6.267145e-01 + 3.224695e+00i
    0 - 7.574665i                 -6.267145e-01 - 3.224695e+00i
    0 + 7.574665i                 -2.323394e-01 + 2.894913e+00i
    ∞                             -2.323394e-01 - 2.894913e+00i

G=1.631020e-02                 dopt=1.008892e+00
```

With the specification given in the program we obtain the attenuation shown in Fig. 2.67.

2.12 Equalizing the Group Delay

By connecting a minimum-phase filter that meets a specification for the magnitude function with an allpass filter, we can, according to Fig. 2.4, equalize the group delay so that the two combined filters meet both a magnitude and group delay specification.

Example 2.11 The program below computes the poles and zeros of an allpass filter for equalizing the group delay of a fifth-order Cauer filter in Example 2.4. The frequency range has been normalized to $\omega_c = 1$.

```
% Determine poles and zeros for the Cauer lowpass filter
Wc=1; Ws=56/40;
Amax=0.28029; Amin=40;
N=CA_ORDER(Wc, Ws, Amax, Amin);
N=5;
[GLP, ZLP, R_ZEROSLP, PLP, Wsnew]=CA_POLES(Wc, Ws, Amax, Amin, N);
W1=0; W2=1;              % Equalization range=passband
Nap=7;                  % 7th-order allpass filter
```

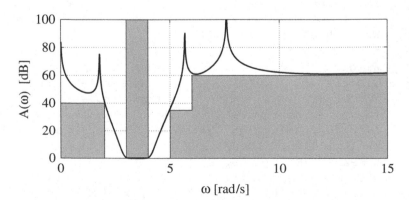

Fig. 2.67 Bandpass filter with a piecewise-constant stopband requirement

```
[PAP, Wpas]=EQ_TG_LP_S(W1, W2, GLP, ZLP, PLP, Nap);

figure(1)
PLOT_PZ_S(-PAP, PAP, Wc, Ws,-1.5, 0.5, 2.5)
PLOT_PZ_S(ZLP, PLP, Wc, Ws, -1.5, 0.5, 2.5)
TgH=PZ_2_TG_S(GLP, ZLP, PLP, Wpas);
TgAP=PZ_2_TG_S(1, -PAP, PAP, Wpas);

figure(2);
subplot('position', [0.08 0.4 0.90 0.5]);
plot(Wpas, TgAP + TgH, 'linewidth', 2)
hold on; grid on; axis([0, 1, 0, 25]);
plot(Wpas, TgH, 'linewidth', 2)
plot(Wpas, TgAP, 'linewidth', 2)
```

The poles and zeros for the allpass filter are

```
ZAP =
0.1834844 - 0.560061i
0.1834844 + 0.560061i
0.1873279 - 0.2830502i
0.1873279 + 0.2830502i
0.1649636 - 0.84411111i
0.1649636 + 0.84411111
0.1921410

PAP =
-0.1834844 - 0.560061i
-0.1834844 + 0.560061i
-0.1873279 - 0.2830502i
-0.1873279 + 0.2830502i
-0.1649636 - 0.84411111i
-0.1649636 + 0.84411111
-0.1921410
```

Note that zeros of an allpass filters lie in the right half-plane and that they are mirror images in the $j\omega$-axis of corresponding poles. That is, poles and zeros of the allpass filter have the same imaginary part and the same real part except for the different signs. Typically, most of the poles of the allpass filter lie closer to the $j\omega$-axis than the poles in the lowpass filter. That is, most of the Q factors are higher than in the lowpass filter.

Figure 2.69 shows the resulting group delays for the lowpass filter in Example 2.4, allpass filter, and resulting overall group delay using a seventh-order allpass filter. Figure 2.68 shows the pole-zero configurations for the lowpass and allpass filters. Synthesis of the allpass filter can be made with the help of the above optimizing program, which yields an approximately equiripple solution. In this case, the minimax error is about 1.7372.

Fig. 2.68 Pole-zero configuration for the lowpass and allpass filters

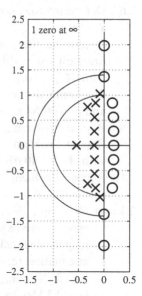

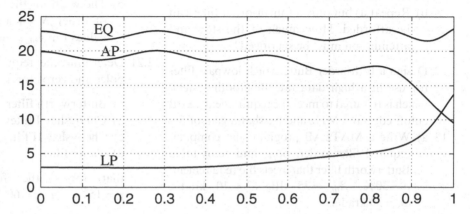

Fig. 2.69 Group delay for a lowpass filter, allpass filter, and the resulting group delay

2.13 Problems

2.1 a) Derive an expression for the required filter order for a Butterworth filter.
     b) Derive an expression for the poles for a Butterworth filter.

2.2 Derive the relation between A_{max}, ε, and ρ.

2.3 The poles in a normalized Butterworth filter are denormalized by multiplying with the factor, r_{p0}, and not with the passband edge, ω_c, which is done with the other approximations. Determine the attenuation at $\omega = r_{p0}$.

2.4 Show that the first $2N-1$ derivatives of the magnitude function squared of a Butterworth filter are zero at $\omega = 0$.

2.5 Determine how the impulse and step responses are affected by a scaling of the frequency with a factor k.

2.6 Determine the extreme values of $T_n(x) = \cos(n\,\mathrm{acos}(x))$.

2.7 Determine the poles and zeros and transfer function for a third-order Butterworth filter with $A_{max} = 0.1$ dB and $\omega_c = 5$ Mrad/s and determine the attenuation at $\omega = 10$ Mrad/s and $\omega = 20$ Mrad/s.

2.8 Repeat Problem 2.7 for a Chebyshev I filter.

2.9 Repeat Problem 2.7 for a Chebyshev II filter with $A_{min} = 40$ dB.

2.10 Repeat Problem 2.7 for a Cauer filter with $A_{min} = 40$ dB.

2.11 a) Determine the required filter order and the poles and zeros for a filter with maximally flat magnitude function that meets the following attenuation requirement: $A_{max} = 0.40959$ dB ($\rho = 30\%$), $A_{min} = 50$ dB, $\omega_c = 3$ Mrad/s, $\omega_s = 12$ Mrad/s.
     b) Repeat a) but only a fourth-order filter can be afforded. How much of the stopband attenuation must be sacrificed?

2.12 Design a fifth-order Butterworth lowpass filter for use in a high data rate Bluetooth system, which is required to meet the requirements: 3 dB cutoff edge at 1 MHz and passband gain of 5.

2.13 a) Write a MATLAB program that computes required filter order, poles and zeros for a Butterworth filter that meets the requirement: $\rho = 30\%$, $A_{min} = 35$ dB, $\omega_c = 10$ krad/s, $\omega_s = 30$ krad/s.

b) Validate the program by plotting the attenuation.
c) Plot the group delay.
d) Plot the poles and zeros in the s-plane.
e) Plot the impulse and step responses in the same diagram.

2.14 Repeat Problem 2.13 for a Chebyshev I filter.

2.15 Repeat Problem 2.13 for a Chebyshev II filter with $A_{min} = 40$ dB.

2.16 Repeat Problem 2.13 for a Cauer filter with $A_{min} = 40$ dB.

2.17 Compare the filters in Problems 2.13, 2.14, 2.15, and 2.16 and plot the magnitudes, group delays, and poles and zeros.

2.18 Determine the constant G in Example 2.2 so that the gain at $\omega = 0$ is 32. Determine also the rate of attenuation increase in dB per octave at high frequencies.

2.19 Determine and compare the required filter order and the poles and zeros for a

a) Butterworth filter
b) Chebyshev I filter
c) Chebyshev II filter
d) Cauer filter

that meets the following requirement: $\rho = 15\%$, $A_{min} = 60$ dB, $\omega_c = 5$ Mrad/s, and $\omega_s = 2.5$ Mrad/s. Mark the poles and zeros in the s-plane and determine the transfer function.

2.20 Compare using MATLAB the Butterworth, Chebyshev I, Chebyshev II, and Cauer approximations with respect to

a) Required order for a given standard lowpass specification.
b) The width of the transition band for filter of order $N = 4$ and $N = 5$. Select typical values for A_{max} and A_{min}.

2.21 Determine the required filter order and the poles and zeros for a

a) Butterworth filter
b) Chebyshev I filter
c) Chebyshev II filter
d) Cauer filter

that meets the following requirements: $\rho = 15\%$, $A_{min} = 60$ dB, $\omega_c = 10$ Mrad/s,

$\omega_s = 25$ Mrad/s. Mark the poles and zeros in the s-plane and determine the transfer function.

2.22 An anti-aliasing filter in front of an analog-to-digital converter (ADC) in a video system is used to bandlimit the input signal before sampling. The passband is up to 8 MHz and the stopband edge is at 12 MHz. The acceptable passband ripple corresponding to $\rho = 5\%$ and the stopband attenuation is at least 40 dB. Select a suitable filter approximation and find the transfer function find and the poles and zeros.

2.23 LP (long-playing) records are engraved with reduced bass levels and increased treble levels because of two main reasons. Low signal frequencies require a larger groove, which has the drawback of shorter recording time and difficulties for the stylus to follow, and, thus, causes distortion. In addition, at high frequencies the stylus has difficulty to accurately follow the groove, which causes high frequency noise. RIAA (Recording Industry Association of America) has standardized a scheme where the high frequencies are amplified during recording to obtain a higher signal-to-noise ratio and attenuated at playback through a filter described by the RIAA frequency curve

$$N = -10\log(1 + \omega^2\tau_1^2) + 10\log(1 + \omega^2\tau_2^2) - 10\log(1 + \omega^2\tau_3^2)$$

where N = level in dB; treble time constant, $\tau_1 = 75$ μs; medium time constant, $\tau_2 = 318$ μs; and bass time constant, $\tau_3 = 3.180$ ms. Find the corresponding transfer function and its poles and zeros. The attenuation should be normalized to 0 dB at 1 kHz.

2.24 Find a second-order allpass transfer function with gain $= -2$ and $\Phi_{AP}(\omega) = -\pi/4$ rad at $\omega = 2\pi$ Mrad/s.

2.25 Determine how the group delay of a lowpass filter is transformed by the LP-HP transformation and mark the positions of the poles and zeros in the s-plane.

2.26 Determine how the group delay of a lowpass filter is transformed by the LP-BS transformation.

2.27 Compute by hand the poles and zeros for a highpass filter of Butterworth type that meets the requirement: $A_{max} = 1$ dB, $\omega_c = 70$ Mrad/s, $A_{min} = 25$ dB, and $\omega_s = 20$ Mrad/s.

2.28 Design an allpass filter that minimizes the overall ripple in the group delay of the cascade of the filter in Problem 2.22 and the allpass filter to less than 5%.

2.29 A filter design program found on the Internet, a so-called shareware program, produces the following transfer function for given specification

$$H(s) = \frac{s^6 - 31s^4 + 175s^2 - 625}{s^3 - 5s^2 + (3 + 4j)s - 15 - 20j}.$$

Give several reasons why you should not buy the program, at least not in this version.

2.30 Design an LP filter with rise time of 3 ms and a delay of 1.5 ms.

2.31 Determine the poles and zeros for a Cauer filter that meets the attenuation requirement: $\rho = 15\%$ $A_{min} = 45$ dB, $\omega_c = 5.5$ Mrad/s, and $\omega_s = 3.5$ Mrad/s.

2.32 Write a MATLAB program that computes the required filter order, poles and zeros, and transfer function, for a

a) Chebyshev I filter
b) Cauer filter that meets the following requirement: $\rho = 30\%$, $\omega_c = 10$ krad/s, $A_{min} = 35$ dB, and $\omega_s = 6$ krad/s.
c) Plot the impulse and step responses.

2.33 Show that lowpass and highpass Butterworth filters can be designed so that: $|H_{LP}(j\omega)|^2 + |H_{HP}(j\omega)|^2 = 1$.

2.34 Determine how the group delay of a lowpass filter is transformed by the LP-BP transformation.

2.35 Determine the poles and zeros for a Chebyshev I filter that meets the specification: $\rho = 30\%$, $A_{min} = 60$ dB, $\omega_{s1} = 2$ krad/s, $\omega_{c1} = 6$ krad/s, $\omega_{c2} = 8.5$ krad/s, and $\omega_{s2} = 25.5$ krad/s.

2.36 Compute by hand the poles and zeros for an analog filter that meets the requirement: $\rho = 30\%$, $\omega_{c1} = 10$ krad/s, $\omega_{c2} = 12$ krad/s, $A_{min} = 35$ dB, $\omega_{s1} = 5$ krad/s, and $\omega_{s2} = 32$ krad/s. The filter approximation should be of Chebyshev I type.

2.37 Write a MATLAB program that computes the required filter order and the poles and zeros as well as the transfer function for a

a) Butterworth filter
b) Chebyshev I filter
c) Chebyshev II filter
d) Cauer filter
   that meets the requirement: $\rho = 30\%$, $\omega_{c1} = 10$ krad/s, $\omega_{c2} = 12$ krad/s, $A_{min} = 35$ dB, $\omega_{s1} = 5$ krad/s, and $\omega_{s2} = 17$ krad/s.

2.38 Add plotting of the impulse and step responses to the MATLAB program developed in Problem 2.37.

2.39 Write a MATLAB program that computes the required filter order and the poles and zeros as well as the transfer function for a Chebyshev I filter that meets the specification:
$$A_{max1} = 0.409586 \text{ dB} \quad \omega_{c1} = 2\pi\, 49.5 \text{ krad/s}$$
$$A_{max2} = 0.409586 \text{ dB} \quad \omega_{c2} = 2\pi\, 50.5 \text{ krad/s}$$
$$A_{min} = 29 \text{ dB} \quad \omega_{s1} = 2\pi\, 48.5 \text{ krad/s}$$
$$\omega_{s2} = 2\pi\, 51.5 \text{ krad/s}$$

2.40 Determine the required filter order and the poles and zeros for a

a) Butterworth filter
b) Chebyshev I filter
c) Chebyshev II filter
d) Cauer filter
   that meets the requirement: $\omega_{s1} = 3$ krad/s, $\omega_{c1} = 5$ krad/s, $\omega_{c2} = 8$ krad/s, and $\omega_{s2} = 15$ krad/s.

Mark the poles and zeros in the s-plane and determine the transfer function.

2.41 Design an equiripple passband filter with $\rho = 30\%$ and $\omega_c = 15$ Mrad/s that meets the following piecewise-constant stopband requirement: $A_{min1} = 65$ dB at $\omega_{s1} = 23$ Mrad/s, and $A_{min2} = 40$ dB at $\omega_{s2} = 30$ Mrad/s.

2.42 Use the POLE_PLACER_BP_EQ_S program to design a BP filter that meets the specification in Problem 2.40 and discuss the pros and cons of using the geometric symmetric LP-BP transformation.

Chapter 3
Passive Filters

3.1 Introduction

The transfer functions that were discussed in Chapter 2 can be realized with so-called *LC* filters, or more correctly *RLC* filters, where *RLC* refer to resistors, inductors, and capacitors. Such filters are normally realized with only passive elements and are implemented with resistors, coils, and capacitors. A passive or lossless circuit element cannot increase the signal energy. Passive *LC* filters belong to the oldest implementation technologies but still play an important role as they are being used in large volumes and are used as prototypes for the design of advanced frequency selective filters.

Passive *LC* filters were first studied by George A. Campbell (USA) and Karl Wagner (Germany) in 1915 in connection with transmission lines and vibrations in mechanical systems. Electrical filter models were also successfully used to improve the frequency response of electromechanical systems such as loudspeakers and phonographs. The most important contributions to filter theory were made in the 1920's and 1930's by Otto J. Zobel, USA, Ronald M. Foster, USA, Wilhem E.A. Cauer, Germany (1900–1945), Otto Brune, South Africa (1901–1982), Hendrik W. Bode, USA (1905–1982), and Sidney Darlington, USA (1906–1997) to mention just a few [30].

A drawback with *LC* filters is that it is difficult to integrate resistors and coils with sufficiently high quality in integrated circuit technology. *LC* filters are for this reason not well suited for systems that are implemented in an integrated circuit.

A more important aspect is that the theory for *LC* filters is used as a basis for realizing high-performance frequency selective filters. This is the case for mechanical, active, discrete-time, and *SC* filters as well as for digital filters. The main reason is that the magnitude function for a well-designed *LC* filter has low sensitivity for variations in the element values.

3.2 Resonance Circuits

An ideal resonator has poles on the $j\omega$-axis. In practice, though, there will always be some losses so the poles will lie in the left half plane. In some literature, the poles are referred to as the natural modes of the network. A measure of quality for a resonance circuit is the Q factor.

Definition 3.1 The Q factor is

$$Q = 2\pi \frac{\text{Maximally stored energy per cycle}}{\text{Energy loss per cycle}}. \quad (3.1)$$

3.2.1 Q *Factor of Coils*

Large inductors are typically implemented with a coil with a magnetic ferrite-based core whereas smaller inductors may be implemented without magnetic cores. The most important problem with an *LC* filter is due to losses in the inductors. These are mainly due to two phenomena. The first is the wire resistance, $r << L$, in the coils, which is

L. Wanhammar, *Analog Filters Using MATLAB*, DOI 10.1007/978-0-387-92767-1_3,
© Springer Science+Business Media, LLC 2009

frequency dependent due to the skin effect. The second component is due to losses in the ferrite core, which exhibits both frequency and current dependence.

A coil can be modeled with a series resistance according to Fig. 3.1. There are also parasitic capacitances, so-called stray capacitances, between the different wires. These stray capacitances can be modeled with a parallel capacitance C_L [26, 138]. The coil is only usable for frequencies where C_L can be neglected, i.e., for frequencies well below the resonance frequency. A coil has the Q vfactor

Fig. 3.1 Model for a coil with ferrite core

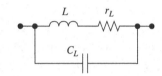

$$Q_L = \frac{\omega L}{r_L}. \tag{3.2}$$

For an ideal coil we have $r_L = 0$. The Q factor should be large for a good coil, typically larger than 100. A too low Q factor impairs the filter's frequency response. The attenuation will increase, especially near the passband edge, and decrease in the stopband. The Q factors of the coils should therefore be sufficiently large at the passband edge. For low frequencies (< 100 Hz), it is difficult to manufacture good coils due to wire losses. The coils also become heavy and take up large space. Good coils can, however, be manufactured for higher frequencies.

Coils with magnetic cores are nonlinear when the current through the coil becomes too large. Coils without magnetic core can be used for frequencies above 10 MHz. Small wire coils for surface assembly in the format 0402, i.e., $1.19 \times 0.64 \times 0.61$ mm, can be manufactured with inductances in the range 1–40 nH and with 5% or 10% tolerances. At high frequencies, wire coils have relatively low Q factors. For example, at 1.8 GHz, a wire coil for surface assembly at 2 nH has Q factors less than 100, which is not enough to implement LC filters that shall meet strict requirements on frequency selectivity.

LC filters for higher frequencies are often implemented directly on the printed circuit board or on a thick-film substrate using coils of spiral type without a magnetic core.

Coils of spiral type are often used in filters and oscillators for high frequencies but are, however, difficult to implement in integrated circuit technologies due to losses that limit the Q factor for coils on a silicon substrate to about 10. Integrated coils have a great potential use of filters and oscillators in radio applications, i.e., in the frequency range 1 to 5 GHz. Because these applications demand considerably higher Q factors, extensive research is for the moment pursued to develop methods of improving the Q factor with integrated coils or replacing these with active components.

3.2.2 Q Factor for Capacitors

A simple model for the losses of a capacitor is a parallel resistance, r, which usually is caused mainly by leakage across the dielectric. The maximum accumulated energy for a sinusoidal voltage over the capacitor is $0.5\,CV^2$ and the losses during a cycle are $\pi V^2/r\omega$ where V is the r.m.s. value. For a capacitor, the Q factor thus becomes

$$Q_C = \omega C r. \tag{3.3}$$

The Q factor for most capacitors is very large. The parallel resistance, r, is usually inversely proportional to the capacitance, i.e.,

$$r = \frac{k}{C} \tag{3.4}$$

where k can be as high as 10^4. Hence, the Q factor for a high-quality capacitor may be $Q_C = \omega k$, which is a very high value.

High-quality discrete capacitors with tolerances 1% and $Q > 1000$ are mica capacitors, which consist of silver-plated mica plates that are encapsulated in a plastic compound. Typically, capacitance values are below 1 nF. They have small sizes and can withstand high temperatures, and the temperature coefficients of these capacitors lie in the range 0 to 10^{-4} 1/°C. Polystyrene and polypropylene capacitors are used in high-quality filters. They typically have $Q's$ >10,000 and typical temperature coefficients of 10^{-4} 1/°C and $2\,10^{-4}$ 1/°C, respectively.

Even resistors have different types of defects, which can be modeled with a series inductor and a parallel capacitor, but their effect on the frequency response is relatively small compared with that of coils. The component's values are affected by the temperature and by aging. It is therefore important at an early stage in the design process to estimate the effect of component tolerances and parasitic elements such as stray capacitances, etc. In Section 3.4, these questions are discussed in detail. In the next section, the cause for the errors in the components in certain types of *LC* filter structures having only a small effect on the magnitude function is discussed.

3.3 Doubly Terminated *LC* Filters

The first usable theory for designing an *LC* filter was the *image parameter method* [48, 76], developed by G.A. Cambell, Otto Zobel, and others (1923). The method does not normally give optimum filters unless advanced methods and skilled designers are used.[1] It has nowadays fully been replaced by a more effective method, which was developed independently by W. Cauer[2] (1931) and S. Darlington and H. Piloty (1939). The method is based on minimizing the mismatch between the load and source resistors and is known as the *insertion loss method*. The insertion loss method, however, requires long and complex computations with high accuracy [26, 29, 112, 124]. Therefore, it did not become the favored approach until computers become widely available in the mid-1960s. The insertion loss method is today the standard method for design of frequency selective filters [100, 146].

3.3.1 Maximum Power Transfer

To explain the good sensitivity properties of correctly designed *LC* filters, we first consider the power transferred from the source to the load in the

circuit shown in Fig. 3.2. We assume that the input signal is $v(t) = V_1 e^{j\omega t}$ where V_1 is the effective value. The dissipated power in the load is $P = Re\{V_2 I_2^*\} = Re\{|V_2|^2/Z_2^*\}$ where V_2 is the r.m.s. value.

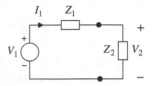

Fig. 3.2 Signal source with source and load impedances

Theorem 3.1: Maximum Power Transfer *The maximum power that can be transferred to the load impedance Z_2 in the circuit shown in Fig. 3.2 is*

$$P_{2max} = \frac{|V_1|^2}{4R_1} \tag{3.5}$$

and it is attained when $Z_2 = Z_1^*$ where $R_1 = Re\{Z_1\}$ and $R_2 = Re\{Z_2\}$.

In the case of $Z_1 = R_1 \neq Z_2 = R_2$, a transformer can be used to match the load to the source. Maximum power is transferred to the load if a transformer with the turns ratio n:1 is placed between the source and load resistors.

The input impedance to the primary side of the transformer is $n^2 R_2$ and the maximum power transfer occurs if we select $R_1 = n^2 R_2$. In this case is an equal amount of power dissipated in the resistor R_1 and R_2, and the ratio of the output and input power is 0.5.

3.3.2 Insertion Loss

Consider the circuit in Fig. 3.2 with real impedances, i.e., resistors. The maximum power transferred to the load, i.e., available from the source, is given by Equation (3.5). If a network is inserted between the source and load impedances, the power dissipated in Z_2 will be equal to or smaller than P_{2max}. Hence, there is a loss in transferred power due to the inserted network, which is referred to as *insertion loss*.

Definition 3.2 The insertion loss is

$$IL = -10 \log\left(\frac{P_2}{P_{in}}\right). \tag{3.6}$$

---

[1] Torben Laurent, Royal Institute of Technology, Sweden.

[2] The civilian Wihelm A.E. Cauer was executed by soldiers of the Red Army in the streets of Berlin in 1945. He became the victim of fate similar to that of Archimedes, who was killed in Syracuse by Roman soldiers in 212 B.C.

Thus, if $IL = 3$ dB, only 50% of the incident power from the source is delivered to the load.

It is common to consider the power that might have been dissipated in the load to be "reflected" back to the source. The *reflected power* is

$$P_r = P_{2max} - P_2. \tag{3.7}$$

In the next few sections, we will show that LC filters that are insensitive to errors in the component values can be designed by minimizing the insertion loss.

3.3.3 Doubly Resistively Terminated Lossless Networks

Consider the doubly resistively terminated network in Fig. 3.3, which is lossless, i.e., it dissipates no power. A lossless reciprocal network can be realized by using only lossless circuit elements, e.g., inductors, capacitors, transformers, and lossless transmission lines. Although other circuit elements can be used, these filters are often referred to as LC filters.

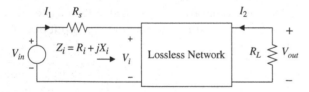

Fig. 3.3 Doubly resistively terminated LC network

We shall later discuss realizations that contain active circuit elements, which enhances the signal energy, but are used in a configuration that makes the overall circuit lossless.

The ratio of the power dissipated in the load resistor and the maximum power, given by Equation (3.5), is

$$\frac{P_{out}}{P_{out\,max}} = \frac{4R_s}{R_L}\frac{|V_{out}(j\omega)|^2}{|V_{in}(j\omega)|^2} \leq 1$$

where $V_{in}(j\omega)$ is the r.m.s. value of the sinusoidal input signal. An important observation is that the

power that the signal source can deliver to the load is limited. The upper bound for the maximal power transfer is the base for the design of filter structures with low element sensitivity.

We define the frequency response[3] as the ratio between input and output voltages, i.e., the relation between signal quantities and corresponding physical signal carrier, according to

$$H(j\omega) = \sqrt{\frac{4R_s}{R_L}\frac{V_{out}(j\omega)}{V_{in}(j\omega)}} \tag{3.8}$$

which means that the frequency response is bounded from above according to

$$\frac{P_{out}}{P_{out\,max}} = |H(j\omega)|^2 \leq 1. \tag{3.9}$$

The corresponding transfer function[4] is

$$H(s) = \sqrt{\frac{4R_s}{R_L}\frac{N(s)}{D(s)}} \tag{3.10}$$

where $D(s)$ is a strictly Hurwitz polynomial as the poles lie inside the left half of the s-plane. The polynomial $N(s)$ is either even or odd, and its zeros are the transmission zeros[5]. For a lossless ladder network, all loss poles (transmission zeros) lie on the $j\omega$-axis.

3.3.4 Broadband Matching

A more difficult and constrained problem is when either the source or load resistor or both are not purely resistive. Another case is when the network between the source and load is fixed, e.g., antenna, loudspeaker, frequency-dependent input and output impedances of an amplifier [23]. The performance can be optimized by placing lossless networks between the source and the amplifier and between the amplifier and load resistor [70]. The approximations discussed in Chapter 2 no longer yields optimal power transfer and selectivity [21, 24, 145].

---

[3]Note that in the passive filter literature, the transfer function, often referred to as the output, is divided by the input signal.

[4]The reciprocal of Equation (3.10) is often referred to as the transducer function.

[5]Often referred to as loss poles.

Formulas for the element values for Butterworth and Chebyshev I filters have been derived for more general terminations [22, 24, 145]. Theoretical limits and performance constraints of wideband matching networks have been derived by Bode-Fano [8, 28, 145].

3.3.5 Reflection Function

Note that the signal source shown in Fig. 3.3 does not deliver maximum power for all frequencies as the input impedance to the reactance network is frequency dependent. This can be interpreted as part of the maximum available power is reflected back to the source. The relationship between the power that is absorbed in R_L and the power that is reflected back to the source can be derived in the following way.

The input impedance to the *LC* network is $Z_i(j\omega) = R_i(\omega) + jX_i(\omega)$. Because the reactance network is lossless, the power into the network will be absorbed in R_L, i.e.,

$$|I_1(j\omega)|^2 R_i(\omega) = \frac{|V_{out}(j\omega)|^2}{R_L}.$$

Furthermore, we have

$$I_1(j\omega) = \frac{V_{in}(j\omega)}{R_s + Z_i(j\omega)}.$$

After some simplifications, we obtain

$$\frac{4R_s}{R_L}|H(j\omega)|^2 = 1 - \left|\frac{Z_i - R_s}{Z_i + R_s}\right|^2$$

and

$$\frac{4R_s}{R_L}|H(j\omega)|^2 = 1 - |\rho_1(j\omega)|^2 \qquad (3.11)$$

where ρ is the *reflection function*[6] for port 1.

Definition 3.3 The reflection function for port 1 is

$$\rho_1(s) = \frac{Z_i - R_S}{Z_i + R_S}. \qquad (3.12)$$

---

[6]In some literature, ρ_1 and ρ_2 are defined with the opposite signs.

The reflection function for port 2 is defined analogously

$$\rho_2(s) = \frac{Z_{i2} - R_L}{Z_{i2} + R_L} \qquad (3.13)$$

where Z_{i2} is the input impedance to port 2. Unless explicitly indicated, $\rho(s)$ will henceforth refer to port 1.

Definition 3.4 The reflection coefficient is

$$\rho = \max\{|\rho_1(j\omega)|\} \quad \text{for} \quad \omega_{c1} \leq \omega \leq \omega_{c2}. \qquad (3.14)$$

The reflection coefficient is related to the passband ripple according to Equation (2.36). The reflection functions at the ports can be written as

$$\rho_1(s) = \frac{F(s)}{D(s)} \qquad (3.15)$$

$$\rho_2(s) = \pm\frac{F(-s)}{D(s)} \qquad (3.16)$$

where the sign in Equation (3.16) is positive (negative) if $Z(s)$ is odd (even). The polynomial $D(s)$, which is equivalent to the denumerator of the transfer function, is either even or odd. In general, the polynomial $F(s)$ has no special constraints on its zeros. We will later discuss the polynomials $F(s)$ and $D(s)$ in more detail.

The voltage standing wave ratio (VSWR) is often used in the high-frequency literature. It is defined as

Definition 3.5 The voltage standing wave ratio (VSWR) is

$$VSWR = \frac{1 + |\rho|}{1 - |\rho|}.$$

3.3.6 Characteristic Function

The characteristic function plays an important role as it tends to simplify the approximation problem and is directly used in the synthesis of *LC* ladder structures.

Definition 3.6 The characteristic function is

$$C(s) = \frac{\rho(s)}{H(s)} = \frac{F(s)}{Z(s)}. \qquad (3.17)$$

$|C(s)|^2$ is proportional the ratio of the reflected power and the transferred power. Hence, the zeros of $C(s)$, i.e., zeros of $F(s)$, are the frequencies at which $P_2 = P_{2max}$, also known as the *reflection zeros*. The poles of $C(s)$ are the frequencies of infinite loss at which $P_2 = 0$, i.e., the transmission zeros of $H(s)$.

A frequency selective filter will have all of the zeros of $C(s)$ in the passband and the poles in the stopband. $F(s)$ is either even or odd, with a degree no greater than that of $P(s)$. The poles and zeros of $C(s)$ will, therefore, lie on the $j\omega$-axis, and hence $C(s)$ is either even or odd.

Sometimes the term *return loss, L_I*, is used, where

$$L_I = -20 \log(|\rho(j\omega)|). \tag{3.18}$$

The maximum return loss, over a frequency band, is a measure of the matching of the load and source in that band. The return loss is infinite when the source and load is perfectly matched.

3.3.7 Feldtkeller's Equation

Feldtkeller's equation (1938) for the lossless two-port is

$$\frac{4R_s}{R_L}|H(j\omega)|^2 + |\rho(j\omega)|^2 = 1 \tag{3.19}$$

which describes how the power is distributed in the filter, i.e.,

$$\frac{P_{out}}{P_{out\,max}} + \frac{P_r}{P_{out\,max}} = 1. \tag{3.20}$$

Alternatively, the Feldtkeller equation can be written as

$$D(s)D(-s) = F(s)F(-s) + Z(s)Z(-s). \tag{3.21}$$

$D(s)D(-s)$ is an even function of s. One of the most difficult problems in filter synthesis is to accurately compute $F(s)$ from $D(s)$ and $Z(s)$, due to unavoidable loss in accuracy in the coefficients in the sum of squared polynomials.

Theorem 3.2: *The transfer function and the reflection function with a doubly resistively terminated reactance network, which is designed for maximum power transfer, have the same poles. The zeros of $\rho(s)$, called reflection zeros, correspond to frequencies with maximum power transfer.*

LC filters, which are designed using the insertion loss method, have *maximum power transfer* at the angular frequencies (*reflection zeros*) ω_k, $k = 1, 2$ and 3, as illustrated in Fig. 3.4.

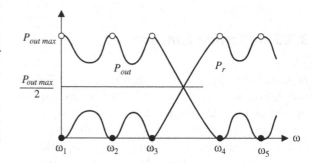

Fig. 3.4 Power transfer to R_L and reflected power as a function of frequency

In a doubly resistively terminated reactance networks the part of the P_{2max} that is not absorbed in R_L will be reflected in the reactance network back to the source. This means that the frequency response and the reflection function are *power complementary magnitude functions*. If the filter is a lowpass filter from the source to R_L, then the reflection function is a highpass filter. However, if the stopband attenuation of the highpass filter is large, the passband ripple of the lowpass filter becomes extremely small. This property with doubly resistively terminated reactance network is used in some applications.

3.3.8 Sensitivity

A doubly resistively terminated LC filter that has been designed for maximum power transfer has a very low sensitivity for errors in the element values. The ripple in the passband and passband edges changes very little if the circuit elements differ some from their nominal values. A measure of sensitivity is *relative sensitivity* of the magnitude function.

Definition 3.7 The relative sensitivity of the magnitude function is

$$S_x^{|H(j\omega)|} = \frac{x}{|H(j\omega)|}\frac{\partial|H(j\omega)|}{\partial x} = \frac{\dfrac{\partial|H(j\omega)|}{|H(j\omega)|}}{\dfrac{\partial x}{x}}. \tag{3.22}$$

The sensitivity equals the relative change in the magnitude function divided by the relative change in the element value.

It is difficult to find a simple and good measure of how the attenuation changes when several element values vary at the same time. The reason for this is that the influence of errors in different element values interacts. In fact, for a doubly resistively terminated reactance network, we will demonstrate below that they tend to cancel. We shall therefore use and interpret sensitivity measures according to Equation (3.22) with care. It is very difficult to compare different filter structures in a fair way. But it is important in an early stage of the design process to estimate the influence of component tolerances. To design a filter, that shall meet a given requirement, we must start from a filter with nominal element values that meets a stricter requirement.

3.3.8.1 Passband Sensitivity

The sensitivity is a function of the angular frequency. The sensitivity in the passband can be determined from the derivative of Feldtkeller's equation

$$\frac{4R_s}{R_L}|H(j\omega)|^2 + |\rho(j\omega)|^2 = 1 \qquad (3.23)$$

with respect to an arbitrary circuit element x. We get

$$\frac{8R_s}{R_L}|H(j\omega)|\frac{\partial|H(j\omega)|}{\partial x} + 2|\rho(j\omega)|\frac{\partial|\rho(j\omega)|}{\partial x} = 0$$

which together with Equation (3.22) yields

$$S_x^{|H(j\omega)|} = \frac{-R_L}{4R_s}\left|\frac{\rho(j\omega)}{H(j\omega)}\right|^2 S_x^{|\rho(j\omega)|}. \qquad (3.24)$$

A. Fettweis showed (1960) that the sensitivity becomes minimal in the whole passband if the filter is designed for maximum power transfer at a number of angular frequencies in the passband. At these angular frequencies, the reflection function $\rho(j\omega)$ is zero, as $Z_{in} = R_s$ and the sensitivity is therefore, according to Equation (3.24), zero. If A_{max} is small, both the reflection coefficient, according to Equation (2.36), and the magnitude of the reflection function $|\rho(j\omega)|$, according to Equation (3.11), will be small. This also leads to that the sensitivity will be

small. If the ripple is increased in the passband, the sensitivity is also increased.

That a doubly resistively terminated *LC* filter has low element sensitivity can also be realized through the following reasoning. Irrespective of if the element value is increased or decreased from its nominal value, P_{out} will decrease, since $P_{out} = P_{out\ max}$ for the nominal element value. Because the derivative is zero where the function has a maximum, i.e.,

$$\frac{\partial P_{out}}{\partial x} = 0. \quad \text{for } \omega = \omega_k \text{ and nominal element values.}$$

If there are many angular frequencies, ω_k, with maximal power transfer in the passband, the sensitivity will be low throughout the passband. This line of reasoning is called *Fettweis-Orchard's*[7] *argument* [86].

Example 3.1 In this example, we study the sensitivity of the frequency response for errors in the element values in the doubly resistively terminated ladder network shown in Fig. 3.5. The filter is a Chebyshev I filter with $A_{max} = 3$ dB. The filter has been chosen with unusually large ripple in the passband to clearly demonstrate the sensitivity for errors in the element values.

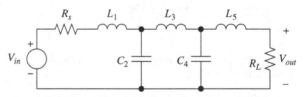

Fig. 3.5 A doubly resistively terminated Chebyshev I filter of ladder type

The element values are varied ±20% around the nominal values. Note that this variation is extremely large and is used only to clearly demonstrate the difference in element sensitivity.

$$R_s = 1, L_1 = L_5 = 3.4813, C_2 = C_4 = 0.7619,$$
$$L_3 = 4.5375, R_L = 1.$$

Because the filter is symmetrical with respect to both structure and element values, i.e., L_1 and L_5 will have the same affect on the magnitude function, which is evident in Fig. 3.6. Fully symmetrical filters, which both symmetrical structure and element values, has, in general, very good sensitivity properties. very good sensitivity properties. Note

---

[7]H.J. Orchard, USA (1922–2004).

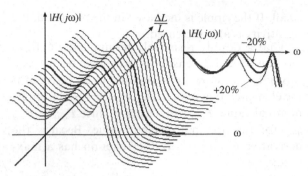

Fig. 3.6 Variation of the magnitude function due to ±20% variation in L_1 or L_5

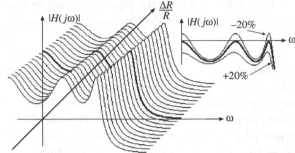

Fig. 3.9 Variation of the magnitude function due to ±20% variation in R_s

that the variation in the magnitude function is convex and that it is relatively small. The peaks in the passband lie essentially on the same level. Thus, the sensitivity is zero at these frequencies. The influence of errors in C_2 and C_4 is also small according to Fig. 3.7. A small reduction and frequency shift can be observed.

The effect of variations in source and load resistance is different even though the elements are symmetrically placed. They mainly affect the filter gain and the ripple size, which is evident from Figs. 3.9 and 3.10. This is often of little importance because we are mainly interested in the frequency selectivity.

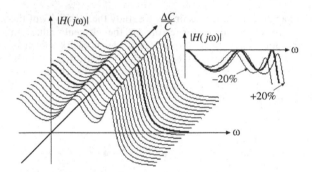

Fig. 3.7 Variation of the magnitude function due to ±20% variation in C_2 or C_4

The filter is symmetrical, independently of errors in the inductor L_3, which is placed in the center of the filter. Experience indicates that errors in such symmetrically placed circuit elements tend to have a small effect on the magnitude function. As shown in Fig. 3.8, a deviation in L_3 mainly causes a frequency shift of the highest pole pair.

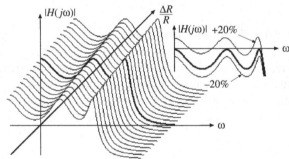

Fig. 3.10 Variation of the magnitude function due to ±20% variation in R_L

A doubly resistively terminated LC filter, which has been designed according to the insertion loss method, is optimal from a sensitivity point of view. Thus, doubly terminated ladder structures are the best filter structures from a sensitivity point of view.

Example 3.2 Compare the sensitivity for errors in the element values in the singly resistively terminated ladder network that are shown in Fig. 3.11 with the filter in Example 3.1. The filter

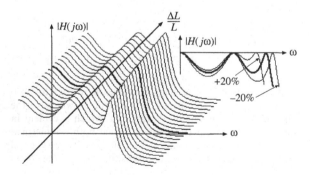

Fig. 3.8 Variation of the magnitude function due to ±20% variation in L_3

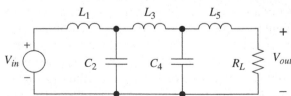

Fig. 3.11 Singly resistively terminated Chebyshev I ladder network

is a Chebyshev I filter with $A_{max} = 3$ dB. The element values vary $\pm 20\%$ around the nominal values

$$L_1 = 2.1491, C_2 = 1.3015, L_3 = 2.6227,$$
$$C_4 = 1.2501, L_5 = 1.7409, R_L = 1.$$

The effect of errors in L_1 is evident from Fig. 3.12. Note that the variation in the magnitude function is very large, especially near the passband edge. The ladder is not symmetrical with respect to element values, and L_1 and L_5 will therefore not have the same sensitivity.

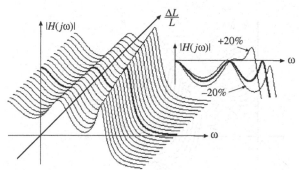

Fig. 3.14 Variation of the magnitude function due to $\pm 20\%$ variation in L_3

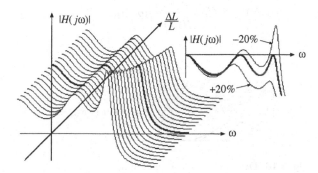

Fig. 3.12 Variation of the magnitude function due to $\pm 20\%$ variation in L_1

According to Figs. 3.13, 3.14, 3.15, 3.16, the effects of errors in the components C_2, L_3, C_4, and L_5 are also very large. More expensive components that have smaller tolerances compared with the doubly terminated filter must therefore be used in order for an implementation to comply with the specification.

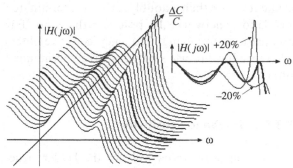

Fig. 3.15 Variation of the magnitude function due to $\pm 20\%$ variation in L_4

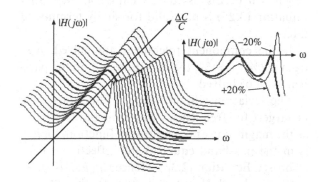

Fig. 3.13 Variation of the magnitude function due to $\pm 20\%$ variation in C_2

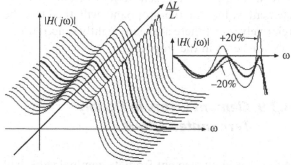

Fig. 3.16 Variation of the magnitude function due to $\pm 20\%$ variation in L_5

The effect of variations in the load resistance is evident from Fig. 3.17. Errors in the load resistor tilt the passband, unlike the doubly resistively terminated ladder network where an error only causes a change of the gain.

A singly resistively terminated ladder for which the principle of maximum power transfer is not valid has, thus, a high sensitivity in the passband. Such *LC* filters should therefore not be used because it would require more expensive components with smaller tolerances.

A singly resistively terminated *LC* ladder network is much more sensitive for variations in the

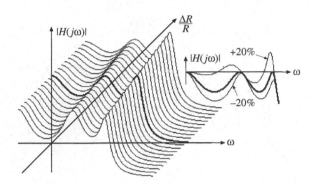

Fig. 3.17 Variation of the magnitude function due to ±20% variation in R_L

element values than a doubly resistively terminated *LC* ladder network. In practice, there are few applications involving frequency selective filters where we have to use singly resistively terminated filters.

3.3.8.2 Stopband Sensitivity

The transmission zeros will, for the ladder structures discussed in Section 3.4.4, depend on only two reactances that affect each zero. If the resonance frequency of the two components is trimmed accurately, which is easy to do, and their temperature dependency matches, then the deviation in the zero will be small. The sensitivity to changes in the element values in the stopband will therefore be relatively low if there are many tightly spaced transmission zeros in the stopband.

3.3.9 Element Errors in Doubly Terminated Filters

Any error in an element from its nominal value will cause a deviation in the attenuation. Moreover, any deviation of an element from its ideal behavior, e.g., losses in an inductor, will also cause a deviation in the attenuation. A naive approach to mitigate this problem is to use better and more accurate components. However, this may become too expensive. A better approach is to select and design a circuit that is less sensitive to errors in its components. Less

expensive components may thereby be used while still meeting the specification.

3.3.9.1 Errors in the Reactive Elements

It can be shown that the deviation shown in Fig. 3.18 in the passband attenuation for a doubly resistively terminated filter is [133]

$$\Delta A(\omega) \leq 8.69\varepsilon \frac{|\rho(j\omega)|}{|H(j\omega)|^2} \omega \tau_g(\omega) \ [\text{dB}] \qquad (3.25)$$

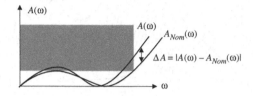

Fig. 3.18 Deviation in the attenuation

where $\varepsilon = |\Delta L/L| = |\Delta C/C|$ represent the uniformly distributed errors in the inductances and the capacitances, i.e., $(1-\varepsilon)L \leq L \leq (1+\varepsilon)L$, etc. It can be shown that ΔA is proportional to the electric and magnetic energy stored in the capacitors and inductors and that Equation (3.25) also holds for commensurate transmission line filters, which will be discussed in Chapter 4. Note that Equation (3.25) is not valid for singly terminated filters.

If the component tolerances are halved, e.g., from ±2% to ±1%, the deviation, ΔA, in the passband will be halved.

The deviation will, according to Equation (3.25), be largest for frequencies where $\omega\tau_g(\omega)$ is largest, as the magnitude of the reflection function, $|\rho(j\omega)|$ is in the passband equal to the reflection coefficient, see Equation (2.36). Hence, $|\rho(j\omega)|$ is small in the passband. Hence, a doubly resistively terminated filter with 3 dB ripple in the passband is significantly more sensitive for element errors than a corresponding filter with smaller passband ripples, e.g., 0.01 dB. Moreover, the Q factors of the poles will be smaller if the passband ripple is smaller. Thus, it is often better to design a filter with a smaller ripple at the expense of a slightly higher filter order.

3.3.9.2 Errors in the Terminating Resistors

The sensitivities with respect to R_s and R_L are [112]

$$S_{R_s}^{H(j\omega)} = -8.69\frac{\rho_1(j\omega)}{2} \qquad (3.26)$$

and

$$S_{R_L}^{H(j\omega)} = -8.69\frac{\rho_2(j\omega)}{2}. \qquad (3.27)$$

The sensitivities $S_{R_s}^{H}(j\omega)$ and $S_{R_L}^{H}(j\omega)$ are small in the passband, as $|\rho(j\omega)| \ll 1$, and equal zero for the frequencies at maximal power transfer.

For a doubly resistively terminated *LC* filter we have

$$|\rho_1(j\omega)| = |\rho_2(j\omega)| = |\rho|_{max} = \sqrt{1 - 10^{-0.1A_{max}}}.$$

The deviation in the attenuation, see Fig. 3.18, due to errors in the terminating resistors is

$$\Delta A(\omega) < 8.69\sqrt{1 - 10^{-0.1A_{max}}}\frac{\Delta R}{R} \qquad (3.28)$$

where we have assumed that the terminating resistors have the same tolerances. Hence, the effect of errors in the resistors can be reduced by selecting a small pass-band ripple and in addition Equations (3.26) and (3.27) indicate that the error becomes zero for the frequencies at maximal power transfer. For example, $A_{max} = 0.5$ dB and resistors with 2% tolerances yield $\Delta A(\omega) < 0.057$ dB. However, this error is an almost constant shift of the frequency response, i.e., essentially an error in the gain of the filter and of little concern.

3.3.9.3 Effects of Lossy Elements

The effect on the attenuation of lossy reactive elements can be estimated in terms of their Q factors where we assume that all inductors have the same Q factor and the same holds for the capacitors [70, 112]

$$\Delta A(\omega) < \frac{8.69}{2}\left(\frac{1}{Q_L} + \frac{1}{Q_C}\right)\omega\,\tau_g(\omega) + \frac{8.69}{2}\left(\frac{1}{Q_L} - \frac{1}{Q_C}\right)Im\left\{\frac{\rho_1 + \rho_2}{2}\right\} \qquad (3.29)$$

and

$$\Delta A(\omega) < \frac{8.69}{2}\left(\frac{1}{Q_L} + \frac{1}{Q_C}\right)\omega\,\tau_g(\omega) + \frac{8.69}{2}\left(\frac{1}{Q_L} - \frac{1}{Q_C}\right)|\rho|_{max}. \qquad (3.30)$$

Also in this case it is favorable to select a small ripple in the passband. The deviation will be largest at the passband edge, i.e., where $\omega\tau_g(\omega)$ is largest.

Example 3.3 Consider the doubly resistively terminated *LC* filters in Example 2.5. We do not consider the sixth-order Cauer filter, because even-order lowpass filters cannot be realized as *LC* filters. However, Cauer type b filters can be realized.

The group delay at the passband edge is for the Chebyshev II, Chebyshev I, and Butterworth filters, 2.7457 s, 7.275 s, and 7.817 s, respectively. The factor $\omega\tau_g(\omega)$ is for the three filters: 5.4914, 14.55, and 15.634, respectively. $A_{max} = 0.01$ dB corresponds to $|\rho|_{max} = 0.048$. Hence, the

first term in Equation (3.30) dominates. We assume that $Q_L > 200$ and $Q_C > 1000$. We get for the four filters:

	$\Delta A(\omega)$ according to Equation (3.30)	$\Delta A(\omega)$ according to Equation (3.25)
Chebyshev II	0.144 dB	0.0457 ε dB
Chebyshev I	0.380 dB	0.1210 ε dB
Butterworth	0.408 dB	0.1300 ε dB

where ε is the tolerances in the reactive elements. Note that the Chebyshev II filter requires only 11 components, which may have 2.8 times larger tolerances, compared with the Butterworth filter, which requires 19 components. In addition, note that it is more important to use reactive components with high Q factors than low-tolerance components.

3.3.10 Design of Doubly Terminated Filters

Instead of using expensive components with low tolerances and large Q factors, we can compensate for an increase in ε, i.e., using components with larger tolerances, using either or all of the following trade-offs so the maximum of $A + \Delta A$ in the passband does not increase.

- Use a doubly resistively terminated reactance network that is designed for maximum power transfer, i.e., Equation (3.25) is valid.
- Reduce $|\rho(j\omega)|$ by reducing the passband ripple, A_{max}, of the filter more than required by the application. However, this requires the filter order to be increased. That is, we can use a few more, but cheaper components to reduce the overall cost of the implementation.
- Use an approximation that has low group delay, i.e., Chebyshev II and Cauer filters are preferred over Butterworth and Chebyshev I filters, see Fig. 2.41.
- Use an approximation with diminishing ripple, as will be further discussed below.

3.3.10.1 LC Filters with Diminishing Ripple

Because of deviations in the attenuation caused by errors in the element values, a part of the allowed ripple in the passband A_{max} must be reserved to allow for the errors in the component values. The filter must therefore be synthesized with a design margin, i.e., with a ripple, which is less than required by the application, A_{max}. According to Equation (3.25), the deviation is smaller for low frequencies and increases toward the passband edge. In practice, however, in order to simplify the synthesis, the design margin is for the standard approximations distributed evenly in the passband even though the margin will not be exploited for lower frequencies.

In order to better exploit the allowed passband ripple in a doubly resistively terminated filter, we

may let the reflection function $\rho(j\omega)$ of the synthesized filter decrease at the same rate as the other factors in $\Delta A(\omega)$ increase. That is, so that $A(\omega) + \Delta A(\omega) \leq A_{max}$. The ripple will decay toward the passband edge and the corresponding LC filter can be implemented with components with larger tolerances, i.e., the filter can be implemented with a lower overall cost. The group delay of a filter with diminishing ripples is slightly smaller than for the original filter.

Figure 3.19 shows the attenuation, $A(\omega)$, in the passband of a fith-order filter where the function $A(\omega) + \Delta A(\omega)$ is equiripple. The attenuation has diminishing ripples in order to compensate for the increasing $\Delta A(\omega)$.

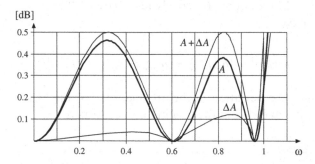

Fig. 3.19 Diminishing ripple

Note that $\Delta A(\omega) = 0$ for the frequencies of maximum power transfer and that $A(\omega)$ also was assigned a small margin at the cutoff edge to allow for the element errors ($\pm 1\%$).

3.4 Lowpass Ladder Structures

The most commonly used type of filter structure lowpass filters are the *ladder network* with alternating series and shunt impedances (reactances) according to Fig. 3.20. Ladder networks for highpass, bandpass, and stopband filters will be discussed in Section 3.5.

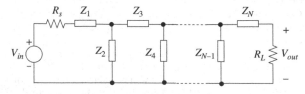

Fig. 3.20 T ladder

Here we focus on doubly resistively terminated ladder networks that are designed for maximum power transfer and therefore have superior sensitivity properties.

For a lowpass filter, the impedances, Z_1, Z_3,.., are inductors and the shunt impedances are series resonance circuits. In the dual network shown in Fig. 3.21, the series impedances are parallel resonance circuits and the shunt branches are capacitors.

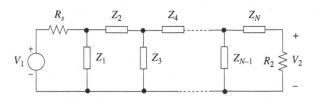

Fig. 3.21 π ladder

Usually, $R_s = R_L$ is chosen in order to get maximum power transfer, i.e., low sensitivity, without needing to use transformers or corresponding circuits. However, for even-order filters, this requires a transformer between the last ladder element and the load resistor. The transformer is not needed if we select $\dfrac{R_L}{R_s} = 2.10^{0.1 A_{max}} - 1 \pm 2\sqrt{10^{0.2 A_{max}} - 10^{0.1 A_{max}}}$
For example: $\rho = 25\% => A_{max} = 0.28028$ dB $=> R_L/R_s = 0.6$ or $R_L/R_s = 1/0.6 = 1.66666$.

In addition to the minimal passband sensitivity, ladder structures also possess advantageous stopband properties, as a ladder structure has a transmission zero, if and only if a shunt admittance or a series impedance is zero for that frequency. That is, a transmission zero is realized by two circuit elements, L and C, in a single branch.

A ladder structure can only realize transfer functions with zeros on the $j\omega$-axis and it cannot realize zeros on the real σ-axis or conjugate complex quads. In fact, not even all transfer functions with only $j\omega$-axis zeros can be realized with a ladder structure with positive elements. If all transmission zeros lie on the $j\omega$-axis and if there is at least one at $s = 0$ or $s = \infty$, then a ladder realization may be possible. Although restricted, ladder structures are used in most practical cases.

The constraints of the transmission zeros and the fact that they depend only on the elements in a single series or shunt impedance causes ladder structures to have low stopband sensitivity to element

variations. This simplifies the tuning of the ladder filters. In practice, it is often enough to tune the inductors in the series and shunt impedances to within a few percent of their nominal values and then tune the frequencies of the resonance circuits.

We stress that the process of determining the element values in any structure from the transfer function is an extremely ill conditioned problem [87]. That is, small errors in the numerical calculations will result in large errors in the element values. It has therefore been necessary to develop special synthesis algorithms based on, for example, the transformed variable technique to alleviate the numerical problems and/or perform all operations using only the roots of polynomials.

On the other hand, the inverse problem, i.e., a small error in an element value in a doubly resistively terminated reactance network, will have a small affect on the magnitude function.

3.4.1 RCLM One-Ports

Conditions for a realizable impedance are given in terms of a *positive-real function*[8], often abbreviated PR function [124, 131] by the following theorem, which is undoubtedly the most important concept in network theory.

Theorem 3.3 *A function $Z(s)$ is realizable as the impedance of a one-port consisting of only R, C, L, and M (mutual inductance) elements with positive values only if it is a rational PR function, i.e., it satisfies Brune's conditions*

- $Z(s)$ is a real rational function of s
- $Re\{Z(s)\} \geq 0$ for $Re\{s\} \geq 0$

or equivalently

- $Z(s)$ is a real rational function of s
- $Re\{Z(j\omega)\} \geq 0$ for all real ω
- All poles of $Z(s)$ are in the closed left half plane and all $j\omega$-axis poles are simple with positive residues.

3.4.2 Generic Sections

Darlington and Piloty showed that a cascade connection of the four basic sections, shown in Figs. 3.22, 3.23, 3.24, 3.25, is sufficient to realize any realizable reactance two-port [7, 124, 145], but more general sections, e.g., E section, are also used [24].

---

[8] Proposed by Otto Brune.

Fig. 3.22 *A* section

Fig. 3.23 *B* section

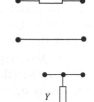

Fig. 3.24 *C* section

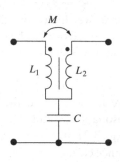

The *A* and *B* sections can only realize a pair of zeros on the $j\omega$-axis, including zeros at $s = 0$ and $s = \infty$. It is convenient to describe the sections with their chain matrices, see Section 5.4. The chain matrix for an *A* section is

$$\begin{bmatrix} V_1 \\ I_1 \end{bmatrix} = \begin{bmatrix} 1 & Z \\ 0 & 1 \end{bmatrix} \begin{bmatrix} V_2 \\ -I_2 \end{bmatrix} \tag{3.31}$$

and for the *B* section

$$\begin{bmatrix} V_1 \\ I_1 \end{bmatrix} = \begin{bmatrix} 1 & 0 \\ Y & 1 \end{bmatrix} \begin{bmatrix} V_2 \\ -I_2 \end{bmatrix}. \tag{3.32}$$

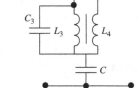

Fig. 3.25 *D* section

The chain matrix for the *C* section is

$$\begin{bmatrix} V_1 \\ I_1 \end{bmatrix} = \frac{1}{CMs^2 + 1} \begin{bmatrix} CL_1s^2 + 1 & (L_1 + L_2 - 2M)s \\ Cs & CL_2s^2 + 1 \end{bmatrix} \begin{bmatrix} V_2 \\ -I_2 \end{bmatrix} \tag{3.33}$$

where $M = \pm\sqrt{L_1 L_2}$. The factor $CMs^2 + 1$ in Equation (3.33) contributes, if $M > 0$, to a pair of finite zeros on the $j\omega$-axis at $s = \pm j/\sqrt{CM}$. In this case, the *C* section is known as the Brune section. If $M < 0$, it is known as the Darlington *C* section, and it realizes a pair of zeros on the real axis. Finally, the *D* section can realize a complex zero quadruplet, i.e., $s = \pm\sigma_z \pm j\omega_z$.

As shown in Figs. 3.24 and 3.25, the *C* and *D* sections contain coupled inductors, which may be difficult to implement. These sections are therefore in practice realized by using electrically equivalent networks [80, 100]. Figure 3.26 shows

four equivalent ladder structures for a *C* section with finite zeros on the $j\omega$-axis.

Note that there is always a negative circuit element. However, these elements can often be removed by performing network transformations on the complete ladder structure, see Section 3.6, in order to arrive at a ladder with only positive elements. Note that the maximum power transfer principle is not valid for a ladder structure with negative element values and it should not be used.

3.4.2.1 Coupled Inductors

Figure 3.27 shows two coupled inductors and their equivalent *T* network. The coupled inductors are described by

$$\begin{cases} V_1 = sL_1I_1 + sMI_2 \\ V_2 = sMI_1 + sL_2I_2 \end{cases} \tag{3.34}$$

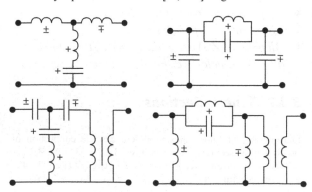

Fig. 3.26 Four possible equivalent circuits for the *C* section with zeros on the $j\omega$-axis

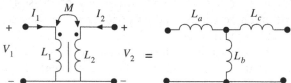

Fig. 3.27 Coupled inductor and their equivalent *T* network

where M is the mutual inductance. The dots in Figs. 3.24 and 3.27 indicate that an increase of the voltage at port 1 results in an increase of the voltage at port 2. The coupling coefficient k, which cannot be greater than unity, is defined by

$$K^2 = \frac{M^2}{L_1 L_2}. \tag{3.35}$$

The chain matrix for the coupled coils is

$$\begin{bmatrix} V_1 \\ I_1 \end{bmatrix} = \begin{bmatrix} \dfrac{L_1}{M} & \left(\dfrac{L_1 L_2 - M^2}{M}\right)s \\[2ex] \dfrac{1}{Ms} & \dfrac{L_2}{M} \end{bmatrix} \begin{bmatrix} V_2 \\ -I_2 \end{bmatrix}. \tag{3.36}$$

The elements in the equivalent T network and the coupled inductors are

$$\begin{cases} L_a = L_1 - L_b \\ L_b = M \\ L_c = L_2 - L_b. \end{cases} \tag{3.37}$$

3.4.2.2 C Section with Tapped Coils

An efficient way to realize a C section is to use a single coil with a tap, as illustrated in Fig. 3.28. The relations between the element values and the elements in T network, which corresponds to a C section, are also shown in Fig. 3.28.

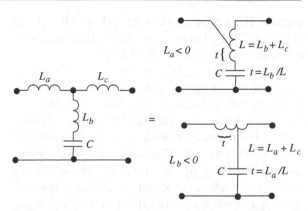

Fig. 3.28 T network and its equivalent tapped coils realization

3.4.3 Lowpass Ladder Structures without Finite Zeros

Figure 3.29 shows ladder networks *of T type* and Fig. 3.30 shows ladder networks *of π type*[9], which can only realize transfer functions of LP type with all zeros at $s = \infty$, i.e., to realize LP filters of Butterworth and Chebyshev I type. T and π refer to the left side of the ladder, i.e., a T type LP ladder always begins with a series (A section) inductor

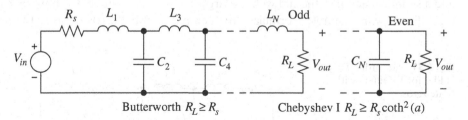

Fig. 3.29 Nth-order T ladder for LP filters without finite zeros

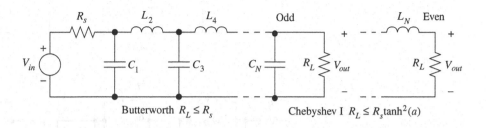

Fig. 3.30 Nth-order π ladder for LP filters without finite zeros

The case when L_c is negative is obtained by interchanging the ports in the upper rightmost circuit.

Theorem 3.4: *A passive, reciprocal ladder network without mutual coupling between branches can only realize a minimum-phase transfer function.*

[9]Also the terms symmetrical and antisymmetrical network are used for odd-order and even-order networks, respectively. The terms refer to the scattering matrix for the corresponding LC two-port, see Chapter 9.

whereas the π type begins with a shunt (B section) capacitor. The π ladder is often preferred because it has fewer inductors than the T ladder.

The nature of a ladder network can directly be obtained by noting that at low angular frequencies ($\omega \approx 0$), the capacitors have no effect and the inductors have low impedance (short circuit). Hence, the network behaves as a resistive voltage divider. At very high frequencies, a series impedance consisting of an inductor behaves as an open circuit; whereas a shunt impedance consisting of a capacitor behaves as a short circuit. This leads to that no signal power is transferred to the load resistance. Hence, each series and shunt impedance create a zero at $s = \infty$.

There are two possible structures for odd-order filters. For odd-order filters with $R_L \leq R_s$, there are shunt capacitors at both ends; whereas in the dual filter with $R_L \geq R_s$, there are series inductors at both ends. Because the ladder network is reciprocal, we can chose to place the voltage source in series with either R_s or R_L. In both cases, we get the same transfer function except for a possible difference in the gain if $R_s \neq R_L$. Even-order filters have in both cases a series inductor at the end with the smaller resistor and a shunt capacitor at the end with the larger resistor. There exist simple formulas to compute

the element values for Butterworth and Chebyshev I filters [21, 33]. For Butterworth and Chebyshev I the terminating resistors are constrained as indicated in Figs. 3.29 and 3.30 where the constant a is

$$a = 0.5 \, \mathrm{asinh}\left(\frac{1}{\varepsilon}\right). \qquad (3.38)$$

In the case of singly terminated ladders, the last element is either a shunt capacitor with an open circuit termination or a series inductor with a short circuit termination.

3.4.4 Lowpass Ladder Structures with Finite Zeros

There are two different types of ladder networks for realization of LP filters with finite zeros. These filters can be used to realize LP filters of Chebyshev II and Cauer type. Figure 3.31 shows a T ladder, which can realize finite zeros because the series resonance circuits work as short circuits at the resonance frequencies.

Figure 3.32 shows a π ladder, which can realize finite zeros because the parallel resonance circuits

Fig. 3.31 Nth-order T ladder for LP filters with finite zeros

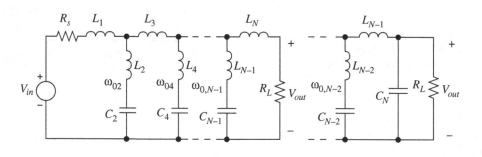

Fig. 3.32 Ladder of π type for LP filters with finite zeros

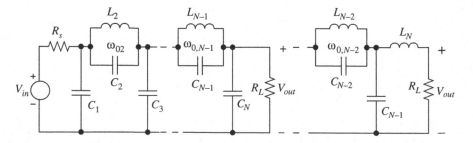

work as open-circuits at the resonance frequencies. In both cases, zeros are realized by reflection of the signal at a resonance circuit to be absorbed in the source resistor. This method results in low sensitivity close to the reflection zeros. The π ladder is often preferred because it has fewer inductors than the T ladder.

The sensitivity with respect to errors in the element values and the spread in the element values depends on in which order the zeros are extracted in the synthesis of the LC ladder. For example, an 11th-order Cauer LP filter, which has five finite zero pairs and one zero at s = ∞, has 6 series and shunt impedances. Hence, there are 6! = 720 different possibilities to assign the zeros to the ladder arms. The designer has to select one of the best among these structures depending on the element sensitivity and spread in the elements. Note that a small spread tends to make all impedances of the same order and, hence, voltages and currents are also of the same order, and the dynamic range will be better. The dynamic signal range is the ratio of the largest input signal that does not cause severe distortion to the thermal noise voltage.

As a rule of thumb, a good ladder network is obtained by beginning at both ends of the ladder and working toward the center and assigning the zeros with the lowest and highest resonance frequency to the first and last branch [48].

Ladder structures can only realize transfer functions of minimum phase type. Thus, they cannot realize transfer functions with zeros in the right half of the s-plane. The transfer function for an LP filter must also have at least one zero in s = ∞ to be realizable with a ladder network. Chebyshev II and Cauer filters of even-order can therefore not be realized without using transformers and often requires unequal terminations. The transfer function can, however, be modified to a b or c type so the filter can be realized with a ladder structure.

An odd-order Cauer filter has a symmetric topology and may have equal termination resistors. The number of resonance circuits is $(N-1)/2$. An even-order Cauer filter of type a requires coupled inductors and therefore type b or c is often preferred.

3.4.5 Design of Lowpass LC Ladder Filters

The synthesis of a ladder structure according to the insertion loss method is complicated and requires high-accuracy computations and is for this reason performed by advanced computer programs. The synthesis of a ladder structure for a realizable transfer function is generally carried out by alternating extracting series impedances and shunt admittances from the input impedance Z_i, using special algorithms [29, 80, 96, 112, 124].

An alternative approach is to first find a feasible ladder, possibly with negative elements, which can by applying certain network transformations be modified into a realizable ladder structure. The latter will be discussed in Section 3.6. The element values are then determined by using a numerical optimization procedure.

The following sections are required in a ladder structure.

- For a transmission zero of order n at $s = 0$, there will be a sequence of n alternating A and B sections, where the A sections will consist of a series capacitor and the B sections will consist of a shunt inductor.
- For a transmission zero of order n at $s = \infty$, there will be a sequence of n alternating A and B sections, where the A sections will consist of a series inductor and the B sections will consist of a shunt capacitor.
- For each conjugate pair of finite, nonzero $j\omega$-axis zeros, there will be one C section.

The sections may be placed in any desired order as long as the alternation of the A and B type sections for the zeros at $s = 0$ and $s = \infty$ is preserved, and the first section (either an A or B section) is compatible with the required input and output impedance of the reactance network at $s = 0$ and $s = \infty$. Finite zeros, or zeros on the negative real axis, requires C sections whereas zero quadruplets requires D sections.

There exists many possible ordering of the sections. For example, consider a 12th-order bandpass filter with four finite zero pairs and double zeros at $s = 0$ and $s = \infty$. The number of possible orders of the $n = 8$ sections is $n!$ divided by the multiplicity of the zeros as these orders appear several times. Hence, we have

$$\frac{8!}{2!2!} = 10,080$$

different configurations. Hence, many different ladders, consisting of the same sections, but in different order, can realize a given transfer function. Some of these may, however, have negative element values. Moreover, the ordering of the sections has a significant effect on the element sensitivity and spreads of the element values. By applying network transformations, the element spread as well as negative element values can often be removed. This further increases the number of potentially good structures. It is therefore a challenging task for the designer to find the "best" realization among the many alternatives.

Here we will use the functions BW_LADDER, CH_I_LADDER and CH_II_LADDER and CA_LADDER, which are a part of the toolbox, to compute the element values in a ladder network that realizes a standard LP approximation. How to use these functions is demonstrated using a few examples.

3.4.5.1 Element Values in Butterworth *LC* Ladders

Simple algorithms for computing the lumped element values in a Butterworth lowpass filter have been derived [33, 96]. A recursive algorithm for a Butterworth lowpass filter with $\omega_c = 1$ and $A_{max} = 3.01$ dB, shown in Fig. 3.29, is

$$\begin{cases} \alpha = \left(1 - \dfrac{4R_s R_L}{(R_s + R_L)^2}\right)^{\frac{1}{2N}} \\[4mm] b_k = 1 + \alpha^2 - 2\alpha\cos\left(k\dfrac{\pi}{N}\right) \qquad k = 1, 2, \ldots, N \\[4mm] a_k = \sin\left(\dfrac{(2k-1)\pi}{2N}\right) \\[4mm] g_1 = \dfrac{2a_1}{1-\alpha}, \qquad g_k = \dfrac{4a_k a_{k-1}}{b_{k-1}g_{k-1}} \qquad k = 2, \ldots, N \end{cases} \qquad (3.39)$$

where g_i are alternating between inductance and capacitance. For $N =$ odd and if $R_L \leq R_s$, g_1 and g_N are shunt capacitors for a π ladder and series inductors for a T ladder. For $N =$ even, there is an inductor at the low-resistance end and a shunt capacitor at the high-resistance end. Because the LC ladder is a reciprocal network, the signal source can be placed at either end. The denormalization is done by multiplying the inductances with R_s/r_{p0} and capacitances with $1/(R_s r_{p0})$ where R_s is the desired source resistor.

Example 3.4 Determine the element values in a doubly resistively terminated LC ladder that realizes a Butterworth filter with $N = 5$, $\omega_c = 20$ krad/s, $\omega_s = 62$ krad/s, $A_{max} = 0.5$, and $A_{min} = 40$ dB. Use a T ladder shown in Fig. 3.33 with $R_s = R_L = 1000\ \Omega$.

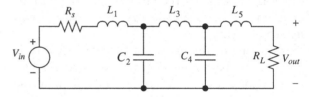

Fig. 3.33 Fifth-order Butterworth filter realized with a T ladder

The element values in the ladder networks can be determined using the function **BW_LADDER**, which implements the algorithm in Equation (3.39). We select $R_s = R_L$ to get low element sensitivity and minimal insertion loss. The attenuation for the filter is shown in Fig. 3.34

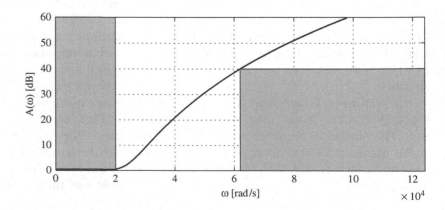

Fig. 3.34 Attenuation for a fifth-order Butterworth filter realized with a T ladder

```
Wc = 20000;
Ws = 62000;
Amax = 0.5;
Amin = 40;
Rs = 1000;
RL = 1000;
Ladder = 1;                          % 1 for a T ladder and 0 for a π ladder
N = BW_ORDER(Wc, Ws, Amax, Amin)
Norder = 5;                          % Must be an integer. We select a 5th-order filter
[L, C, K] = BW_LADDER(Wc, Ws, Amax, Amin, Norder, Rs, RL, Ladder);
Z0 = []; T = 1;                      % Used only for transmission lines
omega = [0:Ws/200:2*Ws];
H = LADDER_2_H(Norder, Z0, L, C, Rs, RL, K, omega, T);
Att = MAG_2_ATT(2*H);               % Normalize attenuation to 0 dB
subplot('position', [0.08 0.4 0.90 0.5]);
PLOT_ATTENUATION_S(omega, Anorm);
PLOT_LP_SPEC_S(Wc, Ws, Amax, Amin);     % Amin = 0 => No stopband spec
axis([0, 2*Ws, 0, 1.5*Amin]);
L'
C'
```

We get for a T ladder

L_1 = 2.503946e-02 C_1 = 0
L_2 = 0 C_2 = 6.55542e-08
L_3 = 8.102939e-02 C_3 = 0
L_4 = 0 C_4 = 6.55542e-08
L_5 = 2.503946e-02 C_5 = 0

We get the following element values for a π ladder

L_1 = 0 C_1 = 2.503946e-08
L_2 = 6.55542e-02 C_2 = 0
L_3 = 0 C_3 = 8.102939e-08
L_4 = 6.55542e-02 C_4 = 0
L_5 = 0 C_5 = 2.503946e-08

Note that the element values for each ladder section are the same, except for a scaling factor. That is, L_1 in the T ladder corresponds to a capacitor $C_1 = L_1/R^2$ in the dual π ladder.

3.4.5.2 Element Values in Chebyshev I LC Ladders

Similar recursive formulas also exists for Chebyshev I filters [33, 96]. Simpler formulas, but involving impedance inverters, are given in [96].

$$
\begin{cases}
\alpha = \begin{cases}
\dfrac{4R_S R_L}{(R_S + R_L)^2} & N = \text{odd} \\[2ex]
\dfrac{4R_S R_L \varepsilon}{(R_S + R_L)^2} & N = \text{even}
\end{cases} \\[4ex]
\eta = \sinh\left[\dfrac{1}{N} a \sinh\left(\dfrac{1}{\varepsilon}\right)\right], \quad \gamma = \sinh\left[\dfrac{1}{N} a \sinh\left(\dfrac{\sqrt{1-\alpha}}{\varepsilon}\right)\right] \\[3ex]
b_k = \eta^2 + \gamma^2 - 2\eta\gamma \cos\left(k\dfrac{\pi}{N}\right) + \sin^2\left(k\dfrac{\pi}{N}\right) \quad k = 1, 2, \ldots, N \\[3ex]
a_k = \sin\left(\dfrac{(2k-1)\pi}{2N}\right) \\[3ex]
g_1 = \dfrac{2a_1}{\eta - \gamma}, \quad g_k = \dfrac{4a_k a_{k-1}}{b_{k-1} g_{k-1}} \quad k = 2, \ldots, N.
\end{cases}
\tag{3.40}
$$

According to Figs. 3.29 and 3.30, g_k is a capacitor for k = odd and inductor for k = even in a π ladder if $R_L \leq R_S \tanh(a)^2$ where a is given by Equation (3.38).

If $R_L \geq R_S \coth(a)^2$, then g_k is an inductor for k = odd and a capacitor for k = even in a T ladder. For Chebyshev I filters, these values are not unique [96].

Example 3.5 Determine the element values in the π ladder shown in Fig. 3.35 to Chebyshev II filter with $\omega_c = 22$ krad/s, $\omega_s = 43$ krad/s, $A_{max} = 0.5$ dB, and $A_{min} = 40$ dB that yields $N = 4.921$. We therefore select $N = 5$ and $R_s = R_L = 1000\ \Omega$.

```
Wc = 22000;
Ws = 43000;
Amax = 0.5;
Amin = 40;
N = CH_ORDER(Wc, Ws, Amax, Amin)
Norder = 5;                         % We select a 5th-order filter
Rs = 1000; RL = 1000;
Ladder = 0;                         % 1 for a T ladder and 0 for a π ladder
[L, C, K] = CH_I_LADDER(Wc, Ws, Amax, Amin, Norder, Rs, RL, Ladder);
Z0=[]; T=1;                         %Used only for transmission lines
omega = [0:Ws/200:2*Ws];
H = LADDER_2_H(Norder, Z0, L, C, Rs, RL, K, omega, T);
Att = MAG_2_ATT(2*H);               % Normalize attenuation to 0 dB
subplot('position', [0.08 0.4 0.90 0.5]);
PLOT_ATTENUATION_S(omega, Anorm);
PLOT_LP_SPEC_S(Wc, Ws, Amax, Amin);   % Amin = 0 => No stopband spec
axis([0, 2*Ws, 0, 1.5*Amin]);
L'
C'
```

The element values are computed using the function CH_I_LADDER, which implements the algorithm in Equation (3.40).

which yields

$C_1 = 7.753501e{-}08$
$L_2 = 5.589212e{-}02$
$C_3 = 1.154921e{-}07$
$L_4 = 5.589212e{-}02$
$C_5 = 7.753501e{-}08$

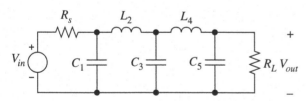

Fig. 3.35 Fifth-order Chebyshev I filter realized with a π ladder

The π ladder has only two inductors, whereas the T ladder has three inductors. For Butterworth and Chebyshev I filters with $R_s = R_L$ and $N =$ odd, the ladder networks become symmetrical with respect to both the structure and element values.

3.4.5.3 Element Values in Chebyshev II *LC* Ladders

Complicated formulas for a few special cases of Chebyshev II filters that involve impedance inverters are given in [96].

Different solutions are obtained, depending on the ordering of the transmission zeros and the ordering of the section, as discussed above. Hence, the element values in Chebyshev II filters are not unique.

Example 3.6 Determine the element values in the T ladder shown in Fig. 3.36 to realize a Chebyshev II type with $\omega_c = 22$ krad/s, $\omega_s = 43$ krad/s, $A_{max} = 0.5$ dB, and $A_{min} = 40$ dB that yields $N = 4.921$. We therefore select $N = 5$ with $R_s = R_L = 1000\ \Omega$.

The element values are computed using the function CH_II_LADDER.

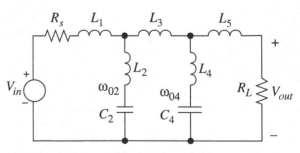

Fig. 3.36 Fifth-order Chebyshev II filter realized with a T ladder

```
Wc = 22000;
Ws = 43000;
Amax = 0.5; Amin = 40;
[G, Z, P] = CH_II_POLES[Wc,Ws,Amax,Amin,N];
Rs = 1000;  RL = 1000;
Norder = 5;                     % We select a 5th-order filter
Ladder = 1;                     % 1 for a T ladder and 0 for a π ladder
[L, C, Rs, RL, Wo, KI] = CH_II_LADDER(G, Z, P, Wc, Ws, Rs, RL, Ladder);
Z0=[]; T=1;                     %Used only for transmission lines
omega = [0: Ws/200: 2*Ws];
H = LADDER_2_H(Norder, Z0, L, C, Rs, RL, K, omega, T);
Att = MAG_2_ATT(2*H);           % Normalize attenuation to 0 dB
subplot('position', [0.08 0.4 0.90 0.5]);
PLOT_ATTENUATION_S(omega, Anorm);
PLOT_LP_SPEC_S(Wc, Ws, Amax, Amin);  % Amin = 0 => No stopband spec
axis([0, 2*Ws, 0, 1.5*Amin]);
L'
C'
Wo'
```

which yields

```
L₁ = 1.87202e-02
L₂ = 3.4900018e-03    C₂ = 5.353957e-08    w₀₂ = 7.315600e+04
L₃ = 6.674866e-02
L₄ = 1.099776e-02    C₄ = 4.448067e-08    w₀₄ = 4.521288e+04
L₅ = 1.255138e-02
```

3.4.5.4 Element Values in Cauer *LC* Ladders

Complicated formulas for a few special cases of Cauer filters that involve impedance inverters are given in [96]. Similar to Chebyshev II filters, the element values in Cauer filters are not unique. A computer-based search among all possible solutions is therefore to be preferred. As selection criteria, the element sensitivity and spread may be used.

Example 3.7 Determine the element values in the π ladder shown in Fig. 3.37 to realize a Cauer filter with $N = 5$, $\rho = 50\%$, $A_{min} = 40.3$ dB, $\omega_c = 10$ krad/s. Use a T ladder with $R_s = R_L = 1000\,\Omega$.

The element values are computed using the function CA_LADDER. The reflection coefficient $\rho = 50\%$ corresponds, according to Equation (2.36), to $A_{max} = 1.2494$ dB.

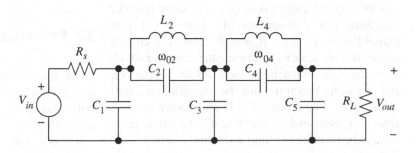

Fig. 3.37 Fifth-order Cauer filter realized with a π ladder

```
Wc = 22000;
Ws = 28000;
r = 0.5;
Amax = -10*log10(1-r^2);
Amin = 40;
Rs = 1000;
RL = 1000;
N = CA_ORDER(Wc, Ws, Amax, Amin);
[G, Z, R_ZEROS, P, Wsnew] = (A_POLES [Wc, Ws, Amax, Amin, N];
Norder = 5;                       % We select a 5th-order filter
Ladder = 0;                       % 1 for a T ladder and 0 for a π ladder
[L, C, Rs, RL, Wo, KI] = CA_LADDER(G, Z, R_ZEROS, P, Wc, Ws, Rs, RL,
Ladder);
Z0 = []; T = 0;                   % Used only for transmission lines
omega = [0:Ws/200:2*Ws];

H = LADDER_2_H(Norder, Z0, L, C, Rs, RL, K, omega, T);
Att = MAG_2_ATT(2*H);         % Normalize attenuation to 0 dB
subplot('position', [0.08 0.4 0.90 0.5]);
PLOT_ATTENUATION_S(omega, Anorm);
PLOT_LP_SPEC_S(Wc, Ws, Amax, Amin);   % Amin = 0 => No stopband spec
axis([0, 2*Ws, 0, 1.5*Amin]);
L'
C'
Wo'
```

We get

$L_2 = 3.763656-02$	$C_1 = 9.076098e-08$	
	$C_2 = 1.839124-08$	$w_{02} = 3.800927e+04$
	$C_3 = 10.17367-08$	
$L_4 = 2.464374e-02$	$C_4 = 5.490668-08$	$w_{04} = 2.718530e+04$
	$C_5 = 6.953684-08$	

For Butterworth and Chebyshev I lowpass filters with $R_s = R_L$, the element values are unique. For the remaining filter types, there are several different sets of element values that have the same transfer function. For even-orders of Chebyshev I and Cauer filters of b type, R_s and R_L cannot be chosen equal. It is possible to determine the ranges in which the resistances must be chosen. As an alternative, the transfer function can be modified so that $|H(0)| = 1$. For Chebyshev II and Cauer lowpass filters of even-order, the transfer function must always be modified so that a double zero at $s = \infty$

is obtained in order for the filter to be realizable as a ladder network.

3.5 Frequency Transformations

In Chapter 2, it was shown how the poles and zeros to an LP filter could be transformed to corresponding poles and zeros of a HP filter or a geometric symmetrical BP or BS filter. The frequency transformation can also be applied directly to the impedances in the

LP filter. This leads to that if all the impedances in, e.g., a ladder network of LP type are replaced with frequency transformed impedances, the same ladder network is obtained as if we instead synthesized it from corresponding frequency transformed transfer function. The principle for frequency transformation of LC filters[10] is illustrated in Fig. 3.38.

The filters that are obtained are based on the same frequency transformations that were used in Chapter 2 and must therefore meet the geometric

and/or load resistance $= 1$ ohm, and the passband edge $\omega_c = 1$ rad/s.

Example 3.8 We show with the means of a simple example that the poles and zeros are not changed if all impedances in a lumped element network are changed with the same factor k. Consider therefore the RLC network shown in Fig. 3.39.

Fig. 3.39 Simple RLC network

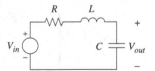

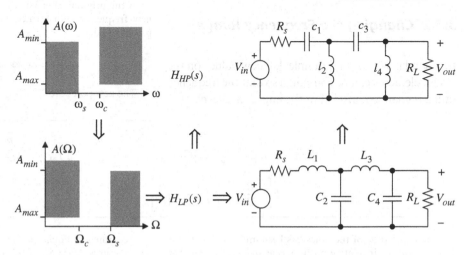

Fig. 3.38 Principle for frequency transformation of an LC filter

symmetry constraint. In general, the LC filters will not be optimal for bandpass and bandstop filters. In this section, we will discuss this technique and its limitations.

3.5.1 Changing the Impedance Level

For the standard approximations for lowpass filters, there exist in the literature many tables with poles, zeros, as well as the element values in the corresponding ladder structure [100]. However, it is only possible to compute such tables for filters with only lumped circuit elements. Both the impedance level and bandwidth of the ladder network can be changed by simple calculations. In most tables, the networks are normalized with source

The frequency response is

$$H(j\omega) = \frac{\dfrac{1}{j\omega C}}{R + j\omega L + \dfrac{1}{j\omega C}} = \frac{1}{1 - \omega^2 LC + j\omega RC}.$$

We multiply all impedances with the factor k.

$$R \to kR$$
$$j\omega L \to kj\omega L \Leftrightarrow L \to kL$$
$$\frac{1}{j\omega C} \to \frac{k}{j\omega C} \Leftrightarrow C \to \frac{C}{k}.$$

Inserting the new element values in the expression for frequency response yields

$$H'(j\omega) = \frac{1}{1 - \omega^2 (kL)\left(\dfrac{C}{k}\right) + j\omega (kR)\left(\dfrac{C}{k}\right)}$$

$$= \frac{1}{1 - \omega^2 LC + j\omega RC} = H(j\omega).$$

The factor k is cancelled in the expression, as all coefficients in the expression consist of products of the type LC and RC.

---

[10]To differentiate between elements in the LP and HP filters, we use capital and lowercase letters, respectively.

Theorem 3.5: *A transfer function corresponding to the ratio V_{out}/V_{in} or I_{out}/I_{in} is invariant with respect to the impedance level of the network whereas a transfer function corresponding to a transfer impedance or admittance is only affected by a change of the gain factor. That is,*

$$F(\alpha R, \alpha L, C/\alpha, s) = \beta F(R, L, C, s) \qquad (3.41)$$

$$where\ \beta = \begin{cases} 1 & F = \text{transfer function} \\ \alpha & F = \text{impedance function} \\ 1/\alpha & F = \text{admittance function.} \end{cases}$$

3.5.2 Changing the Frequency Range

In a similar way as in Example 3.8, new values on the circuit elements can be computed so that the frequency range is changed, instead of the impedance level.

Example 3.9 Show that the frequency scale is changed with the factor k if all the normalized inductances and capacitances are divided with k. Assume that the source filter has the element values R_n, L_n, and C_n, $n = 1, 2,\dots$ at the edge angular frequency ω_0. The new element values are

$$R = R_n \qquad L = \frac{L_n}{k} \qquad C = \frac{C_n}{k}. \qquad (3.42)$$

Insertion of the expression for the frequency response gives

$$H'(j\omega) = \frac{1}{1 - \omega^2 LC + j\omega RC} = \frac{1}{1 - \omega^2 \frac{L_n C_n}{k^2} + j\omega \frac{R_n C_n}{k}}.$$

If the original filter has the same passband edge ω_0, the new frequency transformed filter will have the following passband edge

$$\frac{\omega^2}{k^2} = \omega_0^2 \Rightarrow \omega_{new} = k\omega_0.$$

The new poles are

$$s_{p,1,2} = -\frac{R_n}{2\left(\frac{L_n}{k}\right)} \pm j\sqrt{\frac{1}{\left(\frac{L_n}{k}\right)\left(\frac{C_n}{k}\right)} - \frac{R_n^2}{4\left(\frac{L_n}{k}\right)^2}} = -\frac{kR_n}{2L_n} \pm j\sqrt{\frac{k^2}{L_n C_n} - \frac{k^2 R_n^2}{4L_n^2}}$$

$$= k\left[-\frac{R_n}{2L_n} \pm j\sqrt{\frac{1}{L_n C_n} - \frac{R_n^2}{4L_n^2}}\right].$$

The magnitude of the poles has been multiplied with the factor k, but their relative position remains unchanged. The frequency response of the filter is unchanged when the frequency scale is multiplied with a factor k. In this way, the frequency scale can be changed for filters that only contain *lumped circuit element*, i.e., when the frequencies are so low that the physical dimensions of the circuit elements can be neglected. However, filter structures with distributed circuit elements can in general not be frequency scaled. The network has to be synthesized for the intended frequency range.

Hence, in order to change the impedance level with a factor, R_0, and frequency of a lumped, RLC network with a factor, ω_0, we multiply all

- resistances with R_0
- inductances with R_0/ω_0
- capacitances with $\frac{1}{R_0\omega_0}$.

3.5.3 LP-to-HP Transformation

In Chapter 2, it was shown how the transfer function to an LP filter can be frequency transformed into a HP filter using the mapping

$$S = \frac{\omega_I^2}{s}. \qquad (3.43)$$

According to Equation (3.43), an inductor in the LP filter, with the impedance SL, shall be replaced with an equivalent impedance $\omega_I^2 L/s$, i.e., a capacitor with the capacitance $c = 1/\omega_I^2 L$. In the same way a capacitor in the LP filter with the impedance $1/SC$ shall be replaced with a network with the impedance $s/\omega_I^2 C$, i.e., an inductor with inductance $l = 1/\omega_I^2 C$.

Example 3.10 Determine the element values in a ladder network that meet the filter specification in Example 2.8. Select a T network with $R_s = R_L = 1000\ \Omega$. The requirement on the LP filter was determined in Example 2.7 to $\omega_c = 30$ krad/s, $N = 5$, and $A_{max} = 0.1$ dB.

Figure 3.40 shows the LP filter and corresponding HP filter. In Table 3.1, it is shown how the impedances in the LP filter are transformed.

The element values for the LP filter are computed in the same way as in Example 3.4 using BW_LADDER, which yields

```
L₁ = 1.414535e-02    C₂ = 3.703300e-08
L₃ = 4.577531e-02    C₄ = 3.703300e-08
L₅ = 1.414535e-02
```

The element values in the HP filter are computed according to the relations shown in Table 3.1. R_s and R_L remain unchanged. We get

```
c₁ = 7.854958e-08    l₂ = 3.000327e-02
c₃ = 2.427315e-08    l₄ = 3.000327e-02
c₅ = 7.854958e-08
```

Fig. 3.40 Frequency transformation of a fifth-order LP filter to a HP filter

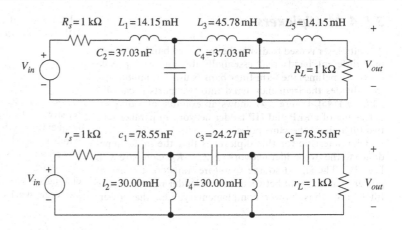

Table 3.1 LP-HP transformation

LP filter	HP filter
$A(\Omega)$ graph with A_{min}, A_{max}, Ω_c, Ω_s $S = \dfrac{\omega_I^2}{s}$	$A(\omega)$ graph with A_{min}, A_{max}, ω_s, ω_c $\Omega_c = \dfrac{\omega_I^2}{\omega_c}$ $\Omega_s = \dfrac{\omega_I^2}{\omega_s}$
R (resistor)	$r = R$ (resistor)
L (inductor)	c (capacitor) $c = 1/\omega_I^2 L$
C (capacitor)	l (inductor) $l = 1/\omega_I^2 C$
L and C parallel	c, l parallel $l = 1/\omega_I^2 C$ $c = 1/\omega_I^2 L$ $\omega_{01} = \omega_I^2 \sqrt{LC}$
L and C series	c, l series $l = 1/\omega_I^2 C$ $c = 1/\omega_I^2 L$ $\omega_{01} = \omega_I^2 \sqrt{LC}$

3.5.4 Multiplexers

A multiplexer is used to divide the frequency band of interest into different bands that essentially do not overlap each other. Sometimes the term filter bank is used. A multiplexer that divides the frequency band into two parts is called a *diplexer* [145]. Figure 3.41 shows an example of a diplexer consisting of an LP and HP ladder network in parallel. The two filters have the same poles.

Characteristic for this diplexer is that the input impedance to the two ladder networks is $Z_i = R_L$. Compare with Fig. 3.2. The signal source therefore transfers a constant power, which is split between the two ladder networks. The filters are, thus, power complementary, i.e., the power,

which is not dissipated in the LP load is dissipated in the HP load and vice versa. Hence, we have

$$P_{2max} = P_{LP} + P_{HP} = (|H_{LP}|^2 + |H_{HP}|^2)P_{in}. \qquad (3.44)$$

Because, $Z_i = R_L$, each of the load resistors can be replaced with an LP-HP pair that divides each of the two frequency bands into two bands. A multiplexer with four bands is thereby obtained. This type of multiplexer has very low sensitivity for errors in the element values.

Figure 3.42 shows an example of a multiplexer with three channels where the sum of the input impedances to the two ladder networks is purely resistive and equals $3R_L$.

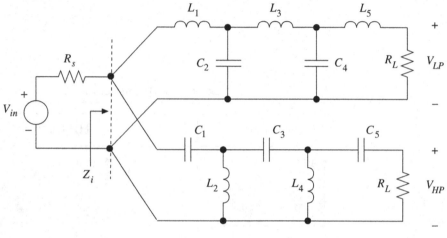

Fig. 3.41 Diplexer of parallel type

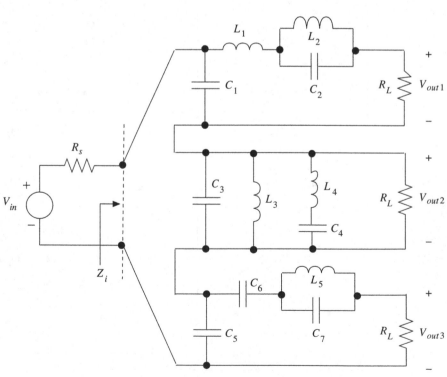

Fig. 3.42 Multiplexer of series type

Note that there exist stringent constraints on the ladder networks, i.e., in the diplexer of parallel type; the ladder networks may not have input impedances, which are zero for any frequency. None of the ladder networks may therefore have a shunt inductor at its input. For the series type, none of the ladder networks may have infinite impedance.

Example 3.11 Figure 3.43 shows a fourth-order branching filter for a speaker system. Because we do not want to waste any signal power in resistor R_s, it has been deleted. Hence, the diplexer has not minimal element sensitivity. However, in this application, this is of less importance as the filters are of low order and not highly frequency selective [72].

of the magnitude function for a geometric symmetrical bandpass filter becomes larger in the lower transition band.

The *relative bandwidth* of a bandpass filter is

$$\frac{\omega_{c2} - \omega_{c1}}{\sqrt{\omega_{c1}\,\omega_{c2}}}. \tag{3.45}$$

A relative bandwidth that is less than 10% is considered as small and if it is less than 1% is very small. The spreads in the element values will become large for BP filters with small relative bandwidths.

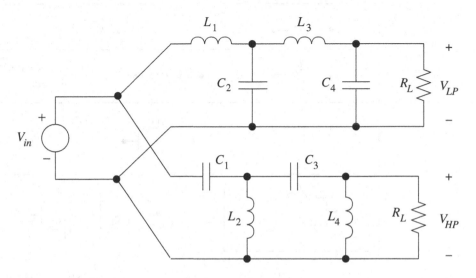

Fig. 3.43 Example of branching filter

With the following element values $L_1 = 800\ \mu H$, $L_2 = 267\ \mu H$, $L_3 = 400\ \mu H$, $L_4 = 1.20\ mH$, $C_1 = 3.52\ \mu F$, $C_2 = 10.6\ \mu F$, $C_3 = 7.03\ \mu F$, $C_4 = 2.34\ \mu F$, and $R_L = 8\ \Omega$, a branching frequency of 3 kHz is obtained. At the branching frequency, half of the power is dissipated in the two load resistors, i.e., the attenuation at the branching frequency is 6 dB.

3.5.5 LP-BP Transformation

Similarly, an LC filter of lowpass type can be transformed to a bandpass filter by using the relations in Table 3.2. Figure 3.44 shows a BP ladder that was derived from an odd-order T ladder with finite zeros. A bandpass filter that is designed through this frequency transformation of a lowpass filter results in a geometric symmetrical bandpass filter. Such a filter is often not optimal because it always has the same number of zeros in both stopbands and the stopbands attenuations are equal. The steepness

In Section 3.6, we will discuss network transformations that will reduce the spread.

An LC filter with a relative bandwidth of less than 1% can in practice not be implemented with RLC structures. In [11, 137] design methods that are suitable for narrow bandpass filters are discussed. Bandpass filters are much more complicated than lowpass filters and the choice of the best configuration often requires considerable skill and experience.

Design of more general types of filter functions requires a direct synthesis from the transfer function [112, 124]. Of importance is the order in which the transmission zeros is realized. For example, in a bandpass filter, it is favorable if a transmission zero in the lower stopband is followed by one in the upper stopband as this usually will yield smaller spreads in the element values. However, all possible orders should in practice be investigated, as there exist orders that yield filters with small spreads and low sensitivity and filters with larger spreads and higher sensitivities.

Table 3.2 LP-BP transformation

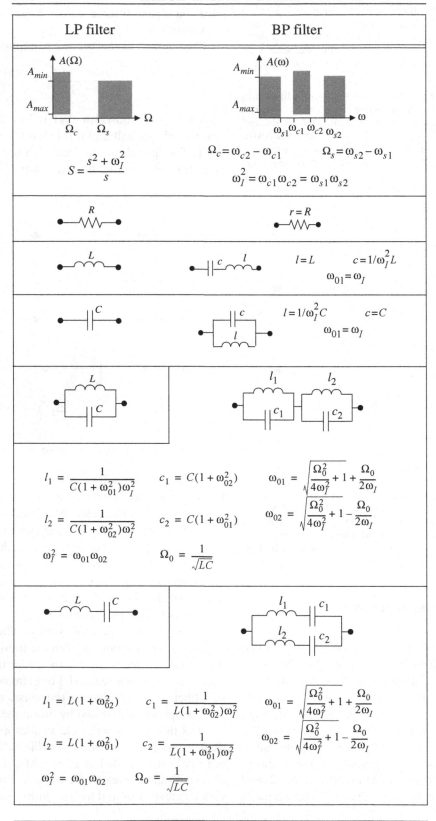

Fig. 3.44 BP filter derived from a T ladder with finite zeros

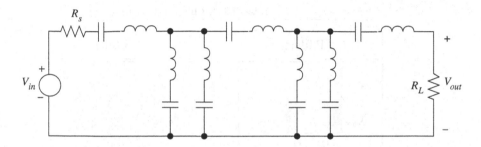

Example 3.12 Determine the element values in a ladder network of Chebyshev I type that meets the specification $A_{max} = 2\,\text{dB}$, $\omega_{c1} = 60\,\text{Mrad/s}$, $\omega_{c2} = 120\,\text{Mrad/s}$, $A_{min1} = 35\,\text{dB}$, $\omega_{s1} = 40\,\text{Mrad/s}$, $A_{min2} = 40\,\text{dB}$ and $\omega_{s2} = 150\,Mrad/s$.

The LP filter specification is computed in the same way as in Example 2.8, which gives $\omega_{clp} = 60\,\text{Mrad/s}$, $\omega_{slp} = 120\,\text{Mrad/s}$. Here ω_{s1} has been increased to $\omega_{s1} = 48\,\text{Mrad/s}$ to meet the symmetry requirement in Equation (2.52).

The element values for the lowpass filter computed in the same way as in Example 3.5 with CH_I_LADDER, which yields for a T network with $R_s = R_L = 1000\,\Omega$

$$L_1 = 4.71836\ \text{e–}05 \qquad C_2 = 1.49744\ \text{e–}11$$
$$L_3 = 6.30454\ \text{e–}05 \qquad C_4 = 1.49744\ \text{e–}11$$
$$L_5 = 4.71836\ \text{e–}05$$

The element values are transformed according to Table 3.2. Figure 3.45 shows the lowpass filter and the corresponding bandpass filter. The inductors are transformed to series resonance circuits

$$
\begin{array}{lll}
l_1 = 4.71836\ \text{e–}05 & l_3 = 6.30454\ \text{e–}05 & l_5 = 4.71836\ \text{e–}05 \\
c_1 = 2.94359\ \text{e–}12 & c_3 = 2.20300\ \text{e–}12 & c_5 = 2.94359\ \text{e–}12 \\
w_{01} = 8.48528\ \text{e+}07 & w_{03} = 8.48528\ \text{e+}07 & w_{05} = 8.48528\ \text{e+}07
\end{array}
$$

The capacitors are transformed to parallel resonance circuits

$$
\begin{array}{ll}
c_2 = 1.49744\ \text{e–}11 & c_4 = 1.49744\ \text{e–}11 \\
l_2 = 9.27511\ \text{e–}06 & l_4 = 9.27511\ \text{e–}06 \\
w_{02} = 8.48528\ \text{e+}07 & w_{04} = 8.48528\ \text{e+}07
\end{array}
$$

With advanced computer software, the element values in a ladder network can be computed for a general transfer function. There must however exist a zero at $s = \infty$ and one at $s = 0$ in order for a BP transfer function to be realizable with a ladder network.

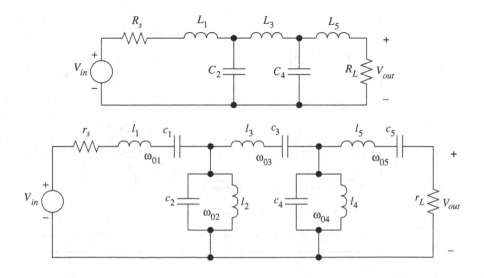

Fig. 3.45 Frequency transformation of an LP filter to a BP filter

Table 3.3 LP-BS transformation

LP filter	BS filter
$$S = \frac{\omega_I^2 s}{s^2 + \omega_I^2}$$	$$\Omega_c = \frac{\omega_I^2}{\omega_{c2} - \omega_{c1}} \qquad \Omega_s = \frac{\omega_I^2}{\omega_{s2} - \omega_{s1}}$$ $$\omega_I^2 = \omega_{c1}\omega_{c2} = \omega_{s1}\omega_{s2}$$
R	$r = R$
L	$l = L \qquad c = 1/(\omega_I^2 L)$ $\omega_{01} = \omega_I$
C	$l = 1/(\omega_I^2 C) \qquad c = C$ $\omega_{01} = \omega_I$
$$l_1 = \frac{L}{(1 + \omega_{01}^2)} \qquad c_1 = \frac{(1 + \omega_{02}^2)}{L\omega_I^2}$$ $$l_2 = \frac{L}{(1 + \omega_{02}^2)} \qquad c_2 = \frac{(1 + \omega_{01}^2)}{L\omega_I^2}$$ $$\omega_I^2 = \omega_{01}\omega_{02} \qquad \Omega_0 = \frac{1}{\sqrt{LC}}$$	$$\omega_{01} = \sqrt{\frac{\omega_I^2}{4\Omega_0^2} + 1} + \frac{\omega_I}{2\Omega_0}$$ $$\omega_{02} = \sqrt{\frac{\omega_I^2}{4\Omega_0^2} + 1} - \frac{\omega_I}{2\Omega_0}$$
$$l_1 = \frac{(1 + \omega_{02}^2)}{C\omega_I^2} \qquad c_1 = \frac{C}{(1 + \omega_{01}^2)}$$ $$l_2 = \frac{(1 + \omega_{01}^2)}{C\omega_I^2} \qquad c_2 = \frac{C}{(1 + \omega_{02}^2)}$$ $$\omega_I^2 = \omega_{01}\omega_{02} \qquad \Omega_0 = \frac{1}{\sqrt{LC}}$$	$$\omega_{01} = \sqrt{\frac{\omega_I^2}{4\Omega_0^2} + 1} + \frac{\omega_I}{2\Omega_0}$$ $$\omega_{02} = \sqrt{\frac{\omega_I^2}{4\Omega_0^2} + 1} - \frac{\omega_I}{2\Omega_0}$$

Fig. 3.46 Example of a ladder network of BS type

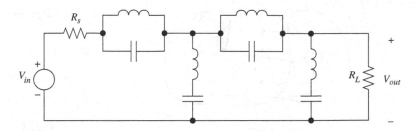

3.5.6 LP-BS Transformation

In a similar way, BS filters can be designed using the relations shown in Table 3.3. For bandstop filters, there are also some restrictions on the ladder network. A bandstop filter does not have any zeros in $s = \infty$ and $s = 0$. Thus, the series arms in the ladder network may not contain a series resonance circuit as it behaves as an open-circuit at $s = \infty$ and $s = 0$. The series branches must therefore be parallel resonance circuits. The shunt branches may not behave as short circuits at $s = \infty$ and $s = 0$. Thus, the shunt branches in the ladder network can only contain series resonance circuits. Figure 3.46 shows an example of a ladder network that can be used for bandstop filters.

3.6 Network Transformations

In this section, we will discuss a number of network transformations that are used to modify the LC network into a more suitable form. A well-designed filter should satisfy the attenuation requirements but should also be easy to manufacture and have the following desired features:

- A minimum number of inductors.
- Small element spreads, i.e., the ratio of the largest and smallest inductor (capacitor) should be small.
- The magnitude of the impedances in the series and shunt branches should be of the same order so that the voltages and currents are of the same order. This will tend to optimize the signal-to-noise ratio.
- Shunt inductors should have a capacitor in parallel to absorb the distributed capacitance of the coils.
- The floating node in a series resonance circuit in a series arm should have a capacitor to ground that absorb any parasitic capacitances.

- Proper distribution of the resonance frequencies in order to make the network less sensitive to deviations in the element values and minimize the element spreads.

3.6.1 Dual Networks

A network and its dual have identical response characteristics and identical voltage transfer functions, $H(s)$, but the input and output impedances may be different. The original network and its dual are described by the same set of differential equations, but voltage has been replaced by currents, resistors by conductors, inductors by capacitors, and capacitors by inductors while the first set of differential equations is derived from the loop and the other from the node equations.

Any planar network[11] can be transformed into its dual by the following rules:

- Place one node in each loop and one outside the network as shown in Fig. 3.47.
- Connect the nodes with a branch through each circuit element in the original network, traversing only one circuit element at a time.
- Place an inverse element with the same numeric element value between the nodes in the dual network. That is, an inductor $L = 3$ H corresponds to a capacitor $C = 3$ F in the dual network. A resistance is replaced with a conductance; that is, $R = 2 \, \Omega$ becomes $G = 2$ S.
- Change voltage sources into current sources and vice versa.

---

[11] A planar network can be drawn on a paper without any crossing branches.

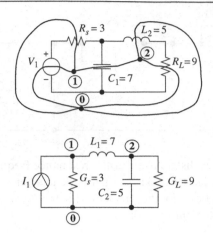

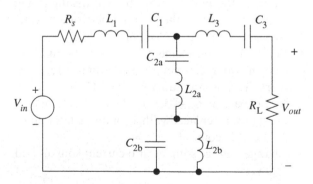

Fig. 3.47 Dual networks

These rules converts series branches into shunt branches and vice versa. The above method will fail for non-planar networks, but it can be modified to include those as well.

Thévenin's theorem, which was derived by von Helmholtz in 1853, and Norton's theorem[12], which was derived by H.F. Mayer in 1926, are examples of dual theorems. As frequently occurs in science and engineering, the name given to a theorem or law may not have been the first or even the main inventor and often more than one person deserve credit for developing a new concept.

Example 3.13 Consider the sixth-order BP ladder shown in Fig. 3.48 that was derived from a third-order Cauer filter, C035025, and derive its dual form. The normalized bandwidth and center frequency are 1 and 10, respectively.

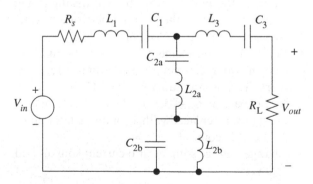

Fig. 3.48 Sixth-order BP ladder

The element values are

$$R_s = R_L = 1,$$

$$L_1 = L_3 = 2.08036 R_s/\omega_c = 2.08036,$$

$$C_1 = C_3 = 0.48069\omega_c/R_s\omega_I^2 = 0.48069\ 10^{-3},$$

$$C_{2a} = 6.2735\omega_c/R_s\ \omega_I^2 = 6.2735,$$

$$L_{2a} = L_{2b} = 1.1617 R_s\omega_c/\omega_I^2 = 11.617\ 10^{-3}, \text{ and}$$

$$C_{2b} = 0.8608/R_s\omega_c = 0.8608.$$

First, we introduce the nodes 0, 1, 2, and 3, each in a separate loop of the network, and the node 0 outside the network. Next, we draw a branch between the nodes through each circuit element. A few of these branches are shown in Fig. 3.49.

Next, we generate the dual network by connecting the dual elements with their numerical values between the corresponding nodes, as shown in Fig. 3.50. Voltage sources are replaced with current sources and impedances with admittances and vice versa. The input current source can be converted to a voltage source using Norton's theorem.

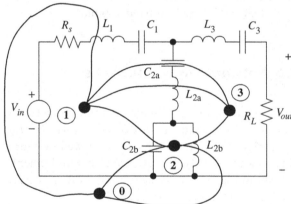

Fig. 3.49 Nodes with a few connected branches through the circuit elements.

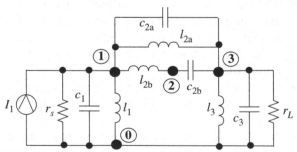

Fig. 3.50 Dual BP ladder

[12] Léon Charles Thévenin (1883) and E.L. Norton (1898–1983).

The element values in the dual network are

$$r_s = G_s = 1, \; r_L = G_L = 1,$$

$$c_1 = c_3 = 2.08036, \; l_1 = l_3 = 0.48069 \, 10^{-3},$$

$$c_{2a} = c_{2b} = 11.617 \, 10^{-3}, \; l_{2a} = 62.735 \, 10^{-3}, \; \text{and}$$

$$l_{2b} = 0.8608.$$

The spread in the element values are proportional to $(\omega_I/\omega_c)^2 = 100$. Hence, the element spread will be very large for BP filters with small relative bandwidths.

3.6.2 Symmetrical and Antimetrical Networks

It is of interest to consider symmetrical two-ports, as they may simplify the design, and experience indicates that symmetrical ladders have slightly better sensitivity properties. However, only a very limited subset of networks has any symmetry properties. We distinguish between electrical and structural symmetrical two-ports.

A structural symmetrical two-port can be divided with respect to the center line into two identical halves that are connected back-to-back as shown in Fig. 3.51 [140]. Unfortunately, only a few standard approximations can be realized by using a structurally symmetrical or antisymmetrical, which often is referred to as antimetrical, ladder network.

Definition 3.8 The relation between the input impedances at ports 1 and 2 for an electrically symmetrical two-port is

$$\frac{Z_1}{R_s} = \frac{Z_2}{R_L} \tag{3.46}$$

and for an electrically antisymmetrical two-port

$$\frac{Z_1}{R_s} = \frac{R_L}{Z_2}. \tag{3.47}$$

The ports in an electrically symmetrical two-port may be interchanged without any change in its electrical behavior. An electrically symmetrical two-port can sometimes be divided into two identical halves, i.e., if it also is structurally symmetrical, but there exist, networks where this is not possible. An example of such an electrically symmetrical network that is not structurally symmetrical is shown in Fig. 3.76.

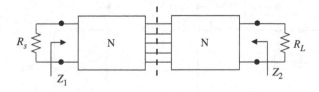

Fig. 3.51 Symmetrical network

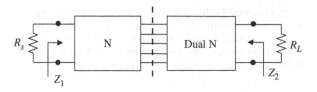

Fig. 3.52 Antimetrical network

A more stricter symmetry requirement is that both the structure and the circuit element values are symmetrical as well. Hence, then the network is also electrically symmetrical. A structurally antimetrical two-port has antisymmetry with respect to the symmetry line. That is, the two halves are dual networks, as shown in Fig. 3.52 [140].

Consider the ladder structures shown in Figs. 3.20 and 3.21. When the characteristic function, $C(s)$, is odd, then Equations (3.15) and (3.16) show that $\rho_1 = \rho_2$, and hence that $Z_1/R_s = Z_2/R_L$. Hence, the two-port networks are symmetric. If $C(s)$ is even, Equations (3.15) and (3.16) show that $\rho_1 = -\rho_2$, and hence that $Z_1/R_s, = R_L/Z_2$ and the two-port networks are antimetrical.

The two structures are dual with respect to the unit terminations and this causes them to have element values that are numerically equal. For example, if $Z_1 = 2.33s$ in the T ladder, then $Z_1 = 1/2.33s$ in the dual π ladder. If $R_s \neq R_L$, then the terminations will be G_s and G_L instead and the remaining element values will be numerically the same.

3.6.3 Reciprocity

Reciprocity is an important property of linear networks, consisting only of linear resistors (R), inductors (L), capacitors (C), and transformers (T) and one independent source. Generally, nonlinear networks and networks with dependent sources are nonreciprocal. Duality and reciprocity are often used to manipulate a network to satisfy termination requirements or to obtain a desired network.

Theorem 3.6: Lord Rayleigh's Reciprocity Theorem [13]
The ratio of a voltage (current) response at one port of a linear RLCT network to a current (voltage) source at another port of the network is the same if the response and source ports are interchanged as illustrated in Fig. 3.53.

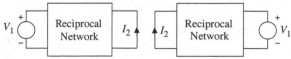

Fig. 3.53 Reciprocal networks

---

[13] Lord Rayleigh (John William Strutt).

In a reciprocal network [21], the response is the same regardless of which direction the signal flows, except for a constant factor. Note that the ratio is voltage to current, or current to voltage, and the theorem does not apply to voltage and current ratios. Note that a symmetrical network is reciprocal.

3.6.4 Bartlett's Bisection Theorem

A useful theorem for modifying a network to obtain suitable terminations is the bisection theorem by A.C. Bartlett (1927).

Theorem 3.7: Bartlett's Bisection Theorem *If a symmetrical network is bisected and one half is impedance scaled including the termination, the frequency response will not change.*

The symmetry requirement includes both the topology and element values. This includes all odd-order Butterworth and Chebyshev I filters, and third-order Chebyshev II and Cauer filters. This result allows many passive filters to be modified for unequal source and load resistances.

Example 3.14 Modify the ladder shown in Fig. 3.50 for a normalized source resistance of 3.

First, we identify the symmetry of the ladder as shown in Fig. 3.54. Next, we multiply the impedances in the left half of the ladder with 3 to increase the source resistance, e.g., $C_1 = c_1/3$, $L_1 = 3l_1$. The resulting structure is shown in Fig. 3.55.

The new element values are $R_s = 3$, $C_1 = c_1/3$, $C_{2a} = c_{2a}/6$, $C_{2b} = c_{2b}/6$, $L_1 = 3l_1$, $L_{2a} = 3l_{2a}/2$, and $L_{2b} = 3l_{2b}/2$. Finally, the inductors and capacitors in the series branches can be combined into simple elements.

3.6.5 Delta-Star Transformations

Delta-star transformations are typically used to modify a network into an alternative network or to reduce the spreads in the component values. The general delta-star transform is shown in Fig. 3.56 and a special case is shown in Fig. 3.57 that can be used to reduce the size of the inductors at the expense of the size of the capacitors.

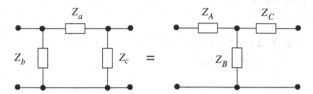

Fig. 3.56 Delta-star equivalence

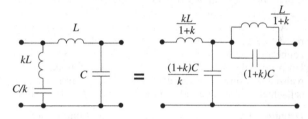

Fig. 3.57 Equivalent two-ports

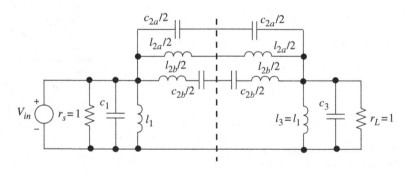

Fig. 3.54 Symmetrical BP ladder structure

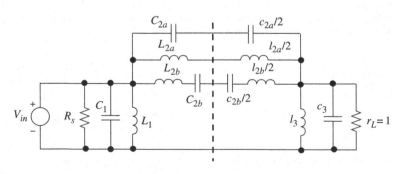

Fig. 3.55 Symmetrical BP ladder structure with unequal termintions

$$\begin{cases} Z_A = \frac{Z_a Z_b}{Z_a + Z_b + Z_c} \\ Z_B = \frac{Z_b Z_c}{Z_a + Z_b + Z_c} \\ Z_C = \frac{Z_a Z_c}{Z_a + Z_b + Z_c} \end{cases} \quad \begin{cases} Z_a = \frac{Z_A Z_C + Z_A Z_B + Z_B Z_C}{Z_B} \\ Z_b = \frac{Z_A Z_C + Z_A Z_B + Z_B Z_C}{Z_C} \quad (3.48) \\ Z_c = \frac{Z_A Z_C + Z_A Z_B + Z_B Z_C}{Z_A} \end{cases}$$

3.6.6 Norton Transformations

The Norton transformations, shown in Figs. 3.58 and 3.59, are useful in the elimination of redundant elements or for the modification of element values [12]. Norton transformations can also be used to reduce the inductance spread and their quality factors in narrow-band filters [55]. Moreover, using several successive network transformations, it is possible to eliminate any redundant elements and obtain equal source and load terminations. We demonstrate the use of Norton transformations by using two examples.

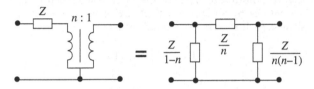

Fig. 3.58 Norton's first two-port equivalence

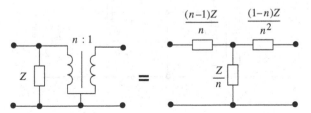

Fig. 3.59 Norton's second two-port equivalence

Note that if $n \neq 1$ in the circuits, then one of the series impedances will be negative. We will later show how to make use of this fact.

A transformer, with turns ratio $n{:}1$, can be removed by multiplying all impedances on the secondary side of a transformer with n^2. Hence, inductances and resistances are multiplied by n^2 and the capacitances by $1/n^2$. A voltage source V and current source I on the secondary side are replaced with a voltage source nV and current source I/n, respectively.

A transformer, with turns ratio $n{:}1$, can be inserted by dividing all impedances on the

secondary side of a transformer with n^2. Hence, inductances and resistances are multiplied by $1/n^2$ and the capacitances by n^2.

One caution must be observed here. The components must be carefully chosen to withstand the high voltages that may occur after applying this transformation.

3.6.7 Impedance Transformations

It is often of interest to modify the impedance level in a network, or part of a network, without changing its frequency response. For example, some element values may be difficult to realize. In other cases, it may be desirable to have a network with impedances of similar magnitudes so that all voltages and currents are of the same order. This will tend to improve the signal-to-noise ratio.

Figure 3.60 shows some equivalent one-ports that can be used to modify the size of the impedances. For example, if $Z_2 = sL_2$ is a too large inductor and Z_1 and aZ_1 are suitable sized capacitors, e.g., $a = 2$, in the left one-port, then the corresponding inductor in the equivalent one-port becomes $2^2 L_2/(2+1)^2 = 4L_2/9$.

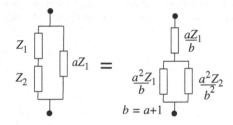

Fig. 3.60 Equivalent one-ports

The Norton transformations may also be used to change the impedance level in a part of the network.

Example 3.15 Consider the bandpass filter shown in Fig. 3.61, which has been derived from a third-order Chebyshev I filter with $A_{max} = 0.1$ dB and the normalized element values $L_{LP1} = L_{LP3} = 1.0316$ and $C_{LP2} = 1.1474$, and $R_s = R_L = 1$. The passband edges of the geometric symmetrical filter are $\omega_{c1} = 100$ Mrad/s and $\omega_{c2} = 105$ Mrad/s.

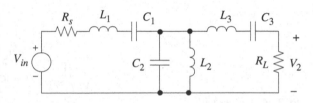

Fig. 3.61 Sixth-order Butterworth bandpass filter

The denormalized element values in Table 3.2 yield

$$L_1 = L_3 = \frac{L_{LP}}{\omega_{c2} - \omega_{c1}} = \frac{1.0316}{5 \cdot 10^6} = 0.20632 \,\mu H$$

$$C_1 = C_3 = \frac{\omega_{c2} - \omega_{c1}}{\omega_I^2 L_{LP}} \tag{3.49}$$

$$= \frac{5 \cdot 10^6}{1.05 \cdot 10^{16} \cdot 1.0316} = 0.4616 \,nF$$

and

$$C_2 = \frac{C_{LP}}{\omega_{c2} - \omega_{c1}} = \frac{1.1474}{5 \cdot 10^6} = 0.22948 \,\mu F$$

$$L_1 = L_3 = \frac{L_{LP}}{\omega_{c2} - \omega_{c1}} = \frac{1.0316}{5 \cdot 10^6} = 0.20632 \,\mu F \tag{3.50}$$

$$C_1 = C_3 = \frac{\omega_{c2} - \omega_{c1}}{\omega_I^2 L_{LP}} = \frac{5 \cdot 10^6}{1.05 \cdot 10^{16} \cdot 1.0316} = 0.4616 \,nF.$$

From Equations (3.49) and (3.50), we find that the element spread is inversely proportional to the relative bandwidth squared. Because the relative bandwidth is about 5%, the element values $C_2 \gg C_1$ and $L_2 \ll L_1$, making the filter difficult to realize. The element spread is about 497 for both the capacitors and inductors.

In order to make the element spread smaller, we insert a transformer on the left side of the shunt branch with turns ratio n_1:1. To compensate for the transformer, we divide all impedances on the secondary side of the transformer with n_1^2. The output voltage will be V_2/n_1. The resulting network is shown in Fig. 3.62.

In the next step, we use Norton's first equivalence, where the capacitor C_1 corresponds to Z in Fig. 3.58, to replace the transformer. We get the network shown in Fig. 3.63.

In the next step, a second transformer with turns ratio n_2:1 is inserted to the right of the inductor L_2, as shown in Fig. 3.64, where $n_3 = n_1 n_2$.

We again use the first Norton equivalence with capacitor $n_3^2 C_3$ as impedance. Note that the turns ratio in Fig. 3.58 is now $n = 1/n_2$, as the equivalence circuit has been flipped over. To retain equal the source and load resistor, we select $n_1 n_2 = 1$.

The new element values shown in Fig. 3.66 are

$$C_{3a} = \frac{(1 - n_2)}{n_2^2} n_3^2 C_3 = \frac{(1 - n_2)}{n_2^2} C_3$$

$$C_{3b} = \frac{n_3^2 C_3}{n_2} = \frac{C_3}{n_1}$$

$$C_{3c} = \frac{(n_2 - 1)}{n_2} n_3^2 C_3 = \frac{(n_2 - 1)}{n_2} C_3.$$

The capacitors, $n_1(n_1 - 1)C_1$ and C_{3a} will be negative if $n_1 < 1$. It is possible to eliminate the effective shunt capacitor, C_x, entirely by an appropriate choice of n_1. The effective shunt capacitor C_x shown in Fig. 3.65 is

$$C_x = n_1(n_1 - 1)C_1 + n_1^2 C_2 + (1 - n_2)n_1^2 C_3.$$

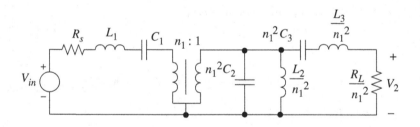

Fig. 3.62 Ladder with inserted transformer

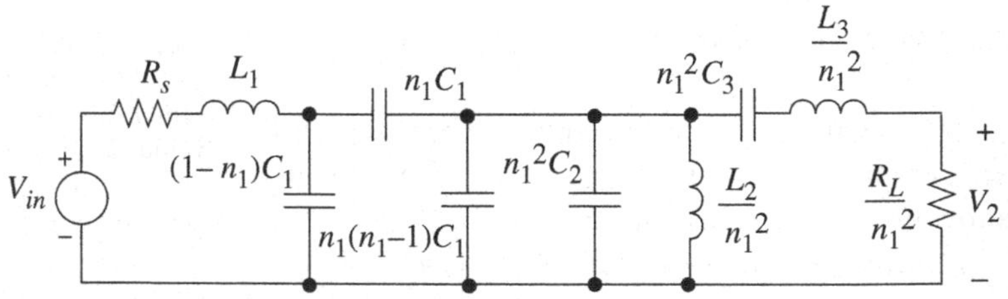

Fig. 3.63 The transformer is removed by using Norton's first equivalence

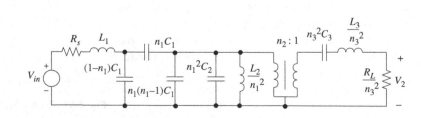

Fig. 3.64 Ladder with a second transformer inserted

Fig. 3.65 The transformer is removed by using Norton's first equivalence

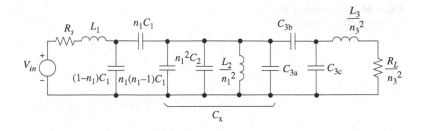

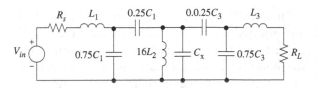

Fig. 3.66 Final structure with less element spread

If we make $C_x = 0$ in order to save one capacitor, then $n_1 = 4.00692 \cdot 10^{-3}$, $n_2 = 249.568$, and the spreads in element values will become $(n_2 - 1)C_3/n_1 n_2 C_1 = 248.6$, which is somewhat large. In practice, however, the shunt inductor will in practice have a parasitic capacitance that may be absorbed into C_x. If we select, for example, $n_1 = 0.25$ and $n_2 = 4$, which yields $C_x = 14.169$ nF, the element spread is reduced to about 122.8 and 16 for the capacitors and inductors, respectively.

3.6.8 Transformations to Absorb Parasitic Capacitance

One problem with the standard synthesis methods is that they result in networks, as shown in Fig. 3.67, that have nodes that are not connected to ground through a shunt branch. Such high-impedance nodes are sensitive to a stray capacitance. In order to remove such sensitive nodes, we may perform one or several successive Norton transformations.

If the LP filter has finite zeros, as shown in Fig. 3.67, the corresponding bandpass filter will have series and shunt branches that contain four circuit elements. This is not good from a sensitivity point of view. Therefore, different types of network transformations should be applied, i.e., Norton transformations, to obtain a BP filter structure without high-impedance nodes [26, 95, 100, 137, 146]. The filter shown in Fig. 3.67 is very sensitive for a stray capacitance, C_{stray}.

Example 3.16 Consider the bandpass filter shown in Fig. 3.67, which has only one critical node between the two parallel resonance circuits in the shunt arm.

First, we place a transformer with turns ratio $n:1$ immediately after the second parallel resonance circuit, $L_{2b}-C_{2b}$ and divide all impedances on the right side of the transformer with n^2. Next, we apply Norton's second equivalence, using the parallel resonance circuit $L_{2b}-C_{2b}$ as impedance, to obtain the ladder network shown in Fig. 3.68.

Fig. 3.67 Frequency transformation of an LP filter with finite zeros to a BP filter

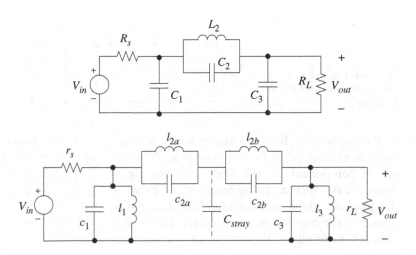

Fig. 3.68 BP structure that is less sensitive to parasitics

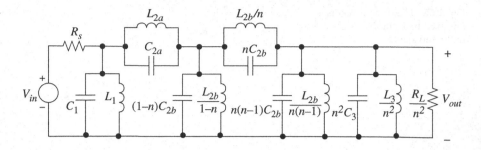

The negative shunt components that result from the transformation are combined in parallel with the positive shunt inductors and capacitors to obtain realizable component values

$$n(1-n)C_{2b} + n^2 C_3 = 0 \Rightarrow n = \frac{C_{2b}}{C_{2b} - C_3} \quad (3.51)$$

or alternatively if we select to eliminate the inductance

$$\frac{L_{2b}}{n(n-1)} \frac{L_3}{n^2 n(n-1)} \frac{L_{2b}}{n^2} + \frac{L_3}{n^2} + \frac{L_{2b} L_3}{n^2 L_{2b} + n(n-1) L_3}$$
$$= 0 \Rightarrow n = \frac{L_3}{L_{2b} + L_3}. \quad (3.52)$$

The value of n may be chosen so that only one additional shunt capacitor or inductor is required. In most cases, we will select to remove an inductor. The two rightmost parallel resonance circuits can be combined to a single circuit. The resulting circuit has four inductors and five capacitors. The Norton transformations will usually not maintain equal source and termination resistances.

3.6.9 Minimum-Inductor Filters

Inductors that are implemented using magnetic cores are bulky, heavy, and expensive compared

In so-called parametric filters, two real zeros are inserted into the characteristic function in order to trade off some of the attenuation in the stopband for a more suitable network structure, or save an inductance, or to have certain relationship between some of the element values, etc. These zeros will either cancel a transmission zero at $s = 0$ or $s = \infty$.

Definition 3.9 A canonical network realizes an impedance with a minimum number of circuit elements.

A transmission zero at $s = 0$ or $s = \infty$ can be obtained using only a series (or shunt) capacitor, but the realization of a finite (nonzero) $j\omega$-axis zero pair, however, requires an inductor. Each pair of real zeros (realized with a C section) also requires one tapped inductor.

Theorem 3.8: *For LP and HP filters, the minimum number of inductors equals the number of finite (nonzero) zero pairs.*

Figure 3.69 shows examples of lowpass and highpass ladders where all inductors are used to realize finite zeros.

Fig. 3.69 Examples of minimum-inductor LP and HP filters

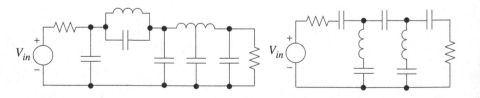

to the capacitors. In applications where a larger number of filters shall be manufactured, the number of components in the filter structure are of economical concern. In addition, in cases when the physical space is restricted, it is essential to minimize the number of components. Hence, it is desirable to minimize the number of inductors.

For BP filters, the situation is more involved. Bandpass filters with at least two zeros at both $s = 0$ and $s = \infty$ can be realized with a minimum-inductor ladder with $N/2$ inductors, where N is the order of the filter. That is, the number of inductors equals the number of zeros, including those at $s = 0$ and $s = \infty$. So-called parametric BP filters require only $N/2 - 1$ inductors [29, 123, 124].

Figure 3.70 shows four additional sections that are useful for realizing BP filters with a minimum number of inductors. The two sections to the left realize transmission zeros below the passband whereas the two on the right realize transmission zeros above the passband.

5 realizes a transmission zero at s = 0 and s = ∞, sections 2 and 4 realize a transmission zero in the upper stopband, and section 3 realizes transmission zeros in the lower stopband.

Example 3.17 Consider the network in Fig. 3.72, which contain, a capacitive loop $C_1 - C_2 - C_3$. The order of the transfer function is therefore one less than the number of reactive components and one capacitor can therefore be eliminated.

Inserting a transformer with turns ratio $n : 1$ in the cut AA'

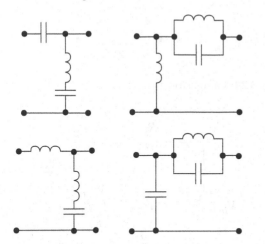

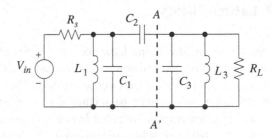

Fig. 3.72 Ladder structure with a capacitive loop

Fig. 3.70 BP sections for realization of finite transmission zeros

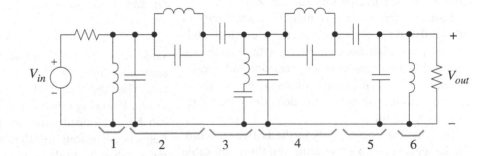

Fig. 3.71 Laurent[14] zig-zag BP filter with minimum number of inductors

Figure 3.71 shows a Laurent[14] zig-zag bandpass filter with double zeros at both $s = 0$ and $s = \infty$, where network transformations have been used to obtain a minimum number of inductors and a small spread in the element values [33, 48, 49, 100]. The Norton transformations can be used to reduce the number of inductors. Note that sections 1 and 6 realize a transmission zero at s = 0, section

and using the first Norton equivalence yields the network shown in Fig. 3.73. It is possible to eliminate either the capacitor resonating with L_1 or with L_2 by inserting the transformer appropriately. Alternatively, if neither of the capacitors is eliminated, instead, the output impedance is modified. To eliminate the capacitors resonating with L_1, we select $C_1 + (1 - n)C_2 = 0$. That is, $n = (C_2 + C_1)/C_2$.

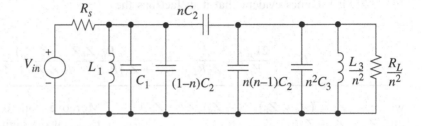

Fig. 3.73 The transformer inserted in the cur A-A' is removed by using Norton's first equivalence

Alternatively, we can eliminate the capacitors resonating with L_2 by selecting

[14]Torben Laurent, Royal Institute of Technology, Sweden.

$$C_3 = 0, \text{ which yields} \qquad n(n-1)C_2 + n^2$$

$$n = C_2/(C_2 + C_3) < 1.$$

The load resistor R_L/n^2 will be larger than R_L. Note that if n is small, then the voltages will be large and the components must be chosen to withstand the high voltages.

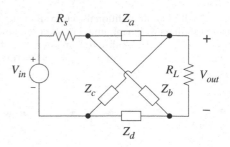

Fig. 3.74 Lattice structure

3.7 Lattice Filters

Passive filters can also be implemented using different types of electromechanical resonators instead of LC resonance circuits, e.g., piezoelectric crystals. These types of filters, which will be further discussed in Section 3.9, often employ so-called *lattice structures*; a name suggested by its schematic shown in Fig. 3.74, with quartz crystals in the branches. The lattice structure can accommodate such resonators much more easily than the corresponding ladder structure.

Lattice filters have extremely high sensitivity in the stopband for errors in the circuit elements, but this is acceptable because the quartz crystals are very stable with respect to temperature and ageing. Several millions of passive filters with crystals or ceramic circuit elements are nowadays manufactured each month.

In principle, the crystals in the lattice structure can be replaced with LC circuits, but this is unusable in practice due to large variations in the element values. Frequency selective lattice filters with regular RLC components are therefore not recommended. Moreover, lattice structures are not canonical and require more than a minimal number of circuit elements.

By redrawing the lattice section, according to Fig. 3.75, it becomes evident that it in fact has the

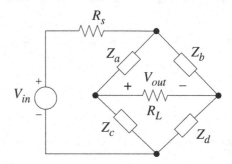

Fig. 3.75 Lattice filter

same structure as a measurement bridge. Using a sinusoidal voltage source, the output voltage becomes zero when the bridge is balanced, i.e., when $Z_a/Z_c = Z_b/Z_d$. Knowing three of the impedances, the fourth can be determined with very high accuracy at that frequency. The four impedances are usually pure reactances when the bridge is used as a filter.

When the bridge is balanced, i.e., at a frequency that corresponds to the transmission zero, the output signal is $V_{out} = 0$. A small variation of any of the reactances will make V_{out} nonzero. Hence, the element sensitivity in the stopband is very large.

The normalized transfer function of the lattice filter in Fig. 3.75 is

$$H_{(s)} = \frac{2V_{out}}{V_{in}} = \frac{R_L(Z_b Z_c - Z_a Z_d)}{Z_1 R_L + (Z_2 + Z_3)R_s R_L + R_s Z_2 Z_3 + Z_3 Z_a Z_b + Z_2 Z_c Z_d} \qquad (3.53)$$

where $Z_1 = (Z_a + Z_c)(Z_b + Z_d)$, $Z_2 = Z_a + Z_b$, and $Z_3 = Z_c + Z_d$.

The lattice structure contrary to ladder structures can realize zeros in the right half of the s-plane. This is useful for realizing allpass filters.

Moreover, one of the best digital filter structures, lattice wave digital filters, is derived from analog lattice structures [135]. In fact, lattice structures are both of practical and theoretical interest in spite of their very high stopband sensitivity.

3.7.1 Symmetrical Lattice Structures

Consider the symmetrical lattice filter shown in Fig. 3.76, where $Z_a = Z_d$, $Z_b = Z_c$, and $R_s = R_L = R$. Equation (3.53) simplifies for the symmetrical lattice to

$$H(s) = \frac{2V_{out}}{V_{in}} = \frac{R(Z_b - Z_a)}{(Z_a + R)(Z_b + R)}$$
$$= \frac{1}{2}\left(\frac{Z_b - R}{Z_b + R} - \frac{Z_a - R}{Z_a + R}\right) \quad (3.54)$$

or

$$H(s) = R\left(\frac{1}{R + Z_a} - \frac{1}{R + Z_b}\right). \quad (3.55)$$

The reflection functions are

$$\rho_1(s) = \rho_2(s) = \frac{1}{2}\left(\frac{Z_b - R}{Z_b + R} + \frac{Z_a - R}{Z_a + R}\right). \quad (3.56)$$

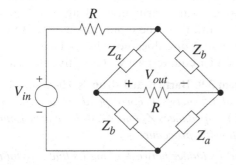

Fig. 3.76 Symmetrical lattice filter

If Z_a and Z_b are pure reactances, the two terms in Equation (3.54) become allpass functions. The impedance Z_a and Z_b can, in principle, be determined by performing a partial fraction expansion of transfer function and assigning the appropriate terms to Z_a and Z_b. There is usually only one assignment among the terms that leads to lattice reactances with positive element values.

It can be shown that lattice filters only can realize odd lowpass and highpass filters and even-order bandpass filters with zeros at both $s = 0$ and $s = \infty$. Another disadvantage of the symmetric lattice structure, besides the very high sensitivity, is the large number of components required, i.e., twice the order of the filter.

3.7.2 Synthesis of Lattice Reactances

In this section, we demonstrate by the means of an example how the lattice reactances of a symmetric lattice can be derived.

The lattice reactances can be computed with the following scheme.

- Find all the transmission zeros and the stopband edge ω_s.
- Compute the characteristic function, which must be an odd function of ω.
- Form a polynomial $X(s)$ by adding the numerator polynomial, multiplied by ε, and denominator polynomial, $p(s)$, of the characteristic function. The roots of $p(s)$ are the transmission zeros.
- Compute the roots of the polynomial $X(s)$ and assign the roots in the left-half s-plane to $h(s)$ (Hurwitz) and the roots in the right-half s-plane to $ah(s)$ (anti-Hurwitz). That is, $X(s) = h(s)\,ah(s)$. A useful program for this step is hurwitz.
- Compute the two lattice reactances, see Theorem 3.10,

$$\text{if } p(s) \text{ is even } Z_a = -R\frac{ah_o(s)}{ah_e(s)} \quad Z_b = -R\frac{h_e(s)}{h_o(s)}$$

and

$$\text{if } p(s) \text{ is odd } Z_a = -R\frac{ah_o(s)}{ah_e(s)} \quad Z_b = -R\frac{h_o(s)}{h_e(s)}.$$

The subscripts e and o indicate polynomials formed by deleting the odd and even terms of the polynomial, e.g., $h(s) = a s^3 + b s^2 + c s + d$ yields $h_o(s) = a s^3 + c s$ and $h_e(s) = b s^2 + d$.

The realization of Z_a and Z_b, can be done by any of the structures discussed in Section 3.7.7.

Apart from a phase shift with π, lattices with any of the reactance pairs (Z_a, Z_b), (Z_b, Z_a), $(R^2/Z_a, R^2/Z_b)$, or $(R^2/Z_b, R^2/Z_a)$ have the same transfer function.

Example 3.18 Determine the lattice reactances for the third-order Cauer filter C031511 when the passband edge is $\omega_c = 1$ rad/s.

Synthesizing the filter, we find that the stopband edge is $\omega_s = 5.2408$ rad/s, $A_{max} = 0.0988324$ dB, $A_{min} = 50.625$ dB, and $\varepsilon = 0.1517165$. The filter has two transmission zeros at $s = \pm j\,6.04657$ rad/s and one at $s = \infty$. The characteristic function is, see Equation (2.49), is

$$C(s) = \varepsilon R_N\left(\frac{s}{\omega_c}, L\right) = 21.68806\frac{s(s^2 + 0.7512366)}{s^2 + 36.561032}.$$

We form the polynomial $X(s) = 21.68806\,s(s^2 + 0.7512366) + (s^2 + 36.56103)$ and solve for its roots

$$X(s) = 21.68806(s - 0.474289 + j1.212363)(s - 0.474289 - j1.212363)(s + 0.994686) =$$
$$= 21.68806(s^2 - 0.948578s + 1.694774)(s + 0.994686).$$

We assign the root in the left half plane to $h(s)$, i.e., $h(s) = s + 0.994686$, and the remaining to $ah(s)$,

i.e., $ah(s) = s^2 - 0.948578\,s + 1.694774$. Because $p(s)$ is odd, we get

$$Z_a = -R\frac{ah_o(s)}{ah_e(s)} = -R\left(\frac{-0.948578}{s^2 + 1.694774}\right) = \frac{RsL}{1 + s^2LC}.$$

Hence, a normalized inductor $L = 0.988578$ in parallel with a capacitor $C = 1.7866$. R is a factor that determines the impedance level, i.e., if $R = 1\,\Omega$ then $L = 0.988578$ H

$$Z_b = -R\frac{h_o(s)}{h_e(s)} R\frac{0.994686}{s}.$$

That is, a capacitor $C = 1.00534$.

The computations above have, of course, been carried out with higher precision than shown.

3.7.3 Element Sensitivity

A major problem of lattice structures is that transmission zeros are realized through a balancing of the branches, i.e., by making $Z_a = Z_b$ at each zero frequency. Furthermore, the impedances must track at all frequencies, which requires a large number of low-tolerance components, making the lattice structure expensive.

Methods to realize transmission zeros, which are based on the principle that two currents are summed in a circuit node, or the difference in voltage between two nodes, as in the lattice structure, are very sensitive for errors in the element values and should therefore be avoided.

The high element sensitivity in the stopband makes, in practice, the lattice structure with coils and capacitors unusable for frequency selective filters. However, lattice structures with crystals in the bridge arms are useable. This works well despite the high element sensitivity in the stopband because the crystals are highly stable and the variations in the element values are very small.

The element sensitivity in the passband is, however, small if the lattice section is designed according to the insertion loss method.

3.7.4 Bartlett and Brune's Theorem

Lattice filters are related to symmetrical ladder filters. A symmetrical ladder filter has always an equivalent lattice structure having equal transfer function and the same input impedance. However, a lattice structure does not always have a symmetrical ladder counterpart with positive elements.

Theorem 3.9: Bartlett and Brune's Theorem *A symmetrical LC network corresponds to a lattice filter with the impedances Z_a and Z_b, which can be found by the following procedure:*

1. *Cut the ladder network along its line of symmetry into two identical parts.*
2. *Short-circuit all nodes that lie on the line of symmetry. The input impedance to this network is Z_a.*
3. *The input impedance of the network (one of the parts) is Z_b.*

Figure 3.37 show some equivalent networks derived by using Bartlett and Brune's theorem. According to Bartlett and Brune's theorem, a symmetrical LC ladder network corresponds to a lattice filter, but the opposite is not always true. Hence, not all minimum-phase transfer functions can be realized by a lattice filter. For example, only lowpass (highpass) filters of odd order and with at least one zero at $s = \infty(s = 0)$ are realizable.

Example 3.19 A symmetrical ladder network is shown in Fig. 3.78 with $L_1 = L_5$ and $C_2 = C_4$. Derive the reactances in the corresponding lattice network.

Fig 3.77 Equivalent
networks based on Bartlett
and Brune's theorem

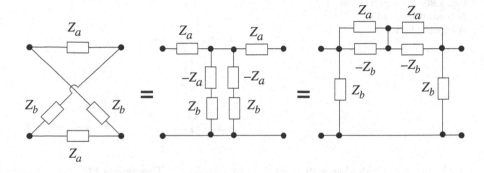

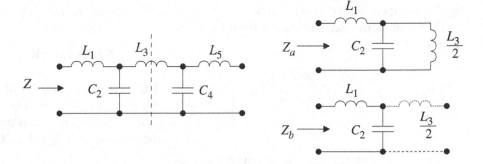

Fig. 3.78 Example of the
use of Bartlett and Brune's
theorem

First, we partition the symmetrical network into two identical parts along the line of symmetry. The inductor L_3 is split into two inductors in series, each with the value $L_3/2$. The reactance Z_a equals the input impedance when the nodes at the line of symmetry are short-circuited. The reactance Z_b equals the input impedance to the remaining part of the network when the nodes are not short-circuited. In this case, the inductor $L_3/2$ is superfluous.

Note that the order of the impedances Z_a and Z_b will always be different.

3.7.5 Bridged-T Networks

A lattice filter can be realized in an equivalent form with the bridged-T network shown in Fig. 3.79. The bridged-T network has the advantage that the source and load have a common ground.

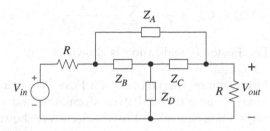

Fig. 3.79 Bridged-T network

The lattice impedances corresponding to the impedances in a *symmetrical bridged-T network*, i.e., $Z_B = Z_C$ and $R_s = R_L = R$, is directly obtain by using Bartlett and Brune's theorem.

We get

$$\begin{cases} Z_a = \dfrac{\dfrac{Z_A}{2} Z_B}{\dfrac{Z_A}{2} + Z_B} = \dfrac{Z_A Z_B}{Z_A + 2Z_B} \\ Z_b = Z_B + 2Z_D. \end{cases} \quad (3.57)$$

An unabridged network is obtained by letting $Z_A \to \infty$. We get

$$\begin{cases} Z_B = Z_C = Z_a \\ Z_D = \dfrac{Z_b - Z_a}{2}. \end{cases} \quad (3.58)$$

3.7.6 Half-Lattices

The number of circuit elements can be reduced by using one of the equivalent unbalanced circuits that are shown in Fig. 3.80. These circuits, called (incorrectly) half-lattices, trade half of the components of

Fig. 3.80 Examples of two-port equivalents of a symmetric lattice

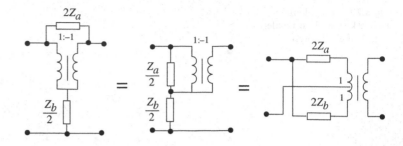

the lattice for an ideal transformer [12]. In addition, these circuits have higher sensitivity in the stopbands than their ladder counterparts.

The chain matrix, see Section 5.4, for the circuits shown in Fig. 3.80 is

$$\begin{bmatrix} V_1 \\ I_1 \end{bmatrix} = \begin{bmatrix} \dfrac{Z_a + Z_b}{Z_b - Z_a} & \dfrac{2Z_aZ_b}{Z_b - Z_a} \\ \dfrac{2}{Z_b - Z_a} & \dfrac{Z_a + Z_b}{Z_b - Z_a} \end{bmatrix} \begin{bmatrix} V_2 \\ -I_2 \end{bmatrix}. \quad (3.59)$$

Lattices and half-lattices are used mostly to realize filters with complex zeros and, especially, filters using electromechanical resonators.

3.7.7 Reactance One-Ports

To obtain a better realization, it may be advantageous to replace an impedance with an equivalent one-port. In this section, we discuss four generic structures for realizing (lossless) reactances and some often used special cases.

Let $F(s)$ denote a reactance, then $F(s)$ is an odd positive-real function with the following properties [124]:

- $F(s)$ is an odd function, i.e., $F(s) = -F(-s)$, with real coefficients
- $F(s) \rightarrow K/s$ or Ks when $s \rightarrow \infty$
- $F(s)$ have only simple poles on the $j\omega$-axis with real and positive residues
- The poles and zeros are alternating on the $j\omega$-axis
- $Re\{F(s)\} = 0$.

Theorem 3.10: *The ratio of the even and odd, or odd and even, parts of a Hurwitz polynomial is an impedance function.*

Theorem 3.11: *The minimum number of reactances necessary to realize a lossless impedance or admittance is equal to its degree.*

3.7.7.1 Foster I and II

The Foster I realization of an arbitrary reactance function $X(s)$ is obtained by a partial fraction expansion of the impedance, which can be performed by the function PART_FRACT_ EXPANSION,

$$X(s) = L_\infty s + \frac{1}{C_0 s} + \sum_{i=1}^{N} \frac{1}{C_i} \frac{s}{s^2 + \omega_i^2} \quad (3.60)$$

The Foster I structure [23] is shown in Fig. 3.81 where $L_i C_i \omega_i^2 = 1$.

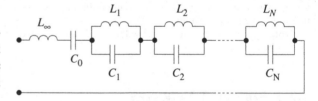

Fig. 3.81 Foster I structure

The Foster II realization of an arbitrary reactance function $X(s)$ is obtained by instead expanding $1/X(s)$

$$Y(s) = C_\infty s + \frac{1}{L_0 s} + \sum_{i=1}^{N} \frac{1}{L_i} \frac{s}{s^2 + \omega_i^2}. \quad (3.61)$$

The Foster II realization is shown in Fig. 3.82 where $L_i C_i \omega_i^2 = 1$.

Two common special cases of a Foster II circuit that may occur in the LP-BP transformation and its corresponding equivalent Foster I circuit are shown in Figs. 3.83 and 3.84.

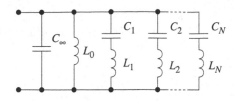

Fig. 3.82 Foster II structure

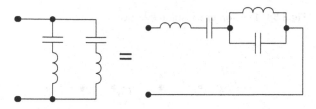

Fig. 3.83 Equivalent Foster I and II structures

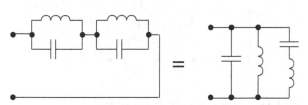

Fig. 3.84 Equivalent Foster I and II structures

3.7.7.2 Cauer I and II

The Cauer I realization [23] is obtained by a continued fraction expansion of $X(s)$ around $s = \infty$ according to

$$X(s) = L_{\infty 1}s + \cfrac{1}{C_{\infty 2}s + \cfrac{1}{L_{\infty 3}s + \cfrac{1}{C_{\infty 4}s + \ldots}}} \quad (3.62)$$

The Cauer I realization is shown in Fig. 3.85. Note that the reactance can either be terminated with an open-circuit or short-circuit for an even-order and odd-order reactance, respectively. For reactances with order ≤ 3, the Cauer I and Foster I structures become identical.

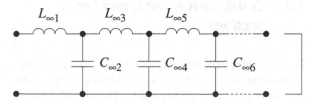

Fig. 3.85 Cauer I structure

The Cauer II realization is obtained by expanding $X(s)$ around $s = 0$ according to

$$X(s) = \frac{1}{C_{01}s} + \cfrac{1}{\cfrac{1}{L_{02}s} + \cfrac{1}{\cfrac{1}{C_{03}s} + \cfrac{1}{\cfrac{1}{L_{04}s} + \ldots}}} \quad (3.63)$$

The Cauer II realization is shown in Fig. 3.86. Note that the reactance can either be terminated with an open-circuit or short-circuit. For reactances with order ≤ 3, the Cauer II and Foster II structures become identical.

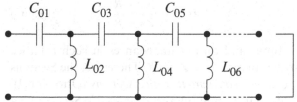

Fig. 3.86 Cauer II structure

Deffinition 3.10 A canonical reactance network has a minimum number of inductors and capacitors.

Both the Foster and Cauer networks are canonical and require a minimum number of inductors and capacitors. Non-canonical networks use more than the minimum number of components to realize an impedance function.

3.8 Allpass Filters

Here we will only discuss lattice structures of allpass type, which are realized with a symmetrical lattice section, i.e., with $R_s = R_L$, $Z_a = Z_d$ and $Z_b = Z_c$, or the corresponding equivalent bridged-T network. The latter network has the advantage of only requiring two impedances and a tapped inductor whereas the lattice structure requires four impedances. Furthermore, the bridged-T network has common ground between input and output.

To equalize the group delay of a minimum-phase filter, $H_m(s)$, i.e., a filter that does not have any zeros in the right-half s-plane, we use an allpass filter in cascade with $H_m(s)$. In fact, a ladder network can only realize transfer functions of minimum-phase type.

An allpass filter, however, has all zeros in the right-half s-plane is are therefore a maximum-phase filter.

3.8.1 Constant-R *Lattice Filters*

If we choose

$$Z_a Z_b = R^2 \qquad (3.64)$$

where Z_a and Z_b are pure reactances, an allpass filter with interesting properties is obtained. The transfer function is

$$H(s) = \frac{Z_b - R}{Z_b + R}. \qquad (3.65)$$

Input and output impedances in such a lattice structure are $Z_i = Z_o = R$. Therefore, the Sections are called *constant-R lattice sections* (Otto Zobel). Note that the multiplexers discussed in Section 3.5.4 also have constant input impedance, $Z_i = R$, but they are not of allpass type.

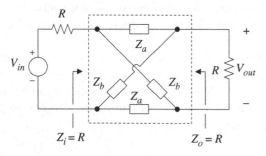

Fig. 3.87 Constant-R lattice

The transfer functions for the two allpass filters shown in Figs. 3.88 and 3.89 are given by Equations (3.64) and (3.65).
We get

$$H_{AP1} = -\frac{s - R/L}{s + R/L} \qquad (3.66)$$

where $C = L/R^2$ and

$$H_{AP2} = \frac{s^2 - \dfrac{s}{RC_2} + \dfrac{1}{L_2 C_2}}{s^2 + \dfrac{s}{RC_2} + \dfrac{1}{L_2 C_2}} \qquad (3.67)$$

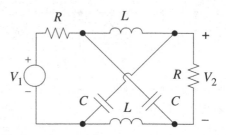

Fig.3.88 First-order constant-R lattice

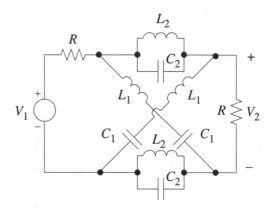

Fig.3.89 Second-order constant-R lattice

where $C_1 = L_2/R^2$ and $L_1 = C_2 R^2$.

3.8.2 Constant-R *Bridged-T Sections*

In order to reduce the number of components compared to the lattice structure, it is common to instead use bridged-T networks. Another advantage is that they have a common ground. A constant-R bridged-T network is obtained using the network shown in Fig. 3.79 with $Z_B = Z_C = R$ and $Z_A = Z_D = R^2$.

The transfer functions for the networks shown in Fig. 3.90 are given by Equations (3.66) and (3.67), respectively. The first-order section, however, requires a negative inductor that can, according to Fig. 3.28, be realized by a tapped coil.

3.8.3 Constant-R *Right-L and Left-L Sections*

The two networks shown in Fig. 3.91 are also constant-R networks, when terminated with a resistor R, and if Equation (3.64) holds. The transfer functions are

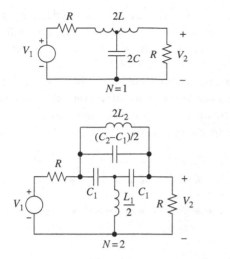

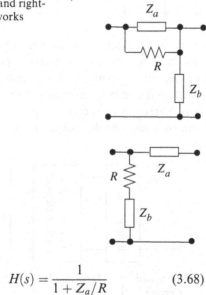

Fig. 3.90 Bridged-T networks

Fig. 3.91 Left-$L$and right-Lconstant-R networks

$$H(s) = \frac{1}{1 + Z_a/R} \qquad (3.68)$$

Constant-R sections are useful for realizing a Hilbert filter, which is a three-port where the two output ports have the same constant magnitude function, but their phase-difference approximates 90° over a prescribed band of frequencies. The structure resembles the diplexer shown in Figure 3.41, but the two branches consist of constant-R sections, i.e., allpass sections.

3.8.4 Equalizing the Group Delay

Theorem 3.12: *The group delay of a reciprocal loss-less two-port is an even nonnegative function of ω.*

Several sections can be cascaded according to Fig. 3.92. Because every section has the load resistor, R, we can replace it with a new lattice structure with the input impedance R, and so on. The lattice sections will not interact.

Figures 3.88 and 3.89 show two examples of first- and second-order lattice structures [11, 26, 146] and the corresponding bridged-T networks are shown in Fig. 3.90.

Example 3.20 Determine the element values in a third-order allpass filter built of cascaded bridged-T networks that equalizes the group delay of the ladder network in Example 2.4 when $R_s = R_L = 1000 \, \Omega$.

The poles of the allpass filter were in Example 2.11 determined to

$$s_{p1} = -7.1328393364267 \pm j\, 6.5766989368398 \text{ krad/s}$$
$$s_{p2} = -7.0319414698343 \pm j\, 19.3771121673633 \text{ krad/s}$$
$$s_{p3} = -6.7508088487487 \pm j\, 32.4760875022130 \text{ krad/s}.$$

From Equation (3.67), we obtain the element values in the lattice structures. We get using $1/RC_2 = -2 \, Re\{s_p\}$ and $1/L_2C_2 = r_p^2$ and $R = 1 \, k\Omega$.

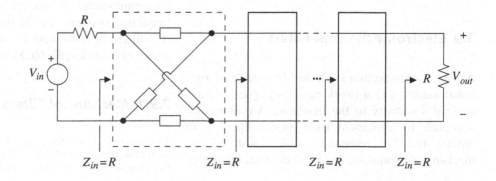

Fig. 3.92 Cascaded constant-R sections

Bridged-T	L_1[mH]	L_2[mH]	C_1[nF]	C_2[nF]
1	70.09831	151.55237	151.5524	70.098312
2	71.10412	33.09767	33.09767	71.10412
3	74.06520	12.27119	12.27119	74.06520

The element values of the LP filter were determined using CA_LADDER. The resulting equalized LP filter is shown in Fig. 3.93.

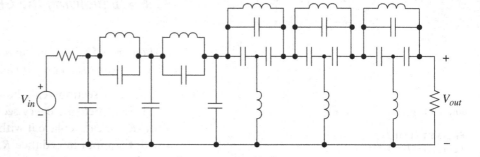

Fig. 3.93 LP filter with equalized group delay

From the lattice structures, the element values can then be computed in the corresponding bridged-T network. In practice, bridged-T networks are used because they have fewer components than lattice structures and they have common ground between the input and output. Note that the allpass sections can be put in any order at the output, or at the input, or a combination thereof, as $R_s = R_L$ in this case.

If a general synthesis program is available, we may instead of using this approach synthesize a ladder with one or several allpass sections inside the ladder as this often lead to a less sensitive overall structure.

3.8.5 Attenuation Equalizing

Constant-resistance lattice sections and bridge-T networks can also be used to equalize the attenuation; for example, to reduce the attenuation in a neighborhood of the passband edge that is deteriorated because of use of coils with too low Q factors. However, in this case the sections are not allpass sections.

3.9 Electromechanical Filters

As discussed in Section 3.3, a doubly resistively terminated lossless LC network can be designed to have optimal sensitivity in the passband. An alternative approach to implement such passive filters is to replace the LC resonance circuits with lossless mechanical resonators. Electromechanical filters that realize a doubly resistively terminated lossless network with maximum power transfer have therefore also optimal sensitivity in the passband. However, electromechanical filters are solely bandpass filters.

In an electromechanical filter, the input signal carrier (voltage or current) is converted into mechanical form, e.g., vibrations, by means of an input transducer, as illustrated in Fig. 3.94. Typically, the signal carrier after the input transducer is a force or velocity. The actual filtering takes place within the mechanical structure, which is composed of resonators and coupling elements. Finally, the filtered signal carrier is then converted back into electrical form by means of the output transducer.

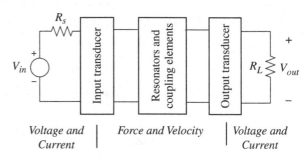

Fig. 3.94 Electromechanical bandpass filter

Electromechanical filters are typically used in telecommunication systems, for example in carrier frequency systems and in intermediate frequency (IF) filters in radio and TV sets as well as in consumer electronics [57, 60, 80].

3.9.1 Mechanical Filters

True mechanical BP filters can be manufactured with the center frequency in the interval 10 kHz to 10 MHz. Mechanical filters, with constant modulus iron-nickel alloy resonators, can be implemented with Q factors up to 10,000, and with better stability than conventional filters built with coils, capacitors, and active elements. For example, it is possible to build a mechanical IF filter with a bandwidth of 500 Hz, centered at 455 kHz, i.e., the relative bandwidth is only 0.11%. Significant rounding of the passband response, due to losses, is avoided due to the high Q factors of the resonators. The center frequency shift may be as low as 50 Hz for a 100°C change in temperature.

Figure 3.95 shows the principle for implementation of a mechanical filter. The transducer converts a voltage to a force using a coil with a magnetorestrictive core, which changes its dimensions proportionally to the magnetic field. The core will therefore vibrate analogously to the input signal. The vibrations in the core are then transferred to the resonator structure by means of thin wires. The wires act as springs when they are less than one-eighth acoustic wavelength long. The frequency response is determined by the resonance frequencies and the coupling between the resonators. The resonators can be shaped in many different ways, i.e., long rods or disks.

The resonators shown in Fig. 3.95 are torodial resonators. Finally, at the output, the vibrations in the last resonator are connected to another magnetorestrictive core. The changes in its dimensions produce a magnetic field that generates a proportional voltage in the output coil.

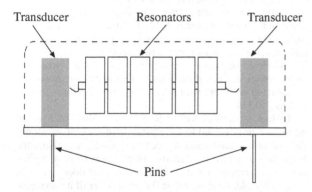

Fig. 3.95 Example of a mechanical filter

Frequency selective filters with mechanical circuit element are characterized by

- Very large Q factors
- Very narrow banded bandpass filters
- Aging and temperature stability
- Low cost
- High reliability
- Cannot be integrated in an IC and have therefore been replaced by integrated circuit compatible technologies.

3.9.1.1 Integrated Microelectromechanical Filters

MEMS (microelectromechanical systems) is an emerging technology that uses the same CMOS processes as for regular integrated circuits to manufacture very small mechanical systems on silicon. Typical subsystems are switches for antennas, IF filters, front-end RF filtering, and demodulation in cellular phones.

The CMOS process that is used has the potential advantage of mass fabrication of the filters, but they do not have sufficient geometrical precision. Typically, the error in the resonator frequency is about 0.5% and may cause significant passband deviation. The resonator structure must therefore, for highly frequency selective filters, be tuned to adjust the filter characteristics. The tuning involves special processes for removing or adding material from the resonators.

Most microelectromechanical filters are implemented using one or several flexural-mode beam elements on top of the silicon substrate, as shown in Fig. 3.96. The beam, which is suspended in free air, is clamped at both ends to the substrate and connected to electrode 2 at one end. Underneath the beam is electrode 1. Both the beam and electrodes are made of conductive materials, such as metal or doped silicon.

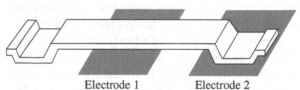

Electrode 1   Electrode 2

Fig. 3.96 Clamped-clamped beam resonator

The micromechanical resonator structure, shown in Fig. 3.96, has two electrical inputs, V_1 and V_2, which are applied to the electrode-1 and the beam through electrode 2, respectively. Hence, a voltage $(V_2 - V_1)$ is applied across the electrode-1-to-beam gap, generating a force between electrode 1 and the movable beam. A variation in the voltage $(V_2 - V_1)$ will therefore produce a proportional vibration in the beam. Several such resonator structures can be electrically or mechanically coupled to implement a higher order filter. At the output, an electrical signal can be obtained, as the vibrating beam will modulate the electrode-to-beam capacitance.

The clamped-clamped resonators are usable for frequencies from about 250 kHz to 100 MHz, and Q factors of up to 10,000 can be obtained. However, at frequencies over 30 MHz, the mechanical energy loss to the substrate becomes large and limits the attainable Q factors. To obtain high Q factors at high frequencies requires beams with both ends free, but this requires additional complex mechanical circuitry.

3.9.2 Crystal Filters

The main difference between a mechanical band-pass filter and a crystal or ceramic filter is that, in the case of mechanical filters, both the resonators and their coupling are mechanical, whereas in crystal and ceramic filters, the resonators are made out of piezoelectric materials and the coupling is accomplished electrically using capacitors and inductors.

Piezoelectricity, which was discovered by the Pierre Curie and Jacques Curie in the 1880s, is found in many crystalline materials, e.g., quartz. One of its first practical usages was in World War I, when sonar systems used quartz crystals to generate directed pulses at audio frequencies to detect submarines.

The piezoelectric effect is stronger in ceramics, but quartz is still widely used, because it has low acoustic losses and good temperature stability.

A resonator consists of a freely supported parallel-sided piezoelectric plate with an electrode on each major face. When a voltage is applied, the impedance is found to show a sharp resonance due to the excitation of acoustic waves, reflected many times between the two faces. Quartz resonators are used as the controlling element of oscillators used in clocks, watches, and in electronic equipment, wherever accurate frequency or timing information is needed.

The resonator structure, shown in Fig. 3.94, consists of piezoelectric crystal or ceramic resonators. The transducers convert between a voltage to pressure or vice versa. Hence, a varying input voltage will make the resonator to vibrate proportionally.

Crystal and ceramic filters are also based on doubly resistively terminated reactance networks that are designed for maximum power transfer, but the conventional resonance circuits consisting of inductors and capacitors are replaced with mechanical resonators. Crystal filters are able to provide narrow bandwidth filtering without excessive loss, temperature drift, and ageing.

Figure 3.97 shows the symbol for a crystal or ceramic resonator and a corresponding electrical model. The capacitance C_p represents the capacitance of the electrodes.

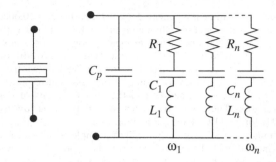

Fig. 3.97 Symbol for a crystal and corresponding model

The equivalent electric circuit contains several (lossy) series resonance circuits, each representing a particular harmonic mode. The resonator can, for example, oscillate in parallel resonance mode, i.e., L_1-C_p, or in series resonance mode, L_1-C_1. In the first case, the impedance of the resonator becomes large whereas in the series resonance case it becomes small. For quartz, the series-resonant circuit, C_1-L_1, has an extremely high Q factor, ranging from about 10^4 to 10^6. Hence, the deterioration of the passband edges due to losses is small.

We may assume that the crystal is operated in the close neighborhood of one of its resonance frequencies, e.g., f_1, and the influence of the other branches (resonances) are neglected. This is, of course, only valid in a narrow frequency range around f_1. This assumption is only valid in the design and implementation of very narrow bandpass filters where it is possible to replace subcircuits in an LC ladder structure with crystals, or crystal-capacitors. This approach, however, is restricted to filters with extremely narrow relative bandwidths $< 0.1\%$.

Ladder structures with relative bandwidths of 5%, or more, are not realizable by crystals and capacitors alone, because of too small capacitance ratios, C_1/C_p. It is therefore better to use a lattice structure, which tends to have higher capacitance ratios than its ladder counterpart does.

In principle, we may realize the whole filter (if it is symmetrical) by a single lattice, but because of sensitivity reasons, it is better to use a cascade of simple lattices and ladder sections. Often, the best structure is a cascade of simple lattice sections with shunt or series capacitors or inductors between the sections. Lattice structures are very sensitive for circuit element errors, but it is still possible to use such structures, as the crystals are very stable [48, 59].

Figure 3.98 shows a *half lattice section* with two crystals, a capacitor, and a transformer. This type of crystal filter is

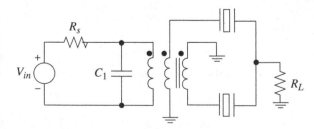

Fig. 3.98 Example of simple crystal filter

often used in the IF part of more expensive receivers. Today, millions of passive filters with crystals or ceramic circuit resonators are manufactured annually and used in most radio and TV receivers.

For filters with relative bandwidths of more than about 10% it is only possible to realize a few selected attenuation peaks by embedding crystals into an *LC* ladder structure. Figure 3.99 shows an example of the implementation of a simple crystal filter. See [80, 123] for a more detailed discussion of electromechanical filters and their design.

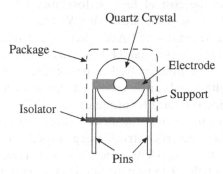

Fig. 3.99 Example of simple crystal

Characteristic of crystal filters, which always are very narrow band BP filters, are that passbands of only a couple of hundred Hz and with center frequencies of tens of MHz can be implemented. Sometimes the term form factor is used, which is defined as the bandwidth at the attenuation 60 dB divided with the bandwidth at the attenuation 6 dB. The form factor can be less than 1.5. Crystals can be manufactured with center frequencies in the range 1.5–400 MHz and with relative bandwidths in the range 0.004–0.05%. Crystal filters have small losses and good signal dynamic.

3.9.3 Ceramic Filters

Ceramic filters are based on piezoelectric ceramics, and the design principles resemble those of crystal filters [123]. Piezoelectric ceramic resonators can only be used to realize narrow bandpass filters. Ceramic resonators have lower Q factors compared to crystal resonators. Typically, Q factors is of the order a few thousand. Ceramic filters are also used in radio and TV sets. A common application is IF filters with, e.g., 10.7 MHz center frequency. The dimension of crystal and ceramic filters is small. For example, a ceramic filter, from Murata™, which is intended for Bluetooth applications, consisting of three resonators, has been implemented in a small package (2 × 4.5 × 4 mm). The filter is a bandpass filter with passband 2400–2500 MHz with a

variation of 2.7–3 dB in the passband and attenuating at least 25 dB at 1950 MHz, 16 dB at 2200 MHz, and 9.5 dB at 2700 MHz.

3.9.4 Surface Acoustic Wave Filters

The first surface acoustic wave filters (SAWs) appeared in the mid-1960's. A SAW device operates by converting electrical energy, through an interdigital transducer (IDT), into a surface acoustic wave on a piezoelectric crystal plate (substrate). The output is typically obtained through a second IDT, as illustrated in Fig. 3.100, but there are resonator structures with a single port. The transducer, which consists of an array of metallic electrodes on the surface of the substrate, determines the frequency response. Another basic SAW component is the reflector, which consists of periodic arrays of either metal strips or grooves.

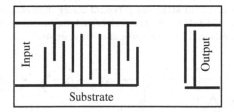

Fig.3.100 SAW filter

By combining IDTs and reflectors, SAW resonators and resonator filters can be implemented for frequencies in the range from about 10 MHz up at a few GHz. The Q factor of SAW resonators is relatively low, typically between 6000 and 12,000. SAW filters can be designed to have linear phase, contrary to analog filters realized with lumped elements. SAWs are relatively easy to manufacture because they require only one or two photoresist and metal deposition steps.

Surface acoustic wave filters (*SAW filter*) are commonly used in radio, TV sets, video players, WLAN, and cellular phones [122, 123]. Diplexers for WCDMA are also implemented in SAW technologies. Surface acoustic wave filters are based on a special type of wave propagation mode of acoustic waves, so-called Rayleigh waves[15], on the surface of a piezoelectric material. However, there exist similar devices that use different types of wave propagation modes. Usually quartz, lithium tantalate (LiTaO3), or lithium niobate are used. Surface acoustic wave filters for cellular phones are

---

[15] Lord Rayleigh,1885.

small, a few mm², but they cannot easily be implemented on top of an integrated circuit.

Surface acoustic wave filters can be manufactured with very narrow transition bands. In principle, surface acoustic wave filters can be designed with linear phase, but in practice, a small phase distortion is obtained. Typically, ±1° to ±5°, and the passband has a constant attenuation of a few dB. Stopband attenuations of 40 to 60 dB can be obtained. Passband ripples as low as ±0.3 dB is achievable. Usable frequency ranges are 10 MHz– 4 GHz and the relative bandwidths are in the range 0.1% to 10%. For example, a SAW filter for GSM with $A_{max} = 2$ dB has a bandwidth of 25 kHz that is centered at 902.5 MHz (transmitter) and 947.5 MHz (receiver). The weight is only 0.15 g and with the dimensions $3.8 \times 3.8 \times 1.6$ mm.

Today, several hundred millions of surface acoustic wave filters are manufactured annually. SAW filters are used in most cellular phone and TV sets. Currently, extensive research is ongoing with aim at replacing SAW filters with other filtering techniques, which can be integrated in a regular integrated circuit, or radio architectures that alleviate the filtering requirements.

Figure 3.101 shows a simplified block diagram for the receiver in a GSM phone with two down converter stages. The first stage is followed by a surface acoustic wave filter (SAW) of bandpass type with a center frequency of approximately 85 MHz and a bandwidth of approximately 6 MHz while the second stage has a typical center frequency in the range 10–20 MHz. This transceiver architecture is expensive, and simpler and less expensive alternative exists.

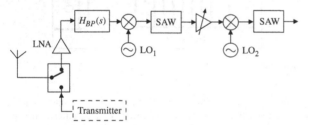

Fig. 3.101 Simplified block diagram for the receiver of a GSM phone

3.9.5 Bulk Acoustic Wave Filters

In the past decade, many other types of acoustic wave filters have been developed in order to achieve ultralow power consumption and greater integration levels, particularly in RF transceivers. For example, devices operating with different types of waves: bulk acoustic waves, Bleustein–Gulyaev waves, Lamb waves, and Love waves. Bulk acoustic wave devices are much smaller than their electromagnetic distributed element counterparts are, as acoustic waves propagate about four to five orders of magnitude slower than electromagnetic waves.

BAW resonators and filters are suitable for wireless communication systems, as they can operate in the range 2–20 GHz with very high Q factors, and have large power capability and low volume. Furthermore, BAW devices have an advantage over technologies such as ceramic or SAW filters, as they can be fabricated on top of integrated circuits. Such a co-integration reduces even further the size and the cost of high-performance RF frontends. Bulk acoustic wave (BAW) and film bulk acoustic resonator (FBAR) filters are determined to replace conventional RF filters in cellular phones as they have now evolved in performance beyond SAW filters and can be manufactured at a very low cost using standard IC process steps.

Thin film bulk acoustic wave resonators (FBARs) convert electrical energy into an acoustic wave, but unlike SAW devices, the wave is directed into the bulk. The primary mode is a longitudinal wave, whereas a SAW employs either a Rayleigh wave or a surface-skimming wave.

There are two main types of resonators, air gap and solidly mounted resonators (SMRs). The first type, which is illustrated in Fig. 3.102, consists of a piezoelectric membrane, sandwiched between two electrodes, that is supported at the edges, but free from the substrate in the resonator region. The electrodes can be directly connected to the top level metal interconnect of the integrated circuit that is used as substrate. The most commonly used piezoelectric material for BAW devices is aluminum nitride (AlN).

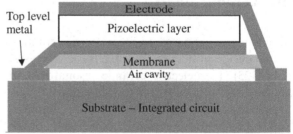

Fig. 3.102 Suspended membrane BAW resonator

An alternative mode of operation is to use Lamb waves (1988). A Lamb wave resonator (LWR) is also composed of a thin piezoelectric layer sandwiched between two thin electrodes on top of the membrane. The piezoelectric layer is small compared to the wavelength. However, whereas the FBAR operates using vertical wave propagation, the LWR employs lateral wave propagation. Lamb wave resonators used in IF filters offers high quality factors (>1000) and on top of IC integration.

The second type of resonator also consists of such a membrane, but the air gap is replaced with a sequence of quarter-wavelength thick layers with alternating high and low acoustic impedances, as illustrated in Fig. 3.103

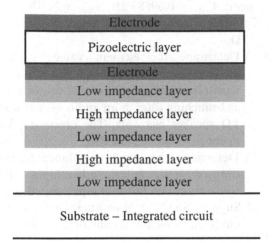

Fig. 3.103 Solidly mounted BAW resonator

The acoustic wave that is generated in the resonator has a component that propagates vertically downwards. Multiple reflections from these layers form a standing wave that approximates a free air surface reflection. Hence, acoustically, the membrane appears to be suspended in free air. In addition, in this case, the electrodes can be directly connected to the top-level metal interconnect of the integrated circuit that is used as substrate.

The resonance frequency of a BAW resonator is determined by the thickness of the piezoelectric layer and the neighboring layers. In an LWR, the resonance appears when the length of the piezoelectric layer $L = \lambda/2$. The BWR and LWR have a resonance and an anti-resonance and can be used as resonators in both ladder and lattice structures as well as in oscillators. However, the special design techniques that are required to effectively use only one of the resonances are beyond the scope of this book.

The required tolerance for the resonance frequency is around 0.1% for filters used in typical communication systems, and the tolerance requirements are in the same range for the piezoelectric layer and the electrodes. These stringent thickness tolerances cannot be met by standard tools for IC processes. Hence, there are still problems regarding thickness tolerances to be solved.

Thin film BAW filters that are based on AlN or ZnO have high Q factors and can handle high signal power and operate at frequencies in the range 2–16 GHz where they are expected to replace SAW filters. In addition, both FBAR and SMR based filters and oscillators have small sizes and their implementation is compatible with standard IC processes. The main advantages of BAW over the SAW filters are intrinsically low insertion loss, the smaller temperature drift, and the possibility to be implemented on top of IC devices.

3.10 Problems

3.1 A coil has the inductance 1 mH and $Q = 500$ at 10 kHz. Determine the size of the equivalent series resistor in the simple model.

3.2 Determine the transfer function of a doubly resistively terminated LC network when the LC part consists of a capacitor and inductor in parallel in series branch.

3.3 Show that the ratio of dissipated power, P_2, and maximally transferred power, P_2max, in a doubly resistively terminated LC ladder network is proportional to $|H(j\omega)|^2$.

3.4 Prove the maximum power transfer theorem.

3.5 Compute the output signal for a doubly resistively terminated LC ladder network with $Rs = RL$ that realizes the transfer function C055030 when the input signal is $V_{in}(t) = 2\sin(\omega t)$ for $\omega = 0$ and $\omega = \omega_c$. Also compute the maximal output power when the filter is terminated with equal resistors.

3.6 a) A filter realized with a T network with $R_s = R_L = 600\ \Omega$, $L_1 = 600$ mH, $C_2 = 3.3\ \mu F$, and $L_3 = 600$ mH has the desired frequency response. Determine the element values if the load resistor is changed to 1 kΩ.

   b) The passband edge, 1 krad/s, but we need a passband edge of 3 MHz. Determine the new element values.

3.7 A lowpass RC section has been normalized so that the attenuation is 3.0102 dB at $\omega_c = 1$,

$R = 1$, and $C = 1$. Denormalize the section so that the 3 dB edge occurs at $f_c = 1.2$ kHz and $R = 10$ kΩ.

3.8 Determine suitable element values using the MATLAB program for ladder network of π type with $R_s = R_L = 50$ Ω that meets the following specification: $A_{max} = 0.01$ dB, $\omega_c = 200$ Mrad/s, $\omega_s = 700$ Mrad/s, and $A_{min} = 50$ dB. The filter shall be of type

a) Butterworth filter
b) Chebyshev I filter
c) Chebyshev II filter
d) Cauer filter

3.9 Determine suitable element values for an LC ladder network of Butterworth type that meets the requirements: $A_{max} = 0.1$ dB, $\omega_c = 100$ krad/s, $\omega_s = 600$ krad/s, and $A_{min} = 60$ dB.

a) T network with $R_s = R_L = 50$ Ω
b) π network with $R_s = R_L = 50$ Ω
c) π network with $R_s = 600$ Ω and $R_L = 300$ Ω

3.10 Determine suitable element values for an LC ladder network of Chebyshev I type that meets the requirements: $A_{max} = 0.5$ dB, $f_c = 500$ kHz, $f_s = 2500$ kHz, and $A_{min} = 30$ dB.

a) T network with $R_s = R_L = 50$ Ω
b) π network with $R_s = R_L = 50$ Ω

3.11 Synthesize a lowpass LC ladder network with a ripple less than $A_{max} = 0.5$ dB in the passband and with $f_c = 22$ MHz, $f_s = 33$ MHz, and $A_{min} = 24$ dB. Use a T ladder with $R_s = R_L = 50$ Ω.

3.12 Determine suitable element values in a T ladder with $R_s = R_L = 50$ Ω that realizes a fifth-order Cauer filter with $A_{max} = 0.1$ dB, $\omega_c = 100$ Mrad/s, $\omega_s = 215$ Mrad/s, and $A_{min} = 40$ dB.

3.13 Synthesize a doubly resistively terminated ladder network of π type that realizes the filter C031534. Use equal termination resistances of 50 Ω. The passband edge is 800π Mrad/s.

3.14 Derive the expressions for the LP-HP transformation of the specification, resistors, inductors, and capacitor.

3.15 Realize an LC ladder of Butterworth type with minimum order that meets the specification: $A_{max} = 0.5$ dB, $A_{min} = 25$ dB, $f_c = 15$ kHz, and $f_s = 6$ kHz. Use a π ladder. $R_s = R_L = 50$ Ω.

3.16 Realize an LC ladder that meets the same specification as in Problem 5.2 but of Chebyshev I type.

3.17 Realize a doubly resistively terminated LC ladder of Cauer type that meets the requirement: $A_{max} = 0.09883$ dB, $A_{min} = 28$ dB, $f_c = 7.5$ MHz, and $f_s = 2$ MHz. Select $R_s = R_L = 50$ Ω.

3.18 a) Determine the element values in the diplexer, shown in Fig. 3.104, with a Butterworth frequency response that can be used as branching filter in an audio system with 8 Ω speaker elements. The crossover frequency should be 2.5 kHz.

b) Determine the input impedance to the diplexer. Discuss any advantages with this particular choice of input impedance.

c) Suggest how the branching filter can be modified to include a subwoofer as well.

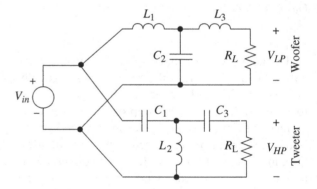

Fig. 3.104 Diplexer

3.19 Derive the expressions for the LP-BP transformation of the specification, resistors, inductors, and capacitors.

3.20 Determine the element values in a Butterworth ladder structure that meets the requirement: $\rho = 15\%$, i.e., $A_{max} = 0.09883$ dB, $A_{min} = 40$ dB, $\omega_{c1} = 40$ Mrad/s, $\omega_{c2} = 90$ Mrad/s, $\omega_{s1} = s$, and/s, and $\omega_{s2} = 3/s$. Use a π ladder network with $R_s = R_L = 100$ Ω. Determine the number of zeros for $s = 0$ and $s = \infty$ and the number of poles for the BP filter.

3.21 Realize a minimal order Chebyshev I LC ladder that meets the following specification: $\rho = 15\%$, i.e., $A_{max} = 0.09883$ dB, $A_{min} = 40$ dB, $\omega_{c1} = 12\pi$ krad/s, $\omega_{c2} = 18\pi$ krad/s, $\omega_{s1} = 8\pi$ krad/s, and $\omega_{s2} = 24\pi$ krad/s. Use a T ladder with $R_s = R_L = 500\ \Omega$.

3.22 Synthesize a Chebyshev I ladder filter of minimal order that meets the requirement: $\rho = 15\%$, i.e., $Amax = 0.09883$, $A_{min} = 22$ dB, $\omega_{c1} = 10\pi$ krad/s, $\omega_{c2} = 20\pi$ krad/s, $\omega_{s1} = 2\pi$ krad/s, and $\omega_{s2} = 40\pi$ krad/s. Use a π ladder with $R_s = 500\ \Omega$ and $R_L = 1\ k\Omega$.

3.23 Realize a Chebyshev I LC ladder filter that meets the requirement: $\rho = 15\%$, i.e., $A_{max} = 0.09883$ dB, $A_{min} = 22$ dB, $\omega_{c1} = 1000\pi$ krad/s, $\omega_{c2} = 2200\pi$ krad/s, $\omega_{s1} = 500\pi$ krad/s, and $\omega_{s2} = 4200\pi$ krad/s. Use a T ladder with $R_s = R_L = 600\ \Omega$.

3.24 Determine the element values in a Cauer ladder that meets the requirement: $\rho = 15\%$, i.e., $A_{max} = 0.09883$ dB, $A_{min} = 40$ dB, $\omega_{c1} = 40$ Mrad/s, $\omega_{c2} = 90$ Mrad/s, $\omega_{s1} = 24$ Mrad/s, and $\omega_{s2} = 150$ Mrad/s. The ladder network shall be of π type and double resistively terminated with $R_s = R_L = 100\ \Omega$.

3.25 Derive the expressions for the LP-BS transformation of the attenuation specification, resistors, inductors, and capacitors.

3.26 Determine the element values in double resistively terminated π $R_s = R_L = 100\ \Omega$. The filter shall be of Chebyshev I type and meet the specification:

$$\omega_{c1} = 2\pi\ 49.5\ \text{rad/s} \qquad \omega_{c2} = 2\pi\ 50.5\ \text{rad/s}$$
$$\omega_{s1} = 2\pi\ 48.5\ \text{rad/s} \qquad \omega_{s2} = 2\pi\ 51.5\ \text{rad/s}$$
$$A_{max1} = A_{max2} = 2\text{dB},\ A_{min} = 35\ \text{dB}$$

3.27 Derive the dual network and its element values to a π ladder that realizes the Cauer filter C051523.

3.28 Find the dual of the network in Fig. 3.45.

3.29 Derive the transfer function for a symmetrical lattice structure and determine necessary constraints on Z_a, Z_b, and R if the network shall realize the transfer function in Equation (3.65) that corresponds to a constant-R lattice section.

3.30 Show how the group delay of a doubly resistively terminated ladder network can be equalized using constant-R lattice sections.

Determine also suitable element values for an allpass filter that realizes the poles: $s_{p0} = -1$ and $s_{p1,2} = -3 \pm j\,4$.

3.31 Show the input impedance of the bridged-T network in Fig. 3.105 is $Z_{in} = R$ and determine the transfer function when $Z_a Z_b = R^2$.

3.32 Find the lattice equivalent to the bridged-T network shown in Fig. 3.106.

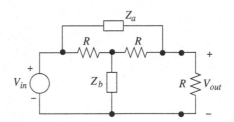

Fig. 3.105 Brided-T network

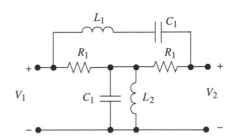

Fig. 3.106 Bridged-T network

3.33 Verify that the bridged-T and the right-L and left-L networks are constant-R networks when terminated with a resistor R and if $Z_a Z_b = R^2$. Verify also that the voltage transfer functions are.

3.34 Validate Equations (3.51) and (3.52) and derive the resulting element values if one inductor is eliminated.

3.35 a) Where in the s-plane can the following structures realize poles and zeros.
  b) Doubly resistively terminated ladder network.
  c) Doubly resistively terminated lattice network.
  d) Why is a doubly resistively terminated ladder preferred over a singly resistively terminated ladder, which has one fewer component?
  e) Give an example of application in which a singly resistively terminated ladder may be an appropriate choice.

Chapter 4
Filters with Distributed Elements

4.1 Introduction

At microwave frequencies, the lumped elements discussed in Chapter 3 tend to become small and therefore difficult to manufacture. In addition, the lumped inductors and capacitors depart from their ideal characteristics due to radiation, loss, and propagation effects. In this chapter, we will discuss the synthesis of analog filter structures that are based on distributed circuit elements. There exist a large number of components that must be modeled by distributed circuit elements. To discuss all filter structures based on commonly used components and associated design techniques is beyond the scope of this book. The interested reader may find a wealth of material in [7, 50, 53, 93, 95–97, 99].

Here we will focus on a subset of components that can be modeled by uniform transmission lines. Moreover, we will focus on lossless transmission lines that can be used to realize doubly resistively terminated reactance networks that are optimal from an element sensitivity point of view if they are designed for maximum power transfer. Uniform lossless transmission lines may be implemented using stripline and microstrip techniques.

These structures are used in many practical applications, but also used to derive digital counterparts, so-called wave digital filters [135]. Because these digital filter structures simulate a doubly resistively terminated reactance network, they will have the same sensitivity properties as the analog filter structures discussed in Chapter 3.

4.2 Transmission Lines

Any distributed component can, in principle, be analyzed by solving Maxwell's equations with the appropriate boundary conditions. However, this general approach does not support synthesis and optimization of filter structures. Here we will therefore discuss a simpler technique that allows us to map the synthesis problem into a form that allows us to utilize the wealth of the lumped-element theory. A disadvantage of this method is that it does not model the electromagnetic field pattern or possible modes of propagation.

An implementation of a transmission line that consists of two or more conductors may support transverse electromagnetic (TEM) waves, characterized by the lack of longitudinal field components. That is, the E and H fields are orthogonal as well as almost orthogonal to the direction of propagation. TEM waves have a uniquely defined voltage, current, and characteristic impedance.

Waveguides, typically consisting of a single tube conductor, support transverse electric (TE) and/or transverse magnetic (TM) waves, which are characterized by the presence of longitudinal magnetic or electric, respectively, field components. A unique definition of characteristic impedance is not possible for such waves. Hence, the underlying assumptions in Section 4.2 are not valid for these waves, and we will henceforth not discuss these structures.

The dimensions and electrical properties of a uniform transmission line are identical at all planes orthogonal to the direction of propagation. A uniform transmission line may be modeled by short line

L. Wanhammar, *Analog Filters Using MATLAB*, DOI 10.1007/978-0-387-92767-1_4,
© Springer Science+Business Media, LLC 2009

segments. Each segment consists of series induc-
tance $l\Delta x$ and resistance $r\Delta x$ as well as shunt capacitance $c\Delta x$ and conductance $g\Delta x$, as shown in
Fig. 4.1.

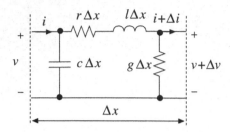

Fig. 4.1 Lumped-element model of a segment of a uniform
transmission line

r = series resistance per unit length of line (Ω/m)
l = series inductance per unit length of line (H/m)
g = shunt conductance per unit length of line (S/m)
c = shunt capacitance per unit length of line (F/m).

The series resistance r is due to the finite conduc-
tivity of the metallic conductors. Because of the skin
and proximity effects, it is also a function of frequency. The conductance g is due to loss in the
insulating material between the conductors. The
series inductance l depends on the magnetic flux
linking the conductors, and c depends on the
charges on the conductors.

Consider a uniform transmission line of length
d that is connected to a voltage source at $x = 0$.
We model a small piece Δx of the line according to
Fig. 4.1. The voltage and current v and i are
approximately

$$\begin{cases} v = \left(ri + l\dfrac{\partial i}{\partial i}\right)\Delta x + v + \Delta v \\ i = \left(gv + c\dfrac{\partial v}{\partial t}\right)\Delta x + i + \Delta i \end{cases} \quad (4.1)$$

or

$$\begin{cases} \Delta v + l\dfrac{\partial i}{\partial t}\Delta x + ri\Delta x = 0 \\ \Delta i + c\dfrac{\partial v}{\partial t}\Delta x + gv\Delta x = 0. \end{cases} \quad (4.2)$$

We get by dividing Equations (4.2) by Δx and
taking the limit as $\Delta x \to 0$

$$\begin{cases} \dfrac{\partial v}{\partial x} + l\dfrac{\partial i}{\partial t} + ri = 0 \\ \dfrac{\partial i}{\partial x} + c\dfrac{\partial v}{\partial t} + gv = 0. \end{cases} \quad (4.3)$$

The derivatives of Equations (4.3) with respect to
x and to t yield

$$\begin{cases} \dfrac{\partial^2 v}{\partial x^2} + l\dfrac{\partial^2 i}{\partial t\partial x} + r\dfrac{\partial i}{\partial x} = 0 \\ \dfrac{\partial^2 i}{\partial x\partial t} + c\dfrac{\partial^2 v}{\partial t^2} + g\dfrac{\partial v}{\partial t} = 0. \end{cases} \quad (4.4)$$

From Equations (4.3) and (4.4), we get the so-
called telegraphist equations for the voltage and
current on the line by changing the order of the
partial derivatives

$$\begin{cases} \dfrac{\partial^2 v}{\partial x^2} - lc\dfrac{\partial^2 v}{\partial t^2} - (lg + rc)\dfrac{\partial v}{\partial t} - rgv = 0 \\ \dfrac{\partial^2 i}{\partial x^2} - lc\dfrac{\partial^2 i}{\partial t^2} - (lg + rc)\dfrac{\partial i}{\partial t} - rgi = 0. \end{cases} \quad (4.5)$$

Assume that the signal source is a sinusoidal voltage
source. Then for the voltage and current at a point x at
the time t, we make use of the complex notation

$$\begin{cases} v(x, t) = Re\{V(x)e^{j\omega t}\} \\ i(x, t) = Im\{I(x)e^{j\omega t}\}. \end{cases} \quad (4.6)$$

Here $V(x)$ and $I(x)$ are the complex amplitudes of
voltages and currents along the line. Equation (4.5)
can be rewritten as

$$\begin{cases} \dfrac{\partial^2}{\partial x^2}V(x) - (r + j\omega l)(g + j\omega c)V(x) = 0 \\ \dfrac{\partial^2}{\partial x^2}I(x) - (r + j\omega l)(g + j\omega c)I(x) = 0. \end{cases}$$

Finally, we get

$$\begin{cases} \dfrac{\partial^2}{\partial x^2}V(x) - \gamma^2 V(x) = 0 \\ \dfrac{\partial^2}{\partial x^2}I(x) - \gamma^2 I(x) = 0 \end{cases} \quad (4.7)$$

where γ is the *propagation constant*

$$\gamma = \sqrt{(r + j\omega l)(g + j\omega c)} = \alpha + j\beta. \quad (4.8)$$

α is the attenuation constant, and β is the phase
constant. Hence, a uniform transmission line is

characterized by the propagation constant and the characteristic impedance, which will be defined in Section 4.2.1.

4.2.1 Wave Description

In this section, instead of using voltages and currents, we will use wave quantities to describe the transmission line. Any waveform can be expressed as two waves: one propagating to the right and the other to the left. The wave quantities are a linear combination of voltages and currents. Hence, this represents only a change of the coordinate system.

Consider the uniform transmission line that is shown in Fig. 4.2. The voltage and current at a position x along the line can be described by an incident and a reflected wave.

Fig. 4.2 Wave description of a transmission line

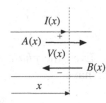

The solution to Equation (4.7) can be written in terms of the voltage waves $A(x)$ and $B(x)$[1] as

$$V(x) = \frac{A(x) + B(x)}{2} \qquad (4.9)$$

where $A(x) = A_1 e^{-\gamma x} = A_1 e^{-(\alpha + j\beta)x}$ is referred to the incident wave that propagates to the right in Fig. 4.2, and $B(x) = B_1 e^{-\gamma x} = B_1 e^{-(\alpha + j\beta)x}$ is the reflected wave that propagates to the left. A_1 and B_1 are constants determined by the boundary conditions for the line. We get from Equation (4.9)

$$\frac{\partial}{\partial x} V(x) = \frac{1}{2}(-\gamma A_1 e^{-\gamma x} + \gamma B_1 e^{\gamma x}) = \frac{\gamma}{2}[-A(x) + B(x)].$$

The expression for the current is obtained from Equations (4.3) and (4.6),

---

[1] We have divided Equations (4.10) and (4.9) with a factor of 2 to adhere to the definitions commonly used in the literature for wave digital filters.

$$\frac{\gamma}{2}[-A(x) + B(x)] + j\omega l I(x) + r I(x) = 0$$

$$\Rightarrow I(x) = \frac{\gamma}{2(r + j\omega l)}[A(x) - B(x)] = \frac{A(x) - B(x)}{2\sqrt{\dfrac{r + j\omega l}{g + j\omega c}}}.$$

Hence, we get

$$I(x) = \frac{A(x) - B(x)}{2Z_0} \qquad (4.10)$$

where

$$Z_0 = \sqrt{\frac{r + j\omega l}{g + j\omega c}} \qquad (4.11)$$

where r, l, g, and c are the primary constants for the line: resistance, inductance, conductance, and capacitance per unit length, respectively. γ is the propagation constant, and Z_0 is the *characteristic impedance*.

4.2.2 Chain Matrix for Transmission Lines

Consider the loaded transmission line shown in Fig. 4.3. The voltage and current at the distance x from the source are

$$\begin{cases} V(x) = \dfrac{A_1 e^{-\gamma x} + B_1 e^{\gamma x}}{2} \\[2mm] I(x) = \dfrac{A_1 e^{-\gamma x} - B_1 e^{\gamma x}}{2Z_0} . \end{cases} \qquad (4.12)$$

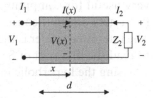

Fig. 4.3 Loaded transmission line

At the source, i.e., $x = 0$, we have $V(0) = V_1$ and $I(0) = I_1$. We get

$$\begin{cases} V(0) = V_1 = \dfrac{A_1 + B_1}{2} \\[2mm] I(0) \ = I_1 = \dfrac{A_1 - B_1}{2Z_0} \end{cases}$$

and

$$
\begin{cases}
A_1 = V_1 + Z_0 I_1 \\
B_1 = V_1 - Z_0 I_1
\end{cases}
$$

$$
\begin{cases}
V(x) = \dfrac{V_1 e^{-\gamma x} + Z_0 I_1 e^{-\gamma x} + V_1 e^{\gamma x} - Z_0 I_1 e^{\gamma x}}{2} = V_1 \cosh(\gamma x) - Z_0 I_1 \sinh(\gamma x) \\[2ex]
I(x) = \dfrac{V_1 e^{-\gamma x} + Z_0 I_1 e^{-\gamma x} - V_1 e^{\gamma x} + Z_0 I_1 e^{\gamma x}}{2 Z_0} = -\dfrac{V_1}{Z_0} \sinh(\gamma x) + I_1 \cosh(\gamma x)
\end{cases}
$$

or in matrix form

$$
\begin{bmatrix} V(x) \\ I(x) \end{bmatrix} =
\begin{bmatrix} \cosh(\gamma x) & -Z_0 \sinh(\gamma x) \\ -\dfrac{\sinh(\gamma x)}{Z_0} & \cosh(\gamma x) \end{bmatrix}
\begin{bmatrix} V_1 \\ I_1 \end{bmatrix}. \quad (4.13)
$$

After inverting the matrix, we get

$$
\begin{bmatrix} V_1 \\ I_1 \end{bmatrix} =
\begin{bmatrix} \cosh(\gamma x) & Z_0 \sinh(\gamma x) \\ \dfrac{\sinh(\gamma x)}{Z_0} & \cosh(\gamma x) \end{bmatrix}
\begin{bmatrix} V(x) \\ I(x) \end{bmatrix}. \quad (4.14)
$$

At the load, $x = d$, we have for $I(x) = I(d) = -I_2$ and $V(x) = V_2$

$$
\begin{bmatrix} V_1 \\ I_1 \end{bmatrix} = K \begin{bmatrix} V_2 \\ -I_2 \end{bmatrix}
$$

where

$$
K = \begin{bmatrix} \cosh(\gamma d) & Z_0 \sinh(\gamma d) \\ \dfrac{\sinh(\gamma d)}{Z_0} & \cosh(\gamma d) \end{bmatrix}. \quad (4.15)
$$

The matrix K, which is called the chain matrix, is very useful for computing the overall chain matrix for a network that consists of cascaded two-ports. The interested reader is therefore urged at this stage to read the first part of Section 5.4.

Using the hyperbolic identity, we get

$$
\cosh(\gamma d) = \frac{\cosh(\gamma d)}{\sqrt{\cosh^2(\gamma d) - \sinh^2(\gamma d)}} = \frac{1}{\sqrt{1 - \tanh^2(\gamma d)}}
$$

and the chain matrix can be written as

$$
K = \frac{1}{\sqrt{1 - \tanh^2(\gamma d)}}
\begin{bmatrix} 1 & Z_0 \tanh(\gamma d) \\ \dfrac{\tanh(\gamma d)}{Z_0} & 1 \end{bmatrix}. \quad (4.16)
$$

4.2.3 Lossless Transmission Lines

Of special interest are lossless lines for which $r = 0$ and $g = 0$. This gives the propagation constant

$$
\gamma = \sqrt{j\omega l \cdot j\omega c} = j\omega \sqrt{lc} = j\beta \quad (4.17)
$$

and the *characteristic impedance*, which now becomes a real positive constant

$$
Z_0 = \sqrt{\frac{r + j\omega l}{g + j\omega c}} = \sqrt{\frac{l}{c}}. \quad (4.18)
$$

Even the *phase constant* β is real. Furthermore, we have

$$
\beta d = \frac{2\pi d}{\lambda} = \frac{2\pi f d}{v} = \frac{\omega d}{v} = \omega \tau \quad (4.19)
$$

where λ is the wavelength, v is the phase velocity, and $\tau = d/v$ is the propagation time from one end to the other of the transmission line[2].

4.2.4 Richards' Variable

Commensurate-length transmission line filters constitute a special case of distributed element networks that can easily be designed by mapping them into a form that resembles a lumped element structure. This mapping involves *Richards' variable*, which is defined as

$$
\Psi \overset{\Delta}{=} \tanh(s\tau) \quad (4.20)
$$

where $\Psi = \Sigma + j\Omega$. Richards' variable is a dimensionless complex variable. Henceforth, we will

---

[2]In the literature for wave digital filters, the sample period is $T = 2\tau$.

assume that Ψ has the same properties and usage as the Laplace variable s.

The real frequencies in the s- and Ψ-planes are related by

$$\Omega = \tan(\omega\tau) \qquad (4.21)$$

where the function $(\omega\tau)$ is periodic with period 2π.

The bilinear transformation between s- and z-planes used in synthesis of digital filters is similar to Richards' variable. In fact, commensurate-length transmission line filters have the same periodic frequency responses as digital filters, except that in digital filters we use the sample period $T = 2\tau$.

4.2.5 Unit Elements

For a lossless transmission line we have $\alpha = 0$, and from Equation (4.19) we get $\gamma d = j\beta d = j\omega\tau$. By replacing $j\omega$ with s, the chain matrix in Equation (4.16) can be written

$$K = \frac{1}{\sqrt{1 - \tanh^2(s\tau)}} \begin{bmatrix} 1 & Z_0\tanh(s\tau) \\ \dfrac{\tanh(s\tau)}{Z_0} & 1 \end{bmatrix}. \qquad (4.22)$$

Substituting Richards' variable into Equation (4.22), we get

$$K = \frac{1}{\sqrt{1 - \Psi^2}} \begin{bmatrix} 1 & Z_0\Psi \\ \dfrac{\Psi}{Z_0} & 1 \end{bmatrix}. \qquad (4.23)$$

The matrix in Equation (4.23) has, similar to the chain matrix for a lumped element network, element values that are rational functions in Richards' variable. The square-root factor can be handled separately during the synthesis, and synthesis procedures (programs) used for lumped element design can therefore be used with small modifications for synthesis of commensurate-length transmission line filters.

Obviously, a transmission line cannot be described by poles and zeros because the elements in the chain matrix are not rational functions in s.

The transmission line filters of interest are often built using only one-ports. At this stage, it is therefore interesting to study the input impedance of the one-port shown in Fig. 4.4. A lossless transmission line described by Equation (4.23) is called a unit element (*UE*).

Fig. 4.4 Terminated transmission line

The input impedance to a unit element, which is terminated by the impedance Z_2, can be derived from Equation (4.23). From Equation (4.23) we get the input impedance to a transmission line, with characteristic impedance Z_0 and loaded with an impedance Z_2, as

$$Z_{in}(\Psi) = \frac{V_1}{I_1} = \frac{Z_2 + Z_0\Psi}{Z_0 + Z_2\Psi}Z_0. \qquad (4.24)$$

We are interested in the input impedance of a lossless transmission line with characteristic impedance Z_0 that is terminated by an impedance in the following three cases

4.2.5.1 Matched Termination ($Z_2 = Z_0$)

$$Z_{in}(\Psi) = \frac{Z_2 + Z_0\Psi}{Z_0 + Z_2\Psi}Z_0 = Z_0. \qquad (4.25)$$

For case $Z_2 = Z_0$, we have a matching between the unit element and the load, and an incident wave reaching the load will not be reflected. The reflected wave amplitude is zero, because at the load we have $V(d) = Z_2I(x) = Z_0I(d)$ and Equation (4.12) yields

$$Ae^{-\lambda d} + Be^{\lambda d} = Ae^{-\lambda d} - Be^{\lambda d}.$$

Identification gives $B = 0$. According to Equation (4.25), the input impedance is purely resistive and equals Z_0.

4.2.5.2 Open-Ended ($Z_2 = \infty$)

$$Z_{in}(\Psi) = \frac{Z_2 + Z_0\Psi}{Z_0 + Z_2\Psi}Z_0 = \frac{Z_0}{\Psi}. \qquad (4.26)$$

Hence, an open-ended unit element can be interpreted as a capacitor, i.e., a Ψ-plane capacitor with the value $1/Z_0$. Note that for a lossless line, Z_0 is a positive real constant.

4.2.5.3 Short-Circuited ($Z_2 = 0$)

$$Z_{in}(\Psi) = \frac{Z_2 + Z_0\,\Psi}{Z_0 + Z_2\,\Psi}\,Z_0 = Z_0\,\Psi. \quad (4.27)$$

A short-circuited unit element can be interpreted as a Ψ-plane inductor with the value Z_0.

4.3 Microstrip and Striplines

Many microwave systems use waveguides for transmission, e.g., radar transceiver to the antenna, because they can handle high power at relative low losses. However, waveguides systems are bulky and expensive. Low-power and cheaper alternatives are stripline, microstrip, slotline, and coplanar waveguides [8, 50, 93, 97, 99]. These transmission lines are compact and may be integrated with active devices to form microwave integrated circuits (*MMICs*). Typically, these transmission lines are used at power levels below 100 W.

We differentiate between symmetrical or asymmetrical geometries. A symmetrical structure is called stripline and an asymmetrical is denoted microstrip.

4.3.1 Stripline

Figure 4.5 shows a cross-sectional view of a symmetrical strip transmission (stripline) structure [70]. Such a stripline can support TEM waves, but it can also support TM and TE waves even at higher modes. With a voltage applied between the center strip and the pair of ground planes, current flows down the center strip and returns via the two ground planes. Although the structure has open sides, it is a non-radiating transmission line. In practice, however, any unbalance in the line causes energy to be radiated out of the sides. To prevent this and suppress higher-mode propagation in the frequency range of interest, the ground planes are connected to each other with screws and by restricting the ground plane spacing to less than $\lambda/4$.

Usually, stripline filters are based on copper-clad printed-circuit boards, where the center conductor thickness, t, is very small in comparison to the other dimensions.

The characteristic impedance depends only on the capacitance per unit length of the stripline. Formulas for the characteristic impedance and formulas for slotlines, bends, junctions, as well as inductors are given in [8, 50]. A program for computing the characteristic impedance can be found in [99].

The characteristic impedance for striplines implemented using a FR4 (epoxy-glass) printed circuit board is typically in the range 30–250 Ω.

4.3.2 Microstrip

Microstrip line is a very popular type of planar transmission line, primarily because it may be fabricated using printed circuit techniques and is easily integrated with other passive and active microwave devices.

The line consists of a thin conductor and a ground plane separated by a low-loss dielectric material as shown in Fig. 4.6. Discrete devices can be directly mounted on top of the microstrip circuit.

The use of high dielectric materials forces the fields into the dielectric and reduces the fields in the air. In most cases, the fields are negligible at a distance $2h$ above the metal conductor. To prevent

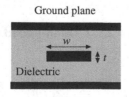

Fig. 4.5 Cross section of stripline transmission line

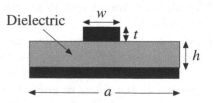

Fig. 4.6 Asymmetric microstrip

radiation losses, the complete microstrip circuit is usually placed in a metal enclosure as shown in Fig. 4.7.

Fig. 4.7 Enclosed microstrip that prevents radiation losses

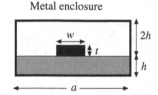

The analysis of microstrip line is complicated because the fields are partly in the dielectric and partly in the air [8]. In fact, the microstrip line cannot supports a pure TEM wave, because the phase velocity of TEM waves in the dielectric is $c/\sqrt{\varepsilon_r}$, but c in the air. The microstrip line therefore supports a hybrid TM-TE wave, which requires an advanced analysis. However, in most practical applications, the dielectric substrate is very thin ($h \ll \lambda$) and the fields are quasi-TEM.

Approximate expressions for the characteristic impedance for microstrips is found in [8, 50]. A program for computing the characteristic impedance can be found in [99]. The characteristic impedance for microstrip implemented using a FR4 (epoxy-glass) printed circuit board is typically in the range 20–120 Ω.

Filters implemented using striplines and microstrip lines are usable in the range 0.1–20 GHz and 0.1–100 GHz, respectively. Filters with suspended microstrip lines may operate in a slightly larger range, i.e., typically 1–200 GHz.

4.3.3 MIC and MMIC Microstrip Filters

The performance of filters based on monolithic microwave integrated circuits (MMICs) is not comparable to standard hybrid MIC and waveguide filters, due to the low Q of filter components. However, depending on system requirements, MMIC techniques allow complete subsystems to be fabricated in a single package, thus leading to high-volume components with low cost. Microwave integrated circuits (MICs) and monolithic microwave integrated circuits (MMICs) may consist of a number of discrete active and passive components, such as transistors, capacitors, inductors, and resistors mounted on a common substrate. The passive elements may even be implemented inside the substrate. The active elements are often used to compensate for the losses in the passive components, which degrade the frequency selectivity. In addition, the active element has the potential for electric tuning. See [123] for active filters with distributed elements.

4.3.3.1 Superconducting Circuits

High-temperature superconducting (HTS) filters are an interesting alternative, especially for very narrowband filters, which otherwise may be degraded due to resistive losses in the transmission lines. Most superconducting filters are implemented as microstrip filters using HTS thin films.

Superconductors exhibit zero intrinsic resistance to direct currents when cooled below a certain temperature. The temperature at which the intrinsic resistance abruptly changes is referred to as the critical temperature. However, for alternating currents, the resistance is not zero, but the resistance of a superconductor is of the order one thousandth of that in the best ordinary conductor. This provides a significant improvement of the Q factors of the components and the performances of highly frequency selective microstrip filters.

4.4 Commensurate-Length Transmission Line Filters

Lumped elements have physical dimensions that are insignificant with respect to the wavelength of the highest operating frequency. One of the great advantages of lumped element networks is that they may be described in terms of a single complex frequency variable. Networks consisting of arbitrary distributed circuit elements are more complex. For example, analysis of a network consisting of transmission lines of different lengths would be very complicated and require more than one complex variable. In general, analysis of such circuits is therefore done by solving Maxwell's equations using, for example, finite element analysis (FEM).

A special case of transmission line networks is when the transmission lines have equal lengths, i.e., they have a common electrical propagation time. Such commensurate-length transmission line networks are a special case of distributed element networks that lend themselves to analysis and synthesis using Richards' variable. In fact, we may often use an analogy between a lumped-element network and the commensurate-length transmission line network counterpart, which is described by Richards' variable to design the later.

4.4.1 Richards' Structures

An arbitrary Nth-order reactance function can be realized by a cascade of N unit elements, a so-called *Richards' structure*, as shown in Fig. 4.8. The far-end is either open or short-circuited [7]. According to Richards' theorem, the unit elements can successively be extracted from the reactance function by the following theorem.[3]

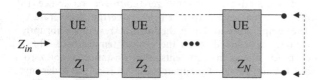

Fig. 4.8 Realization of an Nth-order reactance

Theorem 4.1: Richards' Theorem *Let $Z(\Psi)$ be a positive real impedance function such that $Z(\Psi)/Z(1) \neq \Psi$ and $\neq 1/\Psi$. A unit element with characteristic impedance $Z(1)$ can be extracted and the remaining impedance function is*

$$Z_1(\Psi) = \frac{Z(\Psi) - \Psi Z(1)}{Z(1) - \Psi Z(\Psi)} Z(1). \quad (4.28)$$

Example 4.1 Realize the reactance function, $Z_1(\Psi) = \dfrac{8\Psi^2 + 1}{8\Psi^3 + 3\Psi}$, by a Richards' structure.

We get the characteristic impedance of the first UE:
$$Z_1(1) = \frac{8+1}{8+3} = \frac{9}{11}.$$
The remaining impedance is

$$Z_2(\Psi) = \frac{\dfrac{8\Psi^2 + 1}{8\Psi^3 + 3\Psi} - \Psi \dfrac{9}{11}}{\dfrac{9}{11} - \Psi \dfrac{8\Psi^2 + 1}{8\Psi^3 + 3\Psi}} \frac{9}{11}$$

$$= \frac{9(72\Psi^4 - 61\Psi^2 - 11)}{176\Psi(\Psi^2 - 1)}.$$

We get by factoring[4] out $\Psi^2 - 1$

$$Z_2(\Psi) = \frac{9(72\Psi^2 + 11)(\Psi^2 - 1)}{176\Psi(\Psi^2 - 1)}$$

$$= \frac{9(72\Psi^2 + 11)}{176\Psi}.$$

Now, a second unit element can be extracted,
$$Z_2(1) = \frac{9(72 + 11)}{176} = \frac{747}{176}\Omega.$$
The remaining impedance is

$$Z_3(\Psi) = \frac{\dfrac{9(72\Psi^2 + 11)}{176\Psi} - \Psi \dfrac{747}{176}}{\dfrac{747}{176} - \Psi \dfrac{9(72\Psi^2 + 11)}{176\Psi}} \frac{747}{176}$$

$$= \frac{83}{128\Psi} \text{ and } Z_3(1) = \frac{83}{128}\Omega.$$

$Z_3(1)$ represents a third unit element, with characteristic impedance $= 83/128\Omega$, and that is open-ended (Ψ-capacitor). Hence, the reactance is realized by a cascade of three unit elements that are open-ended at the far end, with characteristic impedances $Z_1 = 9/11\Omega$, $Z_2 = 747/176\Omega$, and $Z_3 = 83/128\Omega$, respectively.

The synthesis of an Nth-order reactance as a cascade of unit elements can be done by using the program RICHARDS_REACTANCE.

4.5 Synthesis of Richards' Filters

Richards' structures may be used as doubly resistively terminated reactance networks, which have low passband sensitivity. However, possible transfer functions are limited to lowpass, highpass, and bandpass filters.

The transfer function is of the form

$$H(\Psi) = \frac{(1 - \Psi^2)^{N/2}}{D_N(\Psi)} \quad (4.29)$$

where $\Psi = \tanh(s\tau)$, and τ is the propagation time through a transmission line. $D_N(\Psi)$ is a Hurwitz polynomial.

$N/2$ transmission zeros are obtained on the real axis at $\Psi = \pm 1$. However, zeros cannot be realized on the imaginary axis. This means that the magnitude response of these filters are similar to an allpole lumped element filter, but the transition

---

[3]Paul I. Richards, USA, 1948.

[4]Solving for the roots of polynomials is a numerically difficult operation.

band is always larger due to the lack of zeros on the imaginary axis. Thus, the filtering effect of a unit element is slightly less than for an inductor or capacitor.

Note also that the frequency response of these filters cannot be changed by frequency normalization. In fact, they have to be designed for the actual passband frequency.

Inserting $\Psi = j\Omega$ into Equation (4.29) yields

$$|H(j\Omega)|^2 = \frac{(1+\Omega^2)^N}{|D_N(j\Omega)|^2}. \qquad (4.30)$$

It can be shown that

$$|H(j\Omega)|^2 = \frac{1}{1+\varepsilon^2 P_N^2(\sin(\omega\tau))} \qquad (4.31)$$

where $\Omega = \tan(\omega\tau)$, and P_N^2 is an even polynomial of order $2N$.

4.5.1 Richards' Filters with Maximally Flat Passband

A maximally flat passband is obtained by making the first $2N-1$ derivatives of the squared magnitude function to vanish at $\omega\tau = 0$, but only the first derivative vanishes at $\omega\tau = \pi$. For a transmission line filter of Richards' type with a maximally flat magnitude function, we have

$$|H(e^{j\omega\tau})|^2 = \frac{1}{1 + \left(\frac{\sin(\omega\tau)}{\alpha}\right)^{2N}} \qquad (4.32)$$

where $\alpha = \varepsilon^{\frac{-1}{N}} \sin(\omega_c\tau)$, and $\varepsilon = \sqrt{10^{0.1 A_{max}} - 1}$. Closed form expressions for the characteristic impedances do not exist, but approximate values can be computed by using the following expressions [96], which have been implemented in the program RICHARDS_MF, for the case $R_s = R_L = 1$:

$$g_n = \frac{2\sin\left(\frac{(2n-1)\pi}{2N}\right)}{\alpha}\left[1 - \frac{\alpha^2 \cos\left(\frac{\pi}{N}\right)}{4\sin\left[\frac{(2n-3)}{2N}\pi\right]\sin\left[\frac{(2n+1)}{2N}\pi\right]}\right] \qquad (4.33)$$

$$Z_n = \begin{cases} g_n, & n = \text{even} \\ 1/g_n, & n = \text{odd} \end{cases} \qquad (4.34)$$

where $n = 1,2,..,N$.

$g_1 = 1.383222$ $Z_1 = 0.722950\,\Omega$
$g_2 = 2.294564$ $Z_2 = 2.294564\,\Omega$
$g_3 = 3.223380$ $Z_3 = 0.310233\,\Omega$
$g_4 = 2.294564$ $Z_4 = 2.294564\,\Omega$
$g_5 = 1.383222$ $Z_5 = 0.722950\,\Omega$

Example 4.2 Synthesize a fifth-order doubly resistively terminated Richards' structure according to Fig. 4.9 with maximally flat passband and $A_{max} = 0.5$ dB, $\omega_c = 2\pi\,300$ Mrad/s, $\tau = 0.5$ ns.

The attenuation for the filter is shown in Fig. 4.10.

The frequency response can be computed with LADDER_2_H. Note that the frequency response is periodic with period $\omega\tau = 2\pi$. The impedance level can be changed by multiplying all characteristic impedances and the terminating resistor with the same factor.

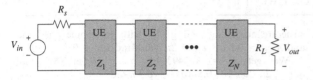

Fig. 4.9 Doubly resistively terminated Richards' structure

4.5.2 Richards' Filters with Equiripple Passband

An algorithm for synthesis of equiripple passband filters is given below [96], which has been implemented in the program RICHARDS_EQ, for the case $R_s = 1$.

Use the formulas above or RICHARDS_MF. We get with $\varepsilon = 0.3493114$, $\alpha = 0.5602789$, and $\omega_c\tau = 2\pi \cdot 300 \cdot 10^6 \cdot 0.5 \cdot 10^{-9} = 0.3\,\pi$ rad

Fig. 4.10 Attenuation for a fifth-order Richards' filter with maximally flat passband

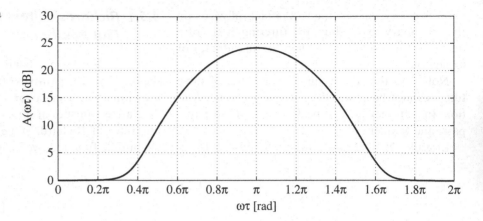

$$A_n = \frac{\left[\eta^2 + \sin^2\left(\frac{(n-2)}{N}\pi\right)\right] \cdot \left[\eta^2 + \sin^2\left(\frac{(n-4)}{N}\pi\right)\right] \cdots}{\left[\eta^2 + \sin^2\left(\frac{(n-1)}{N}\pi\right)\right] \cdot \left[\eta^2 + \sin^2\left(\frac{(n-3)}{N}\pi\right)\right] \cdots} \tag{4.35}$$

and $n = 1, 2, \ldots, n$ with the term $\eta^2 + \sin^2(0)$ replaced by η, i.e.,

$$A_1 = \frac{1}{\eta}, \quad A_2 = \frac{\eta}{\eta^2 + \sin^2\left(\frac{\pi}{N}\right)}, \quad A_3 = \frac{\eta^2 + \sin^2\left(\frac{\pi}{N}\right)}{\left(\eta^2 + \sin^2\left(\frac{2\pi}{N}\right)\right)\eta} \tag{4.36}$$

$$g_n = A_n \left(\frac{2\sin\left(\frac{(2n-1)}{2N}\pi\right)}{\alpha} - \frac{\alpha}{4}\left[\frac{\eta^2 + \sin^2\left(\frac{n}{N}\pi\right)}{\sin\left(\frac{(2n+1)}{2N}\pi\right)} + \frac{\eta^2 + \sin^2\left(\frac{(n-1)}{N}\pi\right)}{\sin\left(\frac{(2n-3)}{2N}\pi\right)}\right] \right) \tag{4.37}$$

where $\alpha = \sin(\omega_c\tau)$ and

$$R_L = \tanh^2\left(\frac{N}{2}\,a\sinh(\eta)\right) = \begin{cases} 1 & N \text{ odd} \\ \frac{\sqrt{1+\varepsilon^2}-\varepsilon}{\sqrt{1+\varepsilon^2}+\varepsilon} & N \text{ even.} \end{cases} \tag{4.38}$$

$$\eta = \sinh\left[\frac{1}{N}a\sinh\left(\frac{1}{\varepsilon}\right)\right] \tag{4.39}$$

$$Z_n = \begin{cases} \frac{1}{g_n}, & n = \text{even} \\ g_n, & n = \text{odd.} \end{cases} \tag{4.40}$$

Example 4.3 Use the formulas or RICHARDS_EQ to synthesize a fifth-order doubly resistively terminated Richards'

structure according to Fig. 4.9 with equiripple passband and $A_{max} = 0.5$ dB, $\omega_c = 2\pi\,300$ Mrad/s, $\tau = 0.5$ ns.

We get $\varepsilon = 0.3493114$, $\alpha = 0.5602789$, and $\omega_c\tau = 2\pi \cdot 300 \cdot 10^6 \cdot 0.5 \cdot 10^{-9} = 0.3\,\pi$ rad

$g_1 = 1.342666A_1$	$A_1 = 2.759994$	$Z_1 = 3.705751\,\Omega$
$g_2 = 3.271357A_2$	$A_1 = 0.759951$	$Z_2 = 0.402241\,\Omega$
$g_3 = 4.114757A_3$	$A_3 = 1.270414$	$Z_3 = 5.227443\,\Omega$
$g_4 = 3.271357A_4$	$A_4 = 0.759951$	$Z_4 = 0.402241\,\Omega$
$g_5 = 1.342666A_5$	$A_5 = 2.759994$	$Z_5 = 3.705751\,\Omega$

Fig. 4.11 Attenuation of a Richards' filter with equiripple passband

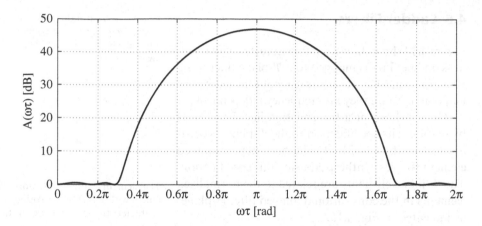

The attenuation for the filter is shown in Fig. 4.11. To get a larger stopband, the value for τ has to be made smaller and the electrical length of the UEs made shorter. This means a larger value of α and a larger spread in the characteristic impedances.

As in the case of LC ladder networks, the dual prototype network could have been used, see Problem 4.5. In this case, the first UE will have a low rather than a high characteristic impedance. The Richards' structure can be implemented using the technique discussed in the next section.

4.5.3 Implementation of Richards' Structures

A Richards' structure can be implemented by using a microstrip as shown in Fig. 4.12. The filter has five UEs. Approximate formulas for computing the size of the microstrip segments can be found in [8].

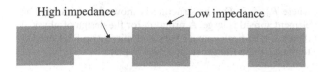

Fig. 4.12 Typical microstrip circuit for a Richards' filter

Alternatively, a coaxial transmission line shown in Fig. 4.13 can be used. The characteristic impedances may be realized as a coaxial line with a stepped inner conductor.

The diameters of the segments can be computed from

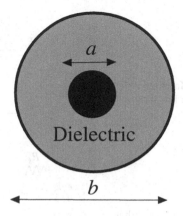

Fig. 4.13 Coaxial transmission line

$$Z_0 = \frac{60}{\sqrt{\varepsilon_r}} \ln\left(\frac{b}{a}\right). \qquad (4.41)$$

These implementations are popular because they take less board space than the ladder structures that will be discussed in the next section. A program for more accurate computation of the characteristic impedance can be found in [99].

4.5.3.1 Stepped Impedance Filters

An alternative design approach exploits that it is easy to implement structures that alternate between high and low characteristic impedances. These filters may be designed by simple approximations [50, 76, 93]. Of course, the approximation may later be improved by using a numerical optimization program. Note that these filters do use transmission lines with different electrical length and their frequency response suffers from spuriousness.

4.6 Ladder Filters

Figure 4.14 shows how a commensurate-length transmission line filter is mapped to a Ψ-plane filter using Richards' variable. Resistors are not affected by the mapping because they are frequency independent.

The synthesis of a transmission line filter starts by mapping its specification to the Ψ-plane according to Equation (4.21). In the next step, a lumped element filter is synthesized using this specification. The Ψ-plane elements are related to the normalized elements in the conventional lumped filter (s-plane) as indicated in Fig. 4.15.

The corresponding lumped element filter shall have a normalized cutoff frequency $\omega_c = \Omega_c$. A third-order lumped element ladder filter of Butterworth has the element values

$$R_s = R_L = 1 \qquad L_1 = 1.290169$$
$$C_2 = 2.580338 \qquad L_3 = 1.290169$$
$$\varepsilon = 0.15262042 \qquad \varepsilon^{-1/3} = 1.8712369$$

The element values are denormalized by multiplying inductances with R_0/Ω_c and dividing capacitances with $\Omega_c R_0$. The characteristic impedances in the transmission line filter,

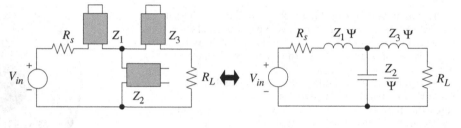

Fig. 4.14 Mapping of a transmission line filter into a Ψ-plane filter

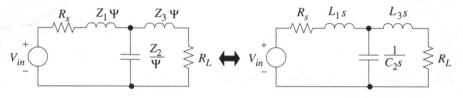

Fig. 4.15 Analogy with a lumped element filter

Finally, the element values for the Ψ-plane filter are obtained from the lumped filter. In general, filters with distributed circuit elements cannot be frequency scaled, i.e., the bandwidth cannot be changed by a simple scaling of the characteristic resistances. However, frequency scaling of commensurate-length transmission line filters can be done according to Equation (4.21) if all transmission lines are used as one-ports.

Example 4.4 Determine the characteristic impedances in a third-order commensurate-length transmission line filter of Butterworth type with a cutoff angle $\omega_c\tau = \pi/4$. The passband ripple is $A_{max} = 0.1$ dB and the terminating resistances are 50 Ω.

The cutoff edge of the Ψ-plane filter, according to Equation (4.21), is

$$\Omega_c = \tan(\pi/8) = 0.4142136.$$

which have the same numerical values as the lumped element filter, are

$$R_s = R_L = 50\,\Omega \qquad Z_1 = \frac{L_1 R_0}{\Omega_c} = 398.645\,\Omega$$
$$Z_2 = \frac{\Omega_c R_0}{C_2} = 3.1356\,\Omega \qquad Z_3 = \frac{L - 3R_0}{\Omega_c} = 398.645\,\Omega$$

where $R_0 = 50\,\Omega$. The structure is shown in Fig. 4.14. The element spread is large. A program for the design of physical dimension of both coaxial lines and microstrip lines needed to realize these filters can be found in [99].

4.7 Ladder Filters with Inserted Unit Elements

The ladder structures discussed in the previous section may be difficult to implement in a planar layout and it may also be necessary to physically separate the branches in a ladder structure, transform series branches into shunt branches, or vice versa, and change impractical characteristic impedances into more realizable ones.

To physically separate the branches in a ladder structure, we may introduce separating UEs between the series and shunt branches. The insertion of unit elements can be done at the synthesis stage, which is most efficient, as their filtering capability can be utilized. Moreover, the element spread becomes smaller. However, we will not discuss this approach here [73, 123]. Another, suboptimal way is to insert unit elements into the ladder structure by using Kuroda-Levy identities.

In the case that the ladder structure shall be used to design a corresponding wave digital filter, it is also advantageous to use UEs that are inserted between the branches in order to avoid delay-free loops.

4.7.1 Kuroda-Levy Identities

Figure 4.16 shows the generic Kuroda[5]-Levy[6] identity. The two networks N_1 and N_2 contain lossless commensurate-length transmission lines. Table 4.1 shows some special cases [70].

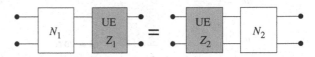

Fig. 4.16 Generic Kuroda-Levy identity

Kuroda-Levy identities can sometimes be used to transform impedances to more practical levels, but they are more often used to remove unwanted series short-circuit stubs from planar designs. We demonstrate the insertion of UEs into a ladder structure by the means of an example.

Table 4.1 Kuroda-Levy identities

$Z_1\Psi$ — UE Z_2	UE Z_3 — $\dfrac{Z_4}{\Psi}$	$\begin{cases} Z_3 = Z_1 + Z_2 \\ Z_4 = Z_2 + Z_2^2/Z_1 \end{cases}$
$Z_1\Psi$ — UE Z_2	UE Z_3 — $Z_4\Psi$ — $n{:}1$	$\begin{cases} Z_3 = nZ_2 \\ Z_4 = nZ_1 \end{cases}\qquad n = \dfrac{Z_1}{Z_1 + Z_2}$
$\dfrac{Z_1}{\Psi}$ — UE Z_2	UE Z_3 — $\dfrac{Z_4}{\Psi}$ — $n{:}1$	$\begin{cases} Z_3 = nZ_2 \\ Z_4 = (n-1)Z_3 \end{cases}\qquad n = \dfrac{Z_1 + Z_2}{Z_2}$
$\dfrac{Z_1}{\Psi}$ — UE Z_2	UE Z_3 — $Z_4\Psi$	$\begin{cases} Z_3 = Z_1 Z_2/(Z_1 + Z_2) \\ Z_4 = Z_2^2/(Z_1 + Z_2) \end{cases}$

---

[5]K. Kuroda, Japan, 1955.

[6]R. Levy, UK.

Example 4.5 Realize a ladder structure of Butterworth type, which satisfies the requirement in Fig. 4.17, where $A_{max} = 0.5\,\text{dB}$, $A_{min} = 20\,\text{dB}$, $\omega_c\tau = \pi/4$, and $\omega_s\tau = \pi/2$. Use a ladder with inserted unit elements.

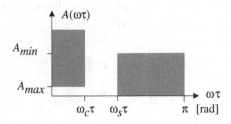

Fig. 4.17 Filter specification

The critical angles in the Ω domain corresponding to $\omega_c\tau$ and $\omega_s\tau$ are

$$\Omega_c = \tan(\omega_c\tau/2) = \tan(\pi/8) \approx 0.4142136$$

$$\Omega_s = \tan(\omega_s\tau/2) = \tan(\pi/4) = 1 \Rightarrow$$

$$\frac{\Omega_s}{\Omega_c} = \frac{1}{0.4142136} \approx 2.414214.$$

The normalized cutoff edge is 2.414214 and $A_{max} = 0.5$ dB, $A_{min} \geq 20$ dB. The specification can be satisfied with a filter of order $N = 3$.

Next, we design a lumped element ladder structure that satisfies the same specification, but with $\omega_c = \Omega_c$ and $\omega_s = \Omega_s$. The element values in the normalized LC filter, shown in Fig. 4.18, are $R_s = R_L = 1$, $L_1 = L_3 = 1.5963$, and $C_2 = 1.0967$.

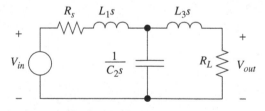

Fig. 4.18 Lumped element ladder structure

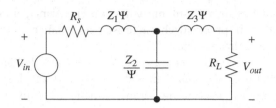

Fig. 4.19 Filter in the Richards' domain

We select to use equal source and load resistance so that the maximal power transfer is obtained for $\omega = 0$. The corresponding filter in the Richards' domain, which is shown in Fig. 4.19, is obtained by an analogy between the lumped and distributed element filters.

We denormalize the element values by multiplying inductances with R_s/Ω_c and dividing capacitances with R_s/Ω. The denormalized characteristic resistances are $Z_1 = Z_3 = L_1R_s/\Omega_c = 3.85381\Omega$ and $Z_2 = R_s\Omega_c/C_2 = 0.3776909\Omega$.

Next, we introduce two unit elements as shown in Fig. 4.20. If these unit elements have the characteristic resistance equal to the load resistor, i.e., matched terminated, then the left-most input impedance to the unit elements is equal to R_L. Thus, the unit elements perform no filtering, they only cause a delay, i.e., increase of the group delay with 2τ.

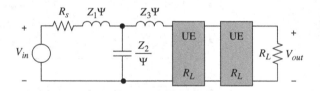

Fig. 4.20 Insertion of non-filtering unit elements

The unit elements can be propagated into the ladder by using the first Kuroda-Levy identity shown in Table 4.1. We get the structures in Figs. 4.21 and 4.22 where $R_3 = R_L + Z_3 = 4.853809\,\Omega$ and $Z_4 = R_L + R_L^2/Z_3 = 1.2594835\,\Omega$.

Using the first identity twice, but in the reverse direction, we get

$$R_4 = \frac{R_3 Z_2}{R_3 + Z_2} = 0.3504233\Omega, \quad Z_5 = \frac{R_3^2}{R_3 + Z_2} = 4.5033858\Omega \text{ and}$$

$$R_5 = \frac{R_L Z_4}{R_L + Z_4} = 0.557421\Omega, \quad Z_6 = \frac{R_L^2}{R_L + Z_4} = 0.442579\Omega.$$

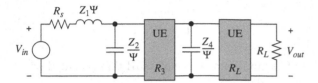

Fig. 4.21 First-time use of Kuroda-Levy identity

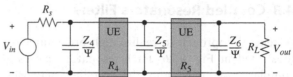

Fig. 4.24 Dual Structure

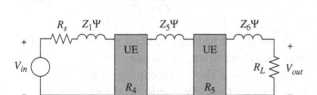

Fig. 4.22 Second-time use of Kuroda-Levy identity

Denormalizing to, e.g., 50 Ω is done by multiplying the characteristic impedances with 50 Ω. Figure 4.23 shows the attenuation when $\tau = 0.5$ ns. Note that the frequency response is periodic like a digital filter with sample frequency $1/\tau = 2$ GHz. Finally, note that the element spread is significantly smaller compared to the filter in Example 4.4.

computation of the characteristic impedances is left as an exercise.

Finally, we scale the impedance level of the circuit by multiplying the normalized characteristic impedances by 50 Ω. The frequency range is determined by the electrical length of the stubs. Figure 4.25 shows a typical microstrip layout for a third-order filter.

Unit elements can also be inserted from the source side, whereby a more symmetric filter is obtained. However, the inserted unit elements do not perform any frequency selective filtering if they are inserted by using the Kuroda-Levy identities.

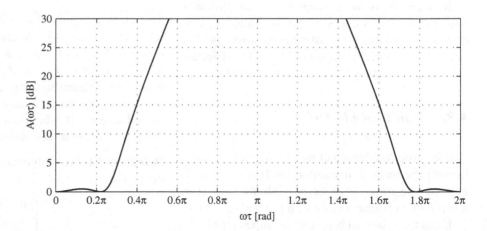

Fig. 4.23 Attenuation for equiripple filter

Example 4.6 Consider the same lowpass filter as in Example 4.5 that should be implemented using microstrip lines. However, the series stubs, i.e., inductors, are difficult to implement as microstrip lines. Redesign therefore the filter so that only shunt capacitors are obtained.

In the first design step in Example 4.5, a series inductor–UE was converted to a UE–shunt capacitor using the first Kuroda-Levy identity. Hence, we may in the same way convert the left-most series inductor by inserting a UE, with characteristic impedance = R_s, between the source resistor and the inductor $Z_1\Psi$. The resulting, dual filter structure is shown in Fig. 4.24. The

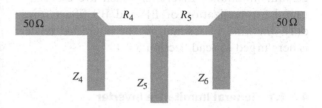

Fig. 4.25 Microstrip layout of the final filter

4.8 Coupled Resonators Filters

A general approach to realize high-order filters is illustrated in Fig. 4.26. Each resonator realizes a complex pole pair, and the coupling networks provide suitable coupling. The resonators may be purely of series or parallel type or a mixture of types, and there are many possible coupling networks that are used in practice [8, 76]. Examples of commonly used coupling networks are capacitors, inductors, transformers, and impedance inverters. However, the resonators and coupling networks should represent a lossless reactance network that is designed for maximal power transfer.

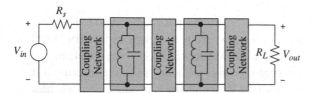

Fig. 4.26 Generic coupled resonators filter

In order to discuss a wider class of coupled resonators structures, we need to briefly discuss the concept of impedance and admittance inverters. The interested reader is recommended to first read Section 5.4 where these two-ports are discussed in more detail.

4.8.1 Immitance Inverters

Higher-order ladder structures consist of resonators in both the series and shunt arms. It is often desirable to use only series or only shunt resonators when implementing a filter with transmission lines.

Earlier we showed how a series impedance could be converted into a parallel impedance using the Kuroda-Levy identities. Another option is to use general immitance inverters, which are especially useful for realization of BP and BS filters using only one kind of resonator. The interested reader is here urged to read Section 5.4.4.

4.8.1.1 General Immitance Inverter

A general immitance inverter, *GII*, is a two-port described by the chain matrix

$$K = \begin{bmatrix} 0 & B(s) \\ C_{(s)} & 0 \end{bmatrix}.$$

The input impedance of a GII that is loaded with an impedance Z_L is

$$Z_{in} = \frac{B(s)}{C(s)} \frac{1}{Z_L}. \qquad (4.42)$$

A series impedance embedded between two impedance inverters, as shown in Example 5.3, is equivalent to a shunt impedance. Two GIIs may, thus, be used to convert series to shunt impedances, and vice versa, and to change the impedance level. Moreover, GII are also used in the synthesis of ladder structures, and the designer can opt to realize these by special circuits or perform network transformations to completely remove them from the final structure.

4.8.1.2 Positive Impedance Inverters

To obtain a positive impedance inverter (*PII*), we select $B(s)/C(s) = K^2$ where K^2 is a real positive constant. The input impedance to a *GII* that is loaded with an inductor, $j\omega L$, is $Z_{in} = K^2/j\omega L = 1/j\omega C$, which is equivalent to a capacitor, $C = L/K^2$.

In the microwave literature, it is common to use a special case of PII, which is referred to as impedance inverter that can be realized with transmission lines.

Definition 4.1 An impedance inverter is (usually) defined as

$$K = \begin{bmatrix} 0 & jK_1 \\ \dfrac{j}{K_2} & 0 \end{bmatrix}. \qquad (4.43)$$

where $K_1 K_2 > 0$. The impedance inverter is also known as a K-inverter. Note that there are several possible ways to select $B(s)$ and $C(s)$ to obtain a chain matrix corresponding to a positive impedance inverter [50, 76]. The input impedance becomes $Z_{in} = K^2/Z_L$.

An alternative selection of $B(s)$ and $C(s)$ yields an impedance inverter that is known in the microwave literature as an admittance inverter. It is the dual of the impedance inverter.

Definition 4.2 An admittance inverter is (usually) defined as

$$K = \begin{bmatrix} 0 & \dfrac{\pm 1}{jJ_1} \\ \mp jJ_2 & 0 \end{bmatrix}. \qquad (4.44)$$

The admittance inverter is also known as a J-inverter. The input admittance is $Y_{in} = J^2/Y_L$, and input impedance is $Z_{in} = 1/(J^2 Z_L)$ where $J_1 J_2 > 0$.

4.8.1.3 Realization of GII

The properties of the K- and J-inverters can be approximated over a limited frequency range by a circuit based on a quarter-wavelength transmission line of characteristic impedance K and characteristic admittance J, respectively [50, 97]. From Equation (4.14) and that $\cosh(jx) = \cos(x)$ and $\sinh(jx) = j\sin(x)$, we get the chain matrix for a quarter-wavelength line

$$K = \begin{bmatrix} \cos(\beta l) & jZ_0 \sin(\beta l) \\ \dfrac{j\sin(\beta l)}{Z_0} & \cos(\beta l) \end{bmatrix} = \begin{bmatrix} 0 & \pm jZ_0 \\ \pm \dfrac{j}{Z_0} & 0 \end{bmatrix} \qquad (4.45)$$

where the length of the transmission lines is $\beta l = \pm \pi/2$. Because a quarter-wavelength line is $\lambda/4$ long only at a single frequency, they are only useful in filters having a relative bandwidth that is less than about 20%.

Figure 4.27 shows a general PII with lumped reactive elements. Figure 4.28 shows a corresponding realization based on transmission lines where $X = \omega L$ with $\phi < 0$ or $X = -1/\omega C$ with $\phi > 0$ and $K = Z_0 \tan\left(\left|\dfrac{\Phi}{2}\right|\right)$, $\phi = -\text{atan}\dfrac{2X}{Y_0}$, and $X = \dfrac{K}{1 - \left(\dfrac{K^2}{Y_0}\right)}$.

Fig. 4.27 K-inverter

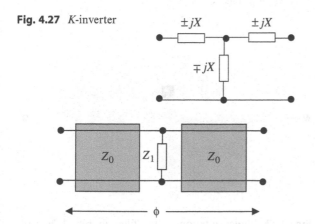

Fig. 4.28 Transmission line K-inverter

The lengths, βl, of the transmission line sections are generally required to be negative for this type of inverter, but often these negative elements can be absorbed into adjacent positive series elements.

Figure 4.29 shows a J-inverter with lumped reactive elements. A corresponding transmission line circuit is shown in Fig. 4.30, where $B = -1/\omega L$ with $\phi > 0$ or $B = \omega C$ with $\phi < 0$ and $J = Y_0 \tan\left(\left|\dfrac{\phi}{2}\right|\right)$, $\phi = -\text{atan}\left(\dfrac{2B}{Y_0}\right)$, and $B = \dfrac{J}{1 - \left(\dfrac{J}{Y_0}\right)^2}$.

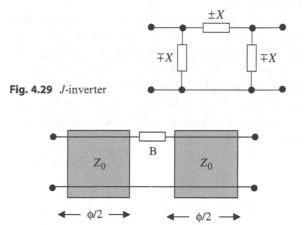

Fig. 4.29 J-inverter

Fig. 4.30 Transmission line J-inverter

Example 4.7 Consider the transmission line circuit in Fig. 4.28 where we assume that X is a transmission line that is short-circuited at the far end. According to Equation (4.27), the transmission line represent a reactance $X = R\Omega$. Now assume that the combined electrical length of the two transmission lines is < 0, e.g., $\beta l_1 = \pi/2$ and $\beta l_2 = -3\pi/2$, the combined circuit has the chain matrix

$$K = \begin{bmatrix} 0 & jZ_0 \\ \dfrac{j}{Z_0} & 0 \end{bmatrix} \begin{bmatrix} 1 & 0 \\ Y & 1 \end{bmatrix} \begin{bmatrix} 0 & -jZ_0 \\ \dfrac{-j}{Z_0} & 0 \end{bmatrix} = \begin{bmatrix} 1 & YZ_0^2 \\ 0 & 1 \end{bmatrix}$$

where $Y = 1/jX$. This corresponds to a series capacitor, $C = R/Z_0^2$. Hence, a quarter-wave open-circuited or short-circuited transmission line stub between two such transmission lines appears as a series inductor or capacitor, respectively. Of course, in the overall network there must be a transmission line with positive electric length to the right of the one with negative length that can absorb the latter.

Quarter-wavelength sections between the stubs act as impedance inverters that effectively convert alternate shunt resonators to series resonators. The stubs and the transmission line sections are $\lambda/4$ long at the center frequency, ω_0.

4.8.2 BP Filters Using Capacitively Coupled Resonators

Bandpass filters can be implemented using microstrip or stripline techniques using a capacitive coupling between the resonators as shown in Fig. 4.31 [50, 97, 123]. An Nth order filter requires N transmission line resonators that are separated by $N-1$ gaps. The gaps can be approximated by series capacitors, and the filter can then be modeled as shown in Fig. 4.31b. The resonators are approximately $\lambda/2$ long at the center frequency, ω_0.

In order to better understand the function of this realization, we redraw the equivalent circuit of Fig. 4.31b with negative-length transmission line sections on either side of the series capacitors. They correspond to an admittance inverter, as seen by comparing Fig. 4.30 and 4.31c. Hence, the gaps correspond to shunt inductors between transmission lines.

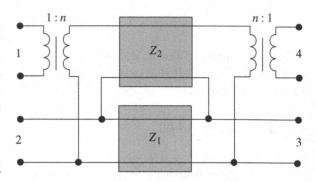

Fig. 4.32 Coupled striplines and corresponding circuit model

Figure 4.32 also shows a corresponding equivalent circuit, where Z_1 is the characteristic impedance for the line (2-to-3) and the second UE, Z_2, is due to the coupling between the two lines [73].

Table 4.2 shows some examples of coupled lines and the corresponding circuit models. The circuit models are directly obtained from Figure 4.32 with the appropriate boundary constraints. Additional cases are found in [73, 123].

Example 4.8 Figure 4.33 shows a possible layout of the filer in Fig. 4.24 using coupled striplines. The structure consists of two copies of the third of the coupled lines in Table 4.2, coupled with a common with a ground (marked with a black square) using a via to the ground plane. The center capacitor is implemented with an open-circuited transmission line (vertical in the figure). The widths of the lines depend on the required characteristic impedances.

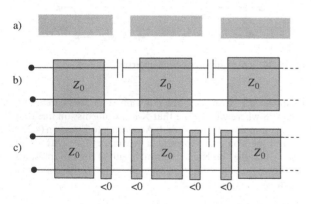

Fig. 4.31 (a) Layout of a capacitive coupled resonator structure, (b) Transmission line model, (c) Transmission line model with negative-length sections forming admittance inverters

4.9 Coupled Line Filters

Figure 4.32 shows two conductors of a stripline that are sufficiently close together so that an electric and magnetic coupling occurs between the two lines.

This configuration can be described by a four-port where the ports are between one end of the line and ground [50, 73, 76, 97, 123].

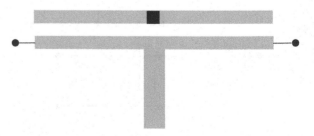

Fig. 4.33 Implementation of the filter in Fig. 4.24 with coupled striplines

Table 4.2 Coupled lines and corresponding circuit models

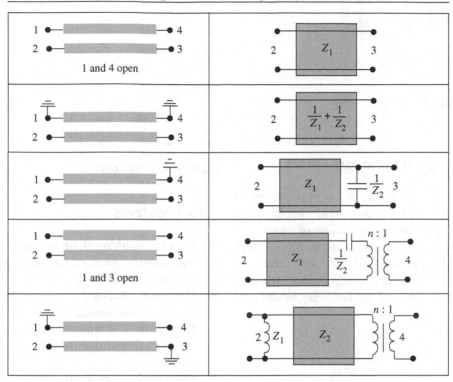

4.9.1 Parallel-Coupled Line Filters

BP and BS filters are particularly easy to implement using parallel-coupled microstrip or striplines for bandwidths less than 20%. Figure 4.34 shows a narrow-band bandpass filter consisting of three cascaded sections [76]. From Table 4.2, we can recognize that this layout corresponds to three UEs interlaced with shunt capacitors. For fractional bandwidths of 15% or less, these filters have practical impedance levels, but for larger fractional bandwidth the structures discussed in Section 4.7 are recommended. A program for the design of physical dimension of microstrip lines needed to realize these filters can be found in [53, 70, 99].

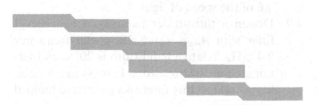

Fig. 4.34 Layout of a BP filter with three coupled lines

In addition, non-commensurate transmission line filters can be designed using coupled lines. Moreover, complex coupled line structures are often used, e.g., interdigital, combline, hairpin-line, and multiplexer structures. Some of these structures will be briefly discussed below. The design of filters based on these types of structures is beyond the scope of this text. See [50, 73, 93, 97, 123] for analysis and synthesis of these types of filters as well as for structures that are suitable for lowpass, highpass, and bandstop filters.

4.9.2 Hairpin-Line Bandpass Filters

Hairpin-line bandpass filters are a variation of parallel-coupled resonator filters where the lines have been bent into a U-shape, so-called hairpin-line resonators, as shown in Fig. 4.35. Hairpin-line bandpass filters have the advantage of compact layouts [50].

Fig. 4.35 Layout of a hairpin-line filter

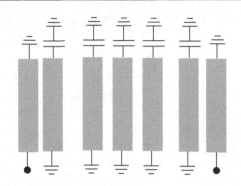

Fig. 4.37 Combline filter

4.9.3 Interdigital Bandpass Filters

Figure 4.36 shows the layout of a typical interdigital bandpass filter that is commonly implemented using microstrips. The layout consists of an array of $\lambda/4$-length transmission line resonators, which alternately are short-circuited at one end and open-circuited at the other end. In general, the physical length and widths of the lines may be different [50, 53, 73].

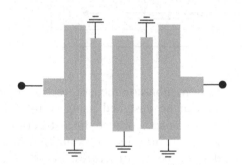

Fig. 4.36 Layout of a fifth-order interdigital filter

4.9.4 Combline Filters

Figure 4.37 shows the layout of a typical combline bandpass filter, which consists of an array of coupled resonators [50, 53, 73, 76]. The resonators consist of transmission lines, which are short-circuited at one end and with a grounded lumped capacitance at the other end. The lumped capacitors may be used for tuning the filter, which may be required particularly for narrow-band filters. The input and output of the filter are through the first and last lines, which are not resonators.

4.10 Problems

4.1 Find the resulting chain matrix of two cascaded transmission lines of length d_1 and d_2. Their characteristic impedance is the same.

4.2 Determine a Richards' structure that corresponds to a series resonance circuit in the Ψ domain.

4.3 Determine a Richards' structure that corresponds to a parallel resonance circuit in the Ψ domain.

4.4 Determine the characteristic impedances of the stubs in Example 4.6.

4.5 Validate that the dual network is obtained by replacing g_i with $1/g_i$ in Equation (4.40).

4.6 Derive the circuit model for a pair of coupled lines where terminals 1 and 3 are used as input and output terminals, terminal 2 is open-circuited, and terminal 4 is grounded.

4.7 Design a lowpass Chebyshev I filter with cutoff frequency 3 GHz, $A_{max} = 0.5$ dB, and $A_{min} > 40$ at 6 GHz. Use commensurate-length transmission lines, and the phase velocity of the dielectric material is 0.6 of the speed of light.

4.8 Design a bandstop commensurate-length transmission line Butterworth filter with 50 Ω resistive terminations. The center frequency is 2.4 GHz, the relative bandwidth is 50%, and the phase velocity of the dielectric material is 0.6 of the speed of light.

4.9 Design a third-order bandpass Chebyshev I filter with $A_{max} = 0.5$ dB, center frequency 2.4 GHz, relative bandwidth is 20%, and terminated in 50 Ω resistors. Discuss the possibility to realize this filter using discrete lumped components.

4.10 A 50 Ω transmission line with length 20 mm is loaded with a complex impedance: $Z_L = 30 + j60$ Ω. The transmission line is operated at 2.4 GHz. Determine the input impedance when the phase velocity of the dielectric material is 0.6 of the speed of light.

4.11 Show that an open-ended loseless transmission line of length $\lambda/4$ behaves as an impedance inverter.

4.12 A lossless transmission line with Z_0 shall be matched to a real load R_L using an impedance inverter. Determine characteristic impedance of the inverter.

Chapter 5
Basic Circuit Elements

5.1 Introduction

Inductors in LC filters for low frequencies require large and heavy ferrite cores, and it is difficult to manufacture inductors with sufficiently large Q factors at low as well as high frequencies. Furthermore, passive LC filters are not compatible with systems that are built using integrated circuit technologies. Therefore, it is desirable to replace inductors with other types of solutions.

It is possible to simulate an inductor with the help of capacitors and active circuit elements such as transistors and operational amplifiers. Because an amplifying element, different from a passive component, can increase the power of the signals, the corresponding filters are called active RC filters.

Active RC filters are often implemented using thin film technology, but it is common that active filters are implemented using monolithic circuit technologies. However, this requires special circuit techniques because we cannot tune the component values in an integrated circuit as easily as when discrete components are used. This is also the case when the filter is implemented inside the PCB.

In this chapter, we will discuss basic circuits that are used as building blocks for active filters. Basic amplifying (active) components are bipolar and FET transistors. A good model for a transistor is, however, very complex and does not correspond to a simple circuit element. To simplify the design of analog filters, it is therefore common to divide the design problem into two main steps.

In the first step, a circuit, with the desired transfer function, consisting of resistors, capacitors, and different types of one-, two- or three-ports (circuit elements), is realized. The latter will in general require amplifying (active) components for their implementation. The choice of passive components is of course important [138], as the cost of a discrete capacitor with high precision is comparable with the cost of an operational amplifier.

The second step consists of realization of the one-, two-, and three-ports with properties as close as possible to the ideal ones.

5.2 Passive and Active n-Ports

In this section, we will discuss some fundamental properties of the n-ports, such as passivity and losslessness.

Theorem 5.1 *A system that initially is at rest and contains no stored energy is passive if the energy, $w(t)$, which is supplied to the system is always nonnegative. That is, for all ports we have*

$$w(t) = \sum_{\text{all ports}} Re\left\{ \int_{-\infty}^{t} i^*(\tau)v(\tau)d\tau \right\} \geq 0 \quad \forall t. \quad (5.1)$$

Theorem 5.2 *A system that initially is at rest and contains no stored energy is lossless if the energy, $w(t)$, which has been supplied to the system ports is zero. That is, the system is lossless if we have for all ports*

$$w(\infty) = \sum_{\text{all ports}} Re\left\{ \int_{-\infty}^{\infty} i^*(\tau)v(\tau)d\tau \right\} = 0. \quad (5.2)$$

L. Wanhammar, *Analog Filters Using MATLAB*, DOI 10.1007/978-0-387-92767-1_5,
© Springer Science+Business Media, LLC 2009

Theorem 5.3

- *An n-port, which contains a finite number of resistors, inductors, capacitors, transformers, gyrators, transmission lines, and dependent voltage and current sources, is time invariant.*
- *An n-port is passive if it also lacks dependent voltage and current sources.*
- *An n-port is reciprocal if it also lacks gyrators.*
- *An n-port is lossless if it only contains inductors, capacitors, transformers, gyrators, and lossless transmission lines.*

5.3 Passive and Active One-Ports

One-ports are characterized by the input current and the voltage across the port. Furthermore, a one-port can be active, passive or lossless, i.e., in the two latter cases it cannot generate any signal power. Below we shall introduce two one-ports, which are active, i.e., they can amplify the signal power, which for analog circuits corresponds to the power of the signal carrier.

5.3.1 Passive One-Ports

Passive circuit elements cannot amplify the power of a signal carrier, i.e., the power of the signal carrier can only be preserved or reduced. Examples of elementary passive circuit elements of one-ports type are resistors, inductors, and capacitors. More complex passive one-ports can be constructed by arbitrary connections of passive circuit elements, i.e., resistors, inductors, capacitors, transformers, gyrators, and lossless transmission lines. We distinguish between passive one-ports, which have losses and dissipate signal energy, and lossless one-ports, which do not dissipate signal energy. A one-port is passive if the energy delivered into it always is non-negative.

Theorem 5.4 *A rational impedance Z is realizable with an RC network that only contains positive resistors and capacitors if and only if*
- *all poles are simple and confined to the negative real axis with positive residues*
- *there are no poles at infinity.*

5.3.2 Active One-Ports

An active one-port can generate signal energy and must contain at least one active (amplifying) circuit element. Examples of active one-ports are resistors, inductors, and capacitors with negative element values, which thus only can be realized using active circuit elements. Sometimes a negative resistor is used to compensate for losses in a passive component, e.g., of a coil to increase its effective Q factor [67].

5.3.2.1 Frequency-Dependent Negative Resistors (FDNRs)

Two active one-ports, which can be used to realize analog filters, are *FDNRs – frequency dependent negative resistors*,[1] which also are called *supercapacitor* and *superinductor* [18]. A supercapacitor has the impedance

$$Z = \frac{1}{s^2 D} \tag{5.3}$$

whereas a superinductor has the impedance

$$Z = s^2 E \tag{5.4}$$

where D [Fs] and E [Hs] are real positive constants. The use of superinductors is not recommended because they are difficult to realize at high frequencies. We will in Section 5.10.4 discuss the realization of these circuits. Figure 5.1 show the symbols used for supercapacitors and superinductors.

Fig. 5.1 Symbols used for supercapacitors and superinductors

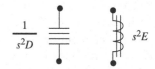

5.4 Two-Ports

Two-ports can also be divided into passive/lossless and active two-ports. Many two-ports are nonreciprocal and require active components to be realized even if the two-port in itself is passive. Figure 5.2 shows the definition of currents and voltages for a

---

[1]Proposed by L.T. Bruton.

Fig. 5.2 Two-port

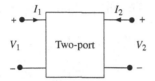

two-port. Note that positive current directions are into the ports.

To describe two-ports, different types of 2×2 matrices are used, i.e., impedance, admittance, and chain matrices together with scattering matrices. The latter are discussed in Chapter 9.

5.4.1 Chain Matrix

Especially useful is the *chain matrix*, $\mathbb{K}$, which is defined as

$$\begin{bmatrix} V_1 \\ I_1 \end{bmatrix} = \mathbb{K} \begin{bmatrix} V_2 \\ -I_2 \end{bmatrix} = \begin{bmatrix} A & B \\ C & D \end{bmatrix} \begin{bmatrix} V_2 \\ -I_2 \end{bmatrix} \quad (5.5)$$

where A, B, C, and D for a two-port, containing only lumped elements, are rational functions in s. The two-port is memoryless if A, B, C, and D are constants.

Theorem 5.5 *A two-port is reciprocal if and only if* $det(\mathbb{K}) = AD - BC = 1$.

A symmetrical two-port has

$$A = D. \quad (5.6)$$

An advantage with the chain matrix is that it can be used to compute the resulting chain matrix for a cascade of two-ports. For example, for a ladder network the resulting chain matrix is the product of the chain matrices for the arms in the ladder network.

Example 5.1 Compute the transfer function of a third-order doubly terminated LC ladder filter using chain matrices when the ladder is of T type.

First we determine the transfer function in terms of the chain matrix of the LC network. We have (note the definition of positive direction of I_2)

$$\begin{cases} V_{in} = R_1 I_1 + V_1 \\ V_{out} = V_2 = -R_2 I_2 \end{cases} \text{ and } \begin{cases} V_1 = AV_2 - BI_2 \\ I_1 = CV_2 - DI_2. \end{cases}$$

After eliminating I_1 and I_2 we get

$$H = \frac{R_2}{AR_2 + B + CR_1R_2 + DR_1}. \quad (5.7)$$

Next we compute the chain matrix for the LC ladder. The chain matrix for a series impedance is

$$\begin{bmatrix} 1 & Z \\ 0 & 1 \end{bmatrix} \quad (5.8)$$

and a shunt admittance

$$\begin{bmatrix} 1 & 0 \\ Y & 1 \end{bmatrix}. \quad (5.9)$$

We assume that the a ladder has no finite transmission zeros. The resulting chain matrix is

$$\begin{bmatrix} A & B \\ C & D \end{bmatrix} = \begin{bmatrix} 1 & sL_1 \\ 0 & 1 \end{bmatrix} \begin{bmatrix} 1 & 0 \\ sC_2 & 1 \end{bmatrix} \begin{bmatrix} 1 & sL_3 \\ 0 & 1 \end{bmatrix} = \begin{bmatrix} C_2L_1s^2 + 1 & C_2L_1L_3s^3 + L_1s + L_3s \\ C_2s & C_2L_3s^2 + 1 \end{bmatrix}.$$

Inserting into Equation (5.7) yields the transfer function.

$$H(s) = \frac{R_2}{C_2L_1L_3s^3 + C_2(R_1L_3 + R_2L_1)s^2 + (R_1R_2C_2 + L_1 + L_3)s + R_1 + R_2}.$$

The magnitude function at $\omega = 0$ is $|H(0)| = \frac{R_2}{R_1+R_2}$. Hence, by multiplying with $(R_1 + R_2)/R_2$, we get a normalized transfer function.

5.4.2 Impedance and Admittance Matrices

In some cases it is convenient to use impedance or admittance matrices. The impedance matrix of a two-port is defined by

$$\begin{bmatrix} V_1 \\ V_2 \end{bmatrix} = \begin{bmatrix} z_{11} & z_{12} \\ z_{21} & z_{22} \end{bmatrix} \begin{bmatrix} I_1 \\ I_2 \end{bmatrix}. \tag{5.10}$$

For a reciprocal two-port, we have $z_{21} = z_{12}$. The admittance matrix of a two-port is defined by

$$\begin{bmatrix} I_1 \\ I_2 \end{bmatrix} = \begin{bmatrix} y_{11} & y_{12} \\ y_{21} & y_{22} \end{bmatrix} \begin{bmatrix} V_1 \\ V_2 \end{bmatrix}. \tag{5.11}$$

For a reciprocal two-port, we have $y_{21} = y_{12}$.

5.4.3 Passive Two-Ports

Passive two-ports can be constructed by arbitrary connections of passive circuit elements, i.e., resistors, inductors, capacitors, transformers, gyrators, and lossless transmission lines. Such two-ports, except for gyrators, are reciprocal. This property allows us to put the signal source at either port of the LC ladder.

Note that reciprocity is a precondition for realization of certain classes of very high-performance analog filters with low element sensitivity. For example, the components, which are used in the LC ladders discussed in Chapter 3, are reciprocal and the whole LC ladder is therefore reciprocal.

5.4.3.1 Transformer

An example of a passive two-port is the *transformer*, which is defined by the relation between currents and voltages for the two ports

$$\begin{cases} V_1 = nV_2 \\ I_1 = \dfrac{-I_2}{n}. \end{cases} \tag{5.12}$$

The chain matrix for a transformer with turns ratio $n : 1$ is

$$\mathbb{K} = \begin{bmatrix} n & 0 \\ 0 & \dfrac{1}{n} \end{bmatrix}. \tag{5.13}$$

The energy that is absorbed by the transformer is

$$\begin{aligned} w &= Re\{I_1^* V_1\} + Re\{I_2^* V_2\} \\ &= Re\{I_1^* V_1\} + Re\left\{-nI_1^* \frac{V_1}{n}\right\} = 0 \end{aligned} \tag{5.14}$$

where $n > 0$ and real. The transformer is, according to Theorem 5.5, a memoryless, lossless, reciprocal two-port that cannot store energy.

The input impedance to a transformer that is loaded on the secondary side with an impedance Z_L is

$$Z_{in} = n^2 Z_L. \tag{5.15}$$

5.4.3.2 Gyrator

A two-port that is described by the relations

$$\begin{cases} V_1 = -r_1 I_2 \\ I_1 = \dfrac{V_2}{r_2} \end{cases} \tag{5.16}$$

where r_1 and $r_2 > 0$ and real is called a *gyrator*[2]. The constants r_1 and r_2 are the gyrator resistances. The chain matrix for the gyrator is

$$\mathbb{K} = \begin{bmatrix} 0 & r_1 \\ \dfrac{1}{r_2} & 0 \end{bmatrix}. \tag{5.17}$$

Thus, the gyrator is, according to Theorem 5.5, a nonreciprocal two-port and therefore is the "direction" of the gyrator essential. The direction is defined, as shown in Fig. 5.3, where a positive input current gives rise to a positive output voltage, according to Equation (5.16), with $r_2 > 0$.

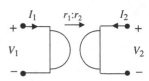

Fig. 5.3 Gyrator

---

[2]Proposed by the inventor of the pentode tube, B.D.H. Tellegen, 1948.

The energy that is absorbed by the gyrator is for $r_1 = r_2$,

$$
\begin{aligned}
w &= Re\{I_1^* V_1\} + Re\{I_2^* V_2\} \\
&= Re\{I_1^* V_1\} + Re\left\{\frac{-r_2 V_1^* I_1}{r_1}\right\} = 0.
\end{aligned}
\tag{5.18}
$$

Thus, the gyrator is a lossless, reciprocal two-port for $r_1 = r_2$, and it cannot store energy.

Example 5.2 Determine the chain matrix for a circuit that consists of two cascaded gyrators according to Fig. 5.4. Notice how the gyrators are connected and their direction.

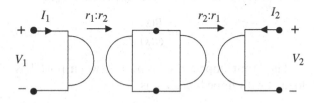

Fig. 5.4 Cascaded gyrators

The resulting chain matrix is obtained by multiplying the chain matrices of the individual gyrators. We get

$$
\mathbb{K} = \begin{bmatrix} 0 & r_1 \\ \frac{1}{r_2} & 0 \end{bmatrix} \begin{bmatrix} 0 & r_2 \\ \frac{1}{r_1} & 0 \end{bmatrix} = \begin{bmatrix} 1 & 0 \\ 0 & 1 \end{bmatrix}.
$$

Comparison with Equation (5.13) shows that the circuit corresponds to a transformer with $n = 1$. The two combined gyrators represent a reciprocal two-port, but internally the circuit is nonreciprocal.

5.4.4 Active Two-Ports

There exist several classes of active two-ports. Here we will discuss the three classes of two-ports: controlled signal sources, immitance converters, and inverters. *Immitance* is a term coined by Bode for denoting an impedance or an admittance.

5.4.4.1 Controlled Signal Sources

There are four different types of *controlled sources* with finite gain:

- VCVS: Voltage controlled voltage source
- VCCS: Voltage controlled current source – transconductance amplifier
- CCVS: Current controlled voltage source – transresistance amplifier
- CCCS: Current controlled current source

The corresponding symbols are shown in Fig. 5.5.

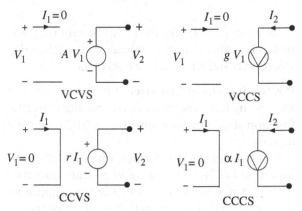

Fig. 5.5 Controlled signal sources with finite gain

5.4.4.2 Generalized Immitance Converters (GICs)

Generalized immitance converters are often used as basic building blocks in high-performance active filters. The symbol for a *generalized immitance converter, GIC*, is shown in Fig. 5.6.

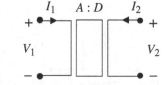

Fig. 5.6 Generalized immitance converter (GIC)

Definition 5.1 A two-port is a generalized immitance converter (GIC) if, when terminated at port 2 with an impedance Z_2, the input immitance at port 1 is $K(s)Z_2^{\pm 1}$, where the converter function, $K(s)$, is independent of Z_2.

The chain matrix for a generalized impedance converter is

$$
\mathbb{K} = \begin{bmatrix} A(s) & 0 \\ 0 & D(s) \end{bmatrix}
\tag{5.19}
$$

where $B(s) = C(s) = 0$. The input impedance to port 1, when port 2 is loaded with an impedance Z_2, is

$$Z_{in1} = \frac{A(s)}{D(s)} Z_2 \qquad (5.20)$$

and the input impedance to port 2, when port 1 is loaded with the impedance Z_1, is

$$Z_{in2} = \frac{D(s)}{A(s)} Z_1. \qquad (5.21)$$

In general, $A(s)$ and $D(s)$ may be rational functions of s. There are two special cases of the GIC, where $A(s)$ and $D(s)$ are real constants.

Positive Impedance Converter (*PIC*): $A(s) = n_1$ and $D(s) = n_2$ have the same signs and the converter function $A/B = n_1/n_2$ is positive. A PIC is active if $n_1 \neq n_2$.

For example, a transformer, which is defined by Equation (5.13), is a positive impedance converter with $A(s) = n$ and $D(s) = 1/n$. The load impedance Z_2 at port 2 is seen from port 1 as an impedance

$$Z_{in1} = n^2 Z_2. \qquad (5.22)$$

Negative Impedance Converter (*NIC*): $A(s) = \pm n_1$ and $D(s) = \mp n_2$ have opposite signs and the converter function is negative. For example, a negative resistance can be realized by using a NIC[3]. Figure 5.7 show the symbols used for impedance converters.

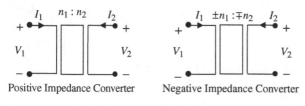

Positive Impedance Converter   Negative Impedance Converter

Fig. 5.7 Positive and negative impedance converters

5.4.4.3 Generalized Immitance Inverters (GIIs)

Generalized immitance inverters are also used as basic building blocks in high-performance active filters. The symbol for a *generalized immitance inverter (GII)* is shown in Fig. 5.8.

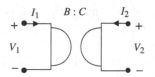

Fig. 5.8 Generalized immitance inverter (GII)

The chain matrix for a generalized immitance inverter is

$$\mathbf{K} = \begin{bmatrix} 0 & B(s) \\ C(s) & 0 \end{bmatrix} \qquad (5.23)$$

where $A = D = 0$. The input impedance to port 1, when port 2 is loaded with an impedance Z_2, is

$$Z_{in1} = \frac{B(s)}{C(s)} \frac{1}{Z_2}. \qquad (5.24)$$

The input impedance to port 2, when port 1 is loaded with impedance Z_1, is

$$Z_{in2} = \frac{B(s)}{C(s)} \frac{1}{Z_1}.$$

In general, $B(s)$ and $C(s)$ may be rational functions of s. The GII has also two special cases, when $B(s)$ and $C(s)$ are real constants.

Positive Impedance Inverter (*PII*): $B(s) = r_1$ and $C(s) = 1/r_2$ have the same signs and the converter function $B/C = r_1 r_2$ is positive. A PII is active if $r_1 \neq r_2$. The gyrator is an example of a positive impedance inverter, i.e., a load impedance Z_2 at port 2 is seen from port 1 as an impedance

$$Z_{in1} = \frac{r_1 r_2}{Z_2}. \qquad (5.25)$$

Hence, gyrator loaded by a capacitor will appear from the input of the gyrator as an inductor.

Negative Impedance Inverter (*NII*): $B(s) = \pm r_1$ and $C(s) = \mp 1/r_2$ have opposite signs and the converter function is negative. Figure 5.9 show the symbols used for impedance invertors

In Section 5.9, we will discuss the realization of these two-ports.

Example 5.3 Determine the converter constants so that the circuit in Fig. 5.10 corresponds to a series impedance.

From Equations (5.9) and (5.23), we get the chain matrix for the circuit

Fig. 5.9 Positive and negative impedance inverters

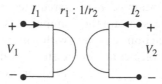

Positive Impedance Inverter, PII

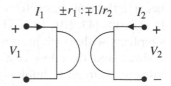

Negative Impedance Inverter, NII

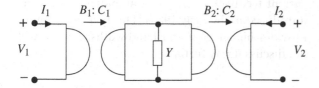

Fig. 5.10 PIIs with embedded admittance Y

$$\begin{bmatrix} 0 & B_1 \\ C_1 & 0 \end{bmatrix} \begin{bmatrix} 1 & 0 \\ Y & 1 \end{bmatrix} \begin{bmatrix} 0 & B_2 \\ C_2 & 0 \end{bmatrix} = \begin{bmatrix} C_2 B_1 & Y B_1 B_2 \\ 0 & C_1 B_2 \end{bmatrix}.$$

A series impedance has the chain matrix in Equation (5.8). Hence, we must have $C_2 B_1 = C_1 B_2 = 1$, which with $B_1 = r_1$ and $C_1 = 1/r_2$ yields $B_2 = r_2$ and $C_2 = 1/r_1$ and the series impedance becomes $Z = r_1 r_2 Y$. For example, if $Y = sC$, i.e., a shunt capacitor, then $Z = sr_1 r_2 C$, i.e., a series inductor $L = r_1 r_2 C$. This is a commonly used circuit to realize inductors in integrated active filters.

5.5 Three-Ports

Three-ports can, in principle, be decomposed into a network consisting of one-ports and two-ports and they are therefore not fundamental circuit elements. However, they are in practice very useful and can be efficiently implemented directly using transistors in integrated circuits. It is therefore convenient to use three-ports, and in some cases four-ports, as basic building blocks.

5.5.1 Passive Three-Ports

A circulator is a lossless nonreciprocal three-port. The symbol for a three-port circulator is shown in Fig. 5.11. A signal incident to port 1 of a circulator is transmitted to port 2. In the same way, a signal incident to port 2 is transmitted to port 3, and a signal incident to port 3 is transmitted to port 1. The arrow in the circulator symbol indicates the sequence in which the incident signals are circulated from port to port.

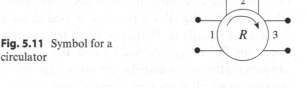

Fig. 5.11 Symbol for a circulator

A circulator can be used between the antenna and the transmitter and receiver. That is, the transmitter is connected to port 1, the antenna to port 2, and the receiver to port 3. A three-port circulator can, as shown in Fig. 5.12, be realized by using a single gyrator.

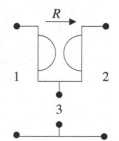

Fig. 5.12 Realization of a circulator using a gyrator

5.5.2 Active Three-Ports

The active elements typically limit the performance of active filters, and we will therefore in the next few sections discuss in more detail the properties of the most common types of active elements and their implementations [109]. Here we are mainly interested in three classes of active three-ports, i.e., operational amplifiers, transconductors, and current conveyors, which basically are generalizations of controlled sources, e.g., VCVS and VCCS.

5.6 Operational Amplifiers

A special case of VCVS, *operational amplifier* (*op-amp*), which has one of the two output terminals grounded and, in principle, infinite gain (A), has become the workhorse for implementation of analog circuits [43, 101, 103]. The main reasons are the simplicity with which most analog functions can be realized and their low cost when implemented in integrated circuit technologies.

The term "operational" comes from the fact that these amplifiers originally were used to realize basic mathematical operations, e.g., addition and integration, in analog simulation machines. The latter, which predated digital computers, was used to solve differential equations. Hence, with the term operational amplifier, we normally mean the physical implementation of an amplifier, but it is often useful to consider an idealized operational amplifier.

An ideal operational amplifier, which is a voltage controlled voltage source, typically with one of the output terminals grounded, is described by

$$V_{out} = A(V_+ - V_-) \qquad (5.26)$$

where $A \to \infty$. Note that here A denotes the transfer function of the amplifier and not the attenuation. The ideal operational amplifier is thus a differential amplifier with infinite gain and infinite input impedance while the output impedance is zero. The symbol used for an operational amplifier is shown in Fig. 5.13.

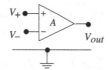

Fig. 5.13 Operational amplifier

A more general version of the operational amplifier, shown in Fig. 5.14, which has both a positive and a negative output terminal is described by

$$V_{out+} - V_{out-} = A(V_+ - V_-).$$

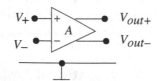

Fig. 5.14 Operational amplifier with differential output

Differential-mode operational amplifiers are normally used to implement active filters in integrated

circuit technology. The advantage with this type of amplifier is that we can realize circuits that operate with differential signals, which tend to reduce the distortion due to nonlinearities. Furthermore, the need for inverters is avoided, as we can choose appropriate output, i.e., inverted or noninverted.

Yet another type of operational amplifier is the current feedback operational amplifier, which was introduced in the early 1980s. This type of amplifier provides higher gain and bandwidths, but we will not discuss it any further.

5.6.1 Small-Signal Model of Operational Amplifiers

A real operational amplifier is, however, not ideal. The gain is finite and frequency dependent. For simple operational amplifiers that do not need to be frequency compensated to be stable, it is often sufficient to model the frequency response with a single pole, i.e.,

$$A(s) = \frac{A_0}{1 + \dfrac{s}{\omega_{3dB}}}. \qquad (5.27)$$

For more advanced operational amplifiers, a higher-order model should be used, i.e., a model with several poles and zeros. Figure 5.15 shows a typical magnitude response for a simple bipolar operational amplifier with $A_0 \approx 10^5$ and $\omega_{3dB} \approx 2\pi$ 10 rad/s. CMOS amplifiers often have a 10-fold smaller A_0. Note that here $|A(j\omega)|$ denotes the magnitude of the amplifier's gain. A more accurate model includes a second pole on the negative real axes at about $s \approx -\omega_t$, which is defined below.

Unity Gain Bandwidth: The angular frequency, ω_t, at which the magnitude function equals 1, i.e., $\omega_t \approx A_0 \omega_{3dB}$, is referred to as the *unity gain bandwidth*. Typically, a simple op amp has $\omega_t \approx 2\pi\ 10^6$ rad/s.

A useful approximation of the frequency response is

$$A(s) \approx \frac{\omega_t}{s}. \qquad (5.28)$$

Gain-Bandwidth Product: We will later use the *gain-bandwidth product* (GB), where $GB = A_0 \omega_{3dB}/2\pi \approx \omega_t/2\pi$, as a measure of the amplifier's finite

Fig. 5.15 Magnitude response of a simple operational amplifier

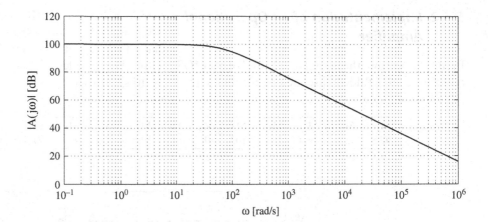

bandwidth. Operational amplifiers with bipolar transistors have often a very high gain-bandwidth product, of the order a few GHz, whereas operational amplifiers with MOS transistors typically have GB in the range 50–100 MHz. The GB varies strongly between different samples of amplifiers (typically ±50%) and it is temperature dependent (±10% from 0 till 70C°).

Typical values for simple operational amplifiers of type 741 are $A_0 \approx 10^5$, $\omega_{3dB} \approx 2\pi \, 10$ rad/s, and $\omega_t \approx 2\pi \, 10^6$ rad/s. GB is typically in the range 1–10 MHz for a simple operational amplifier. A low-power operational amplifier, e.g., MAX416x, has $A_0 \approx 10^4$, $\omega_{3dB} \approx 2\pi \, 10$ rad/s, and $\omega_t \approx 200 \, \pi$ krad/s. Operational amplifiers for higher frequencies have lower gain and much larger GB. For example, the operational amplifier MAX426x has a GB = 450 MHz.

Input Currents: The operational amplifier has a differential input stage. The transistors in this stage require a bias current in order for the transistors to operate in the saturated region. A *DC* path must therefore exist from both of the inputs of the operational amplifier to either ground or to the output of an operational amplifier. The resistance of the DC paths to the positive and negative inputs to the operational amplifier should be the same. Otherwise, the bias currents will cause a nonzero input (offset) voltage to the amplifier. The bias currents are essentially constant but strongly temperature dependent.

Output Voltage: The slope of the output voltage is limited, because charges on capacitances within the operational amplifier cannot be changed arbitrarily fast. A measure of the slope is the *slew rate*, $S = |dV/dt|$ [V/μs], i.e., the derivative of the output

signal with a step as input signal. The slew rate, which essentially is determined by the maximal current in the differential stage, affects signals with large amplitude and high frequency.

Operational amplifiers, like MAX416x and MAX426x, have typical slew rates of 0.1 and 900 V/μs, respectively, whereas simple operational amplifiers typically have $S \approx 0.5$ to 50 V/μs.

Asymmetry: Operational amplifiers are in practice not completely symmetric. Even if the input is zero, the output may be nonzero. A measure of the asymmetry is an equivalent input, *offset voltage*, V_{os}, which is required to make the output voltage zero. Typically, V_{os} is a few mV and somewhat higher for the operational amplifiers with FET transistors in the differential stage.

In an ideal operational amplifier, only the difference between the input voltages is amplified. In practice, however, the mean value of the input voltages, $(V_+ + V_-)/2$, the *common mode voltage*, will be amplified and contribute to the output. As a measure of this contribution, we use the *common mode rejection ratio* (*CMRR*).

Disturbances from the power supply voltages also cause disturbances at the amplifier's output. As a measure of the influence of these disturbances, we used the *power supply rejection ratio* (*PSRR*).

Noise and Distortion: Operational amplifiers are the main cause of noise in active *RC* filters and they have nonlinearities that cause distortion. The design, or selection, of the operational amplifiers is therefore a crucial step in the realization of active filters. Thus, we will henceforth focus on determining the requirements on the operational amplifiers.

5.6.2 Implementation of an Operational Amplifier

Figure 5.16 shows the layout [9, 52, 103] of a CMOS realization of a folded cascode operational ampli-fier in a 0.35-μm CMOS technology. The corre-sponding schematic is shown in Fig. 5.17. The top half of the layout is occupied with a 3.731 pF capa-citor. The gate length is 0.7 μm and the transistor widths (μm) are

M_1	M_2	M_3	M_4	M_5	M_6	M_7	M_8	M_9	M_{10}	M_{11}	M_{12}	M_{13}
92.8	92.8	26.4	26.4	103	61	61	58.8	58.8	41	41	500	158.8

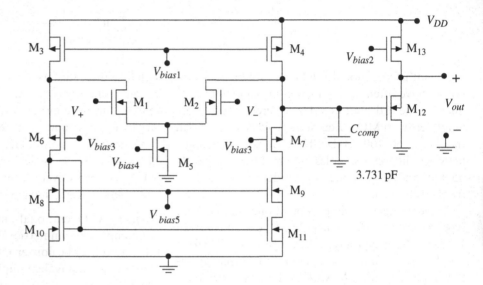

Fig. 5.16 Folded cascode operational amplifier

The amplifier has $A_0 = 70$ dB, $\omega_t = 2\pi\,75$ Mrad/s, phase margin = 60°, PSRRp = 70 dB, PSRRn = 65 dB, CMRR = 119 dB, and power consumption is 7.54 mW with $V_{DD} = 5$ V. The area for the layout is $130 \times 95\ \mu$m. The amplifier is designed to drive a load of 5 pF.

5.7 Transconductors

Transconductors, which also are called *OTA* (*operational transconductance amplifiers*), are often used to implement analog filters in integrated cir-cuits [1, 3, 32, 56, 58, 84, 92, 103, 130]. Figures 5.18 and 5.19 show the symbols used for a single-ended and a differential transconductor, respectively. Note that the definition of the direction of the output cur-rent may vary in literature. It is also common that the transconductor may have a differential output.

Also in this case, it is useful to idealize the compo-nent. The inputs of the ideal transconductor have, similar to an operational amplifier, high impedance. However, a transconductor differs from an opera-tional amplifier by having high output impedance. In fact, the ideal transconductor is a voltage controlled, current source, VCCS. The output current is

$$I = g_m(V_+ - V_-). \qquad (5.29)$$

For an ideal transconductor, g_m is constant $< \infty$ and frequency independent. The transconduc-tance, g_m, is in practice, however, constant over only a small range of input signals. In addition, a transconductor is linear over only a small range of input signals, and differential circuits are therefore used to improve the linearity. To simplify the design and analysis of an analog filter, a *single-ended circuit* is often used. The resulting circuit is

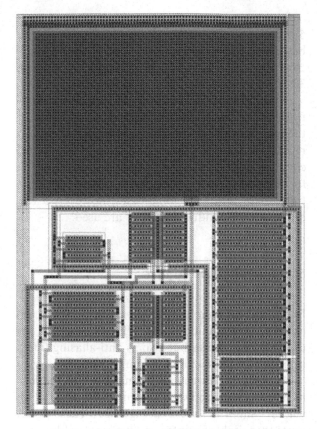

Fig. 5.17 Layout of folded cascode operational amplifier

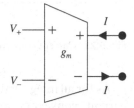

Fig. 5.18 Transconductor

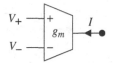

Fig. 5.19 Differential transconductor

at a later stage converted to a differential circuit, which then is implemented. Transconductors suffer from the same limitations as operational amplifiers.

5.7.1 Transconductance Feedback Amplifiers

Wilson [139] proposed an amplifier, the so-called transconductance feedback amplifier (TFA), which can be modeled as shown in Fig. 5.20. This amplifier belongs to the family of constant-bandwidth amplifiers such as the current feedback amplifiers (CFA) [109].

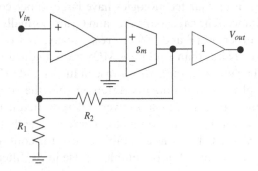

Fig. 5.20 Model of a transconductance feedback amplifier

The TFA has an input stage with high gain, which is followed by a transconductor around which feedback is applied. The voltage at the output of the transconductor is then buffered to yield a low output impedance.

Several differences however, exist between the TFA and the conventional CFA. For example, the CFA cannot be generally configured as an integrator while the TFA can be configured as an integrator making it useful in filter structures. Of less importance is that the CFA can realize a differentiator contrary to the TFA. The use of differentiators is not recommended since they tend to generate high noise.

5.7.2 Small-Signal Model for Transconductors

A simple small-signal model for a transconductor is shown in Fig. 5.21. The transconductance is frequency dependent and is often modeled with only one single real pole, $s_p = -\omega_{3\text{dB}}$, i.e.,

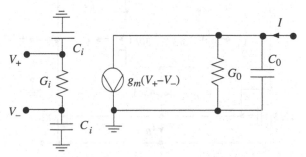

Fig. 5.21 Small-signal model for a transconductor

$$g_m = \frac{g_{m0}}{1 + \dfrac{s}{\omega_{3db}}}. \quad (5.30)$$

Typical values are $g_{m0} = 50\ \mu S$ and $\omega_{3dB} = 2\pi$ 100 Mrad/s. The input conductance G_i is very small for CMOS transconductors and can be neglected except at very high frequencies. Transconductors realized in bipolar technologies have larger input conductance. The capacitances C_i and C_0 are normally of the order 50 fF and 200 fF, respectively, and G_0 is typically less than 1 μS for CMOS transconductors.

The usage of the terms transconductor and OTA for physical components differs in that the transconductance is constant over a larger frequency range for the transconductor compared with the OTA. An OTA is not suitable to use for continuous-time filters, but is suitable for use in *SC* filters.

5.7.3 *Implementation of a Transconductor*

Figure 5.22 shows an example of a simple transconductor realization in a standard digital CMOS process where the width/length (W/L) of the transistor are indicated [1]. Note that the channel length, $L = 3\ \mu m$, is very long compared with the transistors in a digital circuit. This is required to obtain good linearity and low noise. Transconductors can be implemented using CMOS or BiCMOS processes. See [58] for realization of a differential folded cascode transconductor and SPICE models for simulation of transconductor filter and [10] for methods for layout of analog circuits.

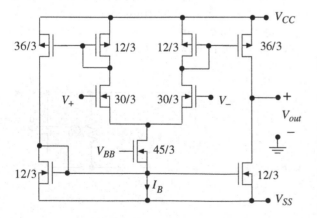

Fig. 5.22 Simple CMOS transconductor

With the power supply voltages $V_{CC} = -V_{SS} = 5$ V and with a control voltage $V_{BB} = -3.24$ V, a bias current $I_B = 336\ \mu A$ is obtained and the transconductance is $g_{m0} = 1.33$ mS and $\omega_{3dB} = 2\pi$ 30 Mrad/s. The value of g_m is determined by V_{BB}.

A complete transconductor filter requires a control circuit, which uses the control voltage V_{BB} to tune the filter parameters to their desired values. To reduce the implementation cost, a common control circuit is often used for all transconductors, which for this reason must be identical and have the same transconductance. In some cases, however, a few different transconductance sizes are used.

5.8 Current Conveyors

Integrated continuous-time filters for high frequencies, i.e., several hundreds of MHz, is needed in for example hard drives but also as anti-aliasing filters in front of ADCs with high sample rates.

A relatively new class of circuit elements that has received great interest is *current conveyors* (*CCs*), whose symbol is shown in Fig. 5.23. These three-ports are suitable for integration of analog filters for high frequencies. Current conveyors can be used to realize most two-ports [34].

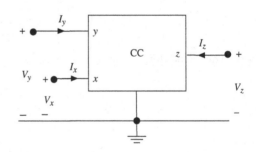

Fig. 5.23 Current conveyor

As for all integrated filters, implementations with current conveyors must be tuned continuously because the circuit element's parameter values vary with temperature, power supply voltage, and from sample to sample [78]. Current conveyors, which can be implemented using both bipolar and CMOS processes, have a circuit complexity of the same order as conventional operational amplifiers.

5.8.1 Current Conveyor I (CCI)

In his diploma work (1968), A.S. Sedra proposed a new type of three-port called current conveyor. A *current conveyor* of type I (*CCI*) is a three-port that is defined by the matrix

$$
\begin{bmatrix} I_y \\ V_x \\ I_z \end{bmatrix} = \begin{bmatrix} 0 & a & 0 \\ 1 & 0 & 0 \\ 0 & b & 0 \end{bmatrix} \begin{bmatrix} V_y \\ I_x \\ V_z \end{bmatrix}. \qquad (5.31)
$$

The first generation of current conveyors, *CCIs*, which no longer are in practical use, is obtained for $a = 1$, which has two variants. A noninverting current conveyor is obtained, CCI+, with $b = 1$, and $b = -1$ yields an inverting current conveyor, CCI−. A current conveyor can be considered as an ideal transistor.

5.8.2 Current Conveyor II (CCII)

A more useful class of current conveyors, type II, was suggested 1970 by Sedra and Smith [114]. A *current conveyor* of type II (*CCII)* is a three-port that is defined by Equation (5.31) where $a = 0$. There are two variants, CCII+ and CCII−, which differ with respect to if the output current I_z is noninverted, $b = 1$, or inverted, $b = -1$, with respect to I_x. Electrically tunable CCIIs have been proposed [78].

Figure 5.24 shows a simple model for a CCII. The voltage at the *y*-port, which has high input impedance, is transferred to the *x*-port, which has low input impedance. The current into the *x*-port is transferred by a current controlled current source (CCCS) between *x* and *z*, which has high output impedance. A current conveyor of the type *CCII−* has an inverting CCCS between *x* and *z*.

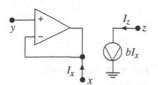

Fig. 5.24 Model for a CCII

Current conveyors can be implemented in both bipolar and CMOS technologies as well as in BiC-MOS technology. Examples of commercial integrated circuits are *CCC*II01 from LTP Electronics, which no longer is manufactured, and *AD*844, which is an operational amplifier with current feedback from Analog Devices, which can be configured as a CCII+. OPA860 from Burr Brown contains a CCII+ and a voltage buffer. OPA2662 is another transconductor that also can be used as a current conveyor of the type CCII+. The circuit package contains two CCII+.

5.8.3 Current Conveyor III (CCIII)

A third-generation current conveyor type III (*CCIII*) is obtained for $a = -1$, which also has two variants. $b = 1$ yields a *CCIII+*, and with $b = -1$ *CCIII−* is obtained.

5.8.4 Small-Signal Model for Current Conveyor II

A simple small-signal model for a current conveyor is

$$
\begin{bmatrix} I_y \\ V_x \\ I_z \end{bmatrix} = \begin{bmatrix} 0 & a & 0 \\ 1 + \varepsilon_y & 0 & 0 \\ 0 & b + \varepsilon_x & 0 \end{bmatrix} \begin{bmatrix} V_y \\ I_x \\ V_z \end{bmatrix} \qquad (5.32)
$$

where $b = \pm 1$ and $|\varepsilon_x| << 1$ and $|\varepsilon_y| << 1$.

Figure 5.25 shows a more accurate small-signal model for CCII, which is suitable for simulation of a analog filters.

R_y and C_y model the input impedance of port *y*. The input impedance for port *x*, which has largest effect on the frequency dependency, is represented by R_{x1}, R_{x2}, L_x, and C_x. The input impedance is small for low frequencies and obtains a maximum at resonance between L_x and C_x. The output impedance for port *z* is modeled with R_z and C_z.

Typical values for an integrated current conveyor are $R_y \approx 10\ \text{M}\Omega$, $C_y \approx 0\ \text{pF}$, $R_{x1} \approx 350\ \Omega$, $R_{x2} \approx 0.5\ \Omega$, $L_x \approx 10\ \mu;\text{H}$, $C_x \approx 100\ \text{pF}$, $R_z \approx 10\ \text{M}\Omega$, $C_z \approx 20\ \text{pF}$, and $k \approx 0.95$ and $h = -1.05$ for CCII+ and $h = 1.05$ for CCII−.

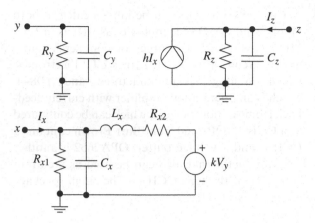

Fig. 5.25 Small-signal model for CCII

*AD*844 CFOA (current feedback operational amplifier) have the following typical values: $R_y \approx$ 10 MΩ, $C_y \approx 0$ pF, $R_{x1} \approx 0.5$ Ω, $R_{x2} \approx 60$ Ω, $L_x \approx 10$ nH, $C_x \approx 2$ pF, $R_z \approx 3$ MΩ, and $C_z \approx 4.5$ pF. The manufacturer also publishes on the Internet a corresponding SPICE model.

5.8.5 CMOS Implementation of a CCII±

Figure 5.26 shows a CMOS implementation of a CCII± with complementary outputs [132]. The circuit, realized using a 0.6-μm CMOS process, has the following parameters when $I_{B1} = 100$ μA and $I_{B2} = 200$ μA: $k = 0.95$, $h_+ = 1.03$, $h_- = -1.03$, $V_{DD} = -V_{SS}$ = 2.5 V.

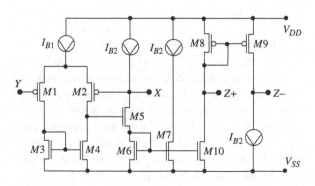

Fig. 5.26 Example of CMOS realization of CCII±

5.9 Realization of Two-Ports

In this section, we will discuss the realization of two-ports using operational amplifiers, transconductors, and current conveyors. These realizations will later be used as building blocks for active filters, but also for the realization of active one-ports.

5.9.1 Realization of Controlled Sources: Amplifiers

Controlled sources are synonyms to amplifiers. For example, the transfer function of a VCCS is of the type I/V; it is often referred to as a transimpedance amplifier, whereas a CCVS is a transresistance amplifier. The four types of amplifiers, i.e., VCVS, VCCS, CCVS, and CCCS, can be realized with the previously discussed active three-ports.

5.9.1.1 Controlled Sources with Operational Amplifiers

Figures 5.27 and 5.28 show two of the most commonly used VCVS that can be realized with operational amplifiers. A CCVS is obtained if the input current is injected at the inverting input of the inverting amplifier and Z_1 is grounded.

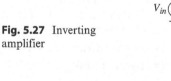

Fig. 5.27 Inverting amplifier

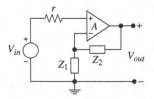

Fig. 5.28 Noninverting amplifier

Inverting Amplifier: The transfer function of the inverting amplifier is

$$H(s) = -\frac{Z_2}{Z_1 + \dfrac{Z_1 + Z_2}{A}}. \quad (5.33)$$

Note that here $A(s)$ denotes the transfer function of the amplifier and not the attenuation. When operated as a regular amplifier, the impedances Z_1 and Z_2 are resistors.

The transfer function with an ideal operational amplifier, i.e., $A \to \infty$,

$$H_{Ideal}(s) = -\frac{R_2}{R_1}. \quad (5.34)$$

The transfer function can therefore be written

$$H(s) = -\frac{R_2}{R_1} \cdot \frac{1}{1 + \dfrac{R_1 + R_2}{R_1 A}} \quad (5.35)$$

where the last term represents an error factor. The phase response is

$$\Phi(\omega) = \text{atan}\{H(j\omega)\} = \pi - \text{atan}\left(\frac{R_1 + R_2}{R_1 \omega_t}\omega\right) \quad (5.36)$$

where we have used the approximation in Equation (5.28). Hence, the finite unity gain bandwidth introduces a phase lag that is significant if the amplifier is used inside a feedback loop.

The inverting amplifier can be used to sum multiple input signals by providing each input source with a resistor connected to the inverting input of the operational amplifier.

In case of realization of an amplifier with very high gain, it may be advantageous to realize the feedback resistor R_2 as a resistive T network [107].

Noninverting Amplifier: The transfer function for the noninverting amplifier is

$$H(s) = \frac{Z_1 + Z_2}{Z_1 + \dfrac{Z_1 + Z_2}{A}} \quad (5.37)$$

and with an ideal operational amplifier, i.e., $A \to \infty$,

$$H(s) = -\frac{Z_1 + Z_2}{Z_1}. \quad (5.38)$$

The transfer function can therefore for resistive impedances be written

$$H(s) = -\frac{R_1 + R_2}{R_1} \cdot \frac{1}{1 + \dfrac{R_1 + R_2}{R_1 A}}. \quad (5.39)$$

The phase response is

$$\Phi(\omega) = \text{atan}\{H(j\omega)\} = -\text{atan}\left(\frac{R_1 + R_2}{R_1 \omega_t}\omega\right). \quad (5.40)$$

For inverting and noninverting amplifiers, we select $r \approx Z_1//Z_2$ at $\omega = 0$, and for the integrator $r \approx R$ in order to reduce the offset voltage due to the bias currents. Note that r does not affect the transfer function. With this choice, the offset voltage at the output of the amplifier is typically reduced with a factor of 4.

Example 5.4 Consider an inverting amplifier with an operational amplifier that can be described by Equation (5.27). The transfer function is given by Equation (5.33). We now replace the operational amplifier with an ideal amplifier with the impedances Z_1 and Z_3. We get: $-\frac{Z_2}{Z_1 + (Z_1 + Z_2)/A} = -\frac{Z_3}{Z_1} \Rightarrow \frac{1}{Z_3} = \frac{1}{Z_2} + \frac{1/Z_1 + 1/Z_2}{A}$. Now, we can interpret this as a correction admittance, $(1/Z_1 + 1/Z_2)/A$, in parallel with the impedance Z_2. For example, with R_1, R_2, and $A = \omega_t/s$, we get $s(1/R_1 + 1/R_2)/\omega_t$, which represents a capacitor $C = \omega_t R_1 R_2/(R_1 + R_2)$ in parallel with R_2.

We have the following general rule for the inverting amplifier:

The correction admittance, parallel to Z_2, is the sum of all admittances connected to the inverting input of the operational amplifier, divided by A.

5.9.1.2 Controlled Sources with Transconductors

A voltage amplifier can be realized with a transconductor and a resistor as shown in Fig. 5.29. We get an inverting amplifier with the gain

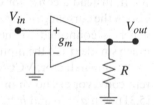

Fig. 5.29 Voltage amplifier

$$\frac{V_{out}}{V_{in}} = -g_m R$$

and a noninverting amplifier if the inputs to the transconductor are interchanged. Several voltages can be added/subtracted by first converting the voltages to currents by using one transconductor for each voltage and then adding their output currents. The voltages can be weighted by selecting different transconductances for the transconductors.

A drawback of this circuit is that the output impedance is R, i.e., neither very low nor very high. Hence, this circuit does not correspond to an ideal voltage nor current source.

Resistors are difficult and expensive to implement in an integrated circuit.

Resistors are therefore often realized with the transconductor circuit shown in Fig. 5.45.

5.9.1.3 Controlled Sources with Current Conveyors

Consider Equation (5.31), which can be rewritten

$$\begin{cases} I_y = aI_x \\ V_x = V_y \\ I_z = bI_x. \end{cases}$$

A current controlled current source, CCCS, is realized with a CCII+ with $a = 0$ and $b = 1$, if the y-port is grounded and the x-port is used as input. Hence, $V_x = V_y = 0$ and $I_z = I_x$. The current gain of the controlled current source is $\alpha = +1$.

If, instead a conveyor of the type CCII−, with $a = 0$ and $b = -1$, is used, a current controlled current source, with the current gain $\alpha = -1$, i.e., $I_z = -I_x$, is obtained. The input impedance to the x-port is low and the circuit is suitable for summing multiple input currents.

Figure 5.30 shows a realization of current controlled current source, CCCS, with the gain $\alpha = R_1/R_2$. If, instead a conveyor of type CCII− is used, we obtain the current gain $\alpha = -R_1/R_2$. The resistor, R_1, which is connected to the y-port, must be chosen much smaller than the input impedance of the y-port.

A tranconductor, VCCS, can be realized by a current conveyor as shown in Fig. 5.31. From Equation (5.31) with $a = 0$ and $b = \pm 1$, we get for the circuit

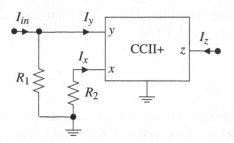

Fig. 5.30 CCCS realized with a current conveyer of type CCII+

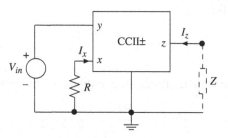

Fig. 5.31 VCCS realized with a CCII±

$$\begin{cases} V_x = V_y = V_{in} \\ V_x = -RI_x \\ I_z = \pm I_x. \end{cases}$$

After elimination of V_y, V_x, and I_x, we obtain

$$\frac{I_z}{V_{in}} = \mp \frac{1}{R}. \tag{5.41}$$

Thus, the circuit realizes a (negative/positive) transconductor using a current conveyor of type CCII±.

5.9.2 Realization of Integrators

Integrators play an important role as basic building blocks in many active filter structures. We will therefore in this section discuss the realization of integrators and their properties in detail [17, 82].

An ideal integrator has the transfer function

$$H(s) = \pm \frac{1}{sRC}. \tag{5.42}$$

An integrator with (+)-sign is called a noninverting integrator or positive integrators and an integrator with (−)-sign is called an inverting integrator or negative integrator.

5.9.2.1 Miller Integrator

Equation (5.33) can be rewritten as

$$H(s) = \frac{-Z_2}{Z_1} \cdot \frac{1}{1 + \dfrac{Z_1 + Z_2}{A(s)Z_1}} \qquad (5.43)$$

where the second term in the denominator represents a deviation from the ideal frequency response of an inverting amplifier.

A *Miller integrator* is obtained by selecting $Z_1 = R$ and $Z_2 = 1/sC$, as shown in the circuit in Fig. 5.32. The transfer function is

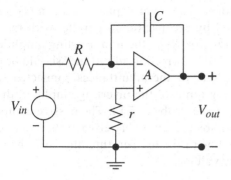

Fig. 5.32 Miller integrator

$$H(s) = \frac{-1}{sRC} \cdot \frac{1}{1 + \dfrac{1 + \dfrac{1}{sRc}}{A(s)}}. \qquad (5.44)$$

Using a simple model of the operational amplifier, i.e., Equation (5.27), and for not too low frequencies, we can approximate $A(s) \approx A_0\omega_{3dB}/s = \omega_t/s$ for $\omega \gg \omega_{3dB}$. Inserting $A(s) = \omega_t/s$ in Equation (5.44) yields

$$H(s) = \frac{-1}{sRC} \cdot \frac{\omega_t}{s + \omega_t + \dfrac{1}{RC}}. \qquad (5.45)$$

The Miller integrator has, thus, a parasitic pole

$$s_p = -\left(\omega_t + \frac{1}{RC}\right) \approx -\omega_t). \qquad (5.46)$$

Now, an ideal integrator should have a pure imaginary frequency response. As a measure of quality of an integrator with the frequency response

$$H(j\omega) = \pm \frac{1}{R(\omega) + jX(\omega)} \qquad (5.47)$$

we use the quality factor

$$Q_I = \frac{X(\omega)}{R(\omega)}. \qquad (5.48)$$

Hence, an ideal integrator should have an infinite Q_I factor. Note that Q_I may be negative. For the Miller integrator, we have according to Equation (5.45)

$$H(j\omega) = \frac{-1}{\dfrac{-\omega^2 RC}{\omega_t} + j\omega RC\left(1 + \dfrac{1}{RC\omega_t}\right)} \qquad (5.49)$$

but as $1/RC \ll \omega_t$, we get

$$H(j\omega) = \frac{-1}{\dfrac{-\omega^2 RC}{\omega_t} + j\omega RC}. \qquad (5.50)$$

The quality factor for the Miller integrator is

$$Q_I = \frac{-\omega_t}{\omega} \approx -|A(j\omega)| \qquad (5.51)$$

where $A(j\omega)$ is the frequency response of the operational amplifier.

A nonideal integrator has according to Equations (5.47), (5.48), and (5.50) the phase response

$$\Phi = \pi - \text{atan}\left(\frac{X(\omega)}{R(\omega)}\right) = \pi - \text{atan}(Q_I)$$
$$= \pi + \text{atan}\left(\frac{\omega_t}{\omega}\right). \qquad (5.52)$$

5.9.2.2 Negative Integrator with Passive Compensation

The effect of finite bandwidth of the operational amplifier can be alleviated by placing a resistor R_0 in the feedback path, as shown in Fig. 5.33. The resistor should be

$$R_0 = 1/(\omega_t C). \qquad (5.53)$$

Alternatively, a capacitor, C_0, in parallel with R, may be used where

$$C_0 = 1/(\omega_t R). \qquad (5.54)$$

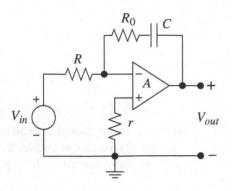

Fig. 5.33 Negative integrator with passive compensation

See Problem 5.23. The problem with passive compensation is that we are trying to match two completely different components, i.e., an RC constant and the unity gain frequency, which depend on different physical processes. Hence, it is difficult to match and achieve good tracking over time, temperature, and power supply voltages.

5.9.2.3 Positive Integrators

A positive integrator has the transfer function $H(s) = +1/sRC$, unlike the (negative) Miller integrator, which has the transfer function $H(s) = -1/sRC$. In many applications, noninverting integrators are needed.

A simple solution that is shown in Fig. 5.34 is to use a Miller integrator that is followed by an inverting amplifier. The measure of quality for a Miller integrator in series with an inverting amplifier is [112]

$$Q_I = \frac{-1}{\omega\left(\dfrac{1}{\omega_{t1}} + \dfrac{2}{\omega_{t2}}\right)} \quad (5.55)$$

where ω_{t1} and ω_{t2} are the unity gain frequencies for the integrator and the inverting amplifier, respectively. Hence, this version of positive integrator has about three times as large phase lag compared with the Miller integrator.

5.9.2.4 Noninverting Integrator with Passive Compensation

The parasitic poles in the integrator and the inverting amplifier causes excess phase in the overall transfer function. This excess phase shift is a major problem when the integrator is used inside a feedback loop, because it will change the pole positions. The excess phase can, however, be reduced by a capacitor in parallel with the feedback resistor R_2 in the noninverting amplifier. In general, the correcting admittance should be equal to the sum of the admittances connected to the inverting input of the inverting amplifier, divided by ω_t/s. See Problem 5.14. The passive compensation is, however, not very efficient because ω_t varies both with the temperature and with the power supply voltage.

5.9.2.5 Noninverting Integrators with Active Compensation

A more efficient approach is to place a unity gain amplifier in the feedback path, as shown in Fig. 5.35. In this case, it is required that the two operational amplifiers are matched and track over time and temperature [107].

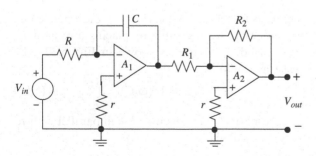

Fig. 5.34 Positive integrator

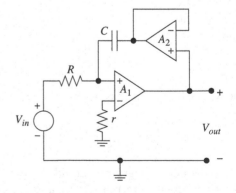

Fig. 5.35 Noninverting integrator with active compensation

The quality factor for this circuit is

$$Q_I = -|A(j\omega)|^3 < 0.$$

However, to obtain this high quality factor required very tightly matched amplifiers.

Phase-Lead Integrator: A better way of reducing the error in the phase function is to use an integrator with active compensation, as shown in Fig. 5.36. The transfer function is

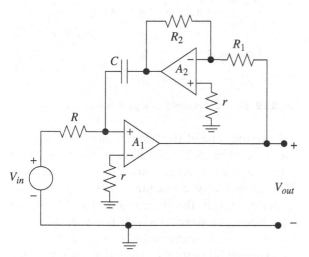

Fig. 5.36 Noninverting integrator with active compensation

$$H(s) \approx \frac{R_1}{R_2} \frac{1}{sRC}. \qquad (5.56)$$

The quality factor of the noninverting integrator is

$$Q_I = \frac{1}{\omega\left(\dfrac{2}{\omega_{t2}} - \dfrac{1}{\omega_{t1}}\right)} \qquad (5.57)$$

and with matched operational amplifiers, i.e., $\omega_{t1} = \omega_{t2}$, is

$$Q_{Imatched} = \frac{\omega_t}{\omega}. \qquad (5.58)$$

Note that the quality factor is positive. Hence, the excess phase

$$\Phi = -\text{atan}\,(Q_I) = -\text{atan}\left(\frac{\omega_t}{\omega}\right) \qquad (5.59)$$

is positive for this integrator. This circuit is therefore known as phase-lead integrator. Comparing with the uninverting integrator in Fig. 5.34 and Equation (5.55), the quality is three times better for matched amplifiers.

We will later use an inverting and a noninverting integrator in a feedback loop. High performance is obtained if a Miller integrator, which has a phase lag, $-\text{atan}(Q_I)$, is compensated for by a phase-lead integrator with a phase lead of $+\text{atan}(Q_I)$ [18, 112].

Improved Miller-Inverter Integrator: An alternative circuit is shown in Fig. 5.37 [17]. The positive input to the second amplifier has been lifted off ground and connected to the virtual ground of the integrator. By choosing $R_0 = 1/\omega_{t1}C$ and $\alpha = R_2/R_1 = \omega_{t2}/\omega_{t1}$, the transfer function becomes

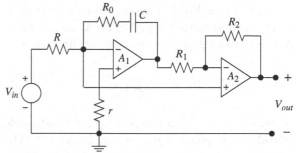

Fig. 5.37 Noninverting integrator

$$H(s) = \frac{\alpha}{sRC}.$$

The quality factor of this integrator is the same as for the Miller integrator. Hence, we obtain a three-fold reduction of the phase lag compared with the circuit shown in Fig. 5.34.

An even better positive integrator is obtained if the integrator in Fig. 5.37 is replaced by the integrator shown in Fig. 5.35, but this requires an additional amplifier [112].

5.9.2.6 Transconductor-Based Integrators

As mentioned previously, many techniques to build higher order filters use integrators as the basic component. This is particularly true in integrated active filters.

Figure 5.38 shows a so-called g_m-C integrator. The output voltage is

Fig. 5.38 g_m-C integrator
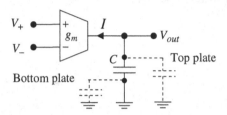

$$V_{out} = \frac{-I}{sC} = -\frac{g_m}{C}\frac{1}{s}V_{in}$$

where $V_{in} = (V_+ - V_-)$. The integrator's transfer function is

$$H(s) = -\frac{g_m}{C}\frac{1}{s}. \quad (5.60)$$

A circuit realizing a transconductor, however, will have finite output impedance, i.e., nonzero output conductance G_0, and the transfer function is

$$H(s) = -\frac{g_m}{C}\cdot\frac{1}{s+a} \quad (5.61)$$

where $a = -G_0/C$ is a real pole. This parasitic pole introduces a phase error of the integrator. The phase of an ideal integrator is 90° for all frequencies. A small deviation from 90° affects strongly the frequency response of an integrator-based filter.

Example 5.5 Determine the effect of finite input and output impedances of the transconductor for the integrator shown in Fig. 5.38.

The integrator's output voltage is

$$V_2 = \frac{g_m V_{in}}{(G_0 + s(C_0 + C))}$$

and the transfer function is

$$H(s) = \frac{g_m}{(C_0 + C)}\frac{1}{s + \dfrac{G_0}{C_0 + C}}.$$

The transconductor's output capacitance, C_0, which is nonlinear and depends on the signal level, is added to the load capacitance. Depending on how the integrator is used, the input admittance of the following circuit will also be added.

With typical values, $G_0 = 1\ \mu S$, $C_0 = 200\ fF$, and the load capacitance $C = 4.8\ pF$, the real pole becomes $s_p = -200$ krad/s

Parasitic elements, which will appear in an implementation of an analog filter, will affect the frequency response. Typically, the resistive parasitic can be neglected and often also inductive coupling to adjacent components and wires. The dominating parasitic elements are stray capacitances between different electrical nodes and ground. In an integrated circuit, an implementation of a capacitor will have two major parasitic capacitors; the top-plate-to-ground and the bottom-plate-to-ground, as illustrated in Fig. 5.39.

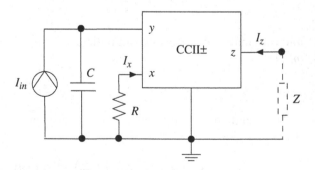

Fig. 5.39 Stray capacitance in a g_m-C integrator

Note that one of the stray capacitances (*bottom plate*), see Fig. 5.39, will have a constant (zero) voltage and will therefore not affect the integrator. The second stray capacitance (*top plate*) can be estimated during the design and absorbed in the capacitor C. However, the effective capacitance at the output of the transconductor will be affected.

It is essential that all capacitors in an integrated circuit are grounded and that floating capacitors are avoided. The later introduces parasitic capacitances that affect the frequency response.

5.9.2.7 Current Conveyor-Based Integrators

Figure 5.40 shows a (negative/positive) integrator that operates in current mode. With an ideal current conveyor, we have the following relation for the circuit

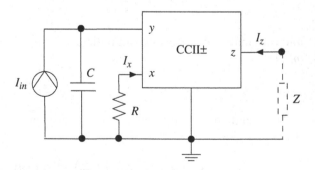

Fig. 5.40 Integrator

$$\begin{cases} V_y = \dfrac{I_{in}}{sC} \\ V_y = V_x = -RI_x \\ I_z = \pm I_x. \end{cases}$$

After elimination of V_y, V_x, and I_x, we get the transfer function

$$H(s) = \frac{I_z}{I_{in}} = \frac{\mp 1}{sRC}. \qquad (5.62)$$

Note that the transfer function is represented by the ratio between two currents. An integrator, which operates in voltage mode, can be realized with a transconductor, according to Fig. (5.66), and with a capacitor as load.

Filter structures, which are based on integrators, i.e., two-integrator loops and leapfrog filters, which operate in current mode can easily be realized with these integrators.

5.9.3 Realization of Immitance Inverters and Converters

Generalized immitance inverters and converters are flexible two-ports that are used in many active filter realizations. Both the GII and GIC can be realized using operational amplifiers, transconductors, or current conveyors.

5.9.3.1 Antoniou's GIC

A useful two-port, which is shown in Fig. 5.41, is Antoniou's *GIC* (*generalized impedance converter*) [18]. The input impedance with ideal amplifiers is

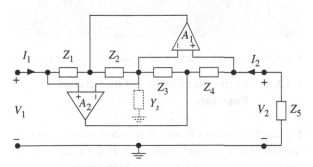

Fig. 5.41 Antoniou's GIC

$$Z_{in} = \frac{V_1}{I_1} = \frac{Z_1 Z_3}{Z_2 Z_4} Z_5 = K(s) Z_5. \qquad (5.63)$$

Hence, if we view Z_5 as the load at port 2, the circuit realizes a positive impedance converter

(PIC). We will in Sections 5.10.4 and 5.10.5 discuss in detail how this circuit can be used to realize inductors and FDNRs.

The chain matrix with ideal operational amplifiers is

$$\mathbb{K} = \begin{bmatrix} 1 & 0 \\ 0 & \dfrac{Z_2 Z_4}{Z_1 Z_3} \end{bmatrix}. \qquad (5.64)$$

Y_s is a compensation admittance that will be discussed later. We may also consider Z_2 or Z_4 as the load. In this case, however, the two ports do not have a common ground. In this case, the circuit realizes a positive impedance inverter (PII).

5.9.3.2 Transconductor-Based Gyrator

Equation (5.16), which describes two voltage controlled current sources (*voltage controlled current source*; VCCS), can be realized with two transconductors according to Fig. 5.42 [18]. A simple analysis of the circuit yields $r_1 = 1/g_{m1}$ and $r_2 = 1/g_{m2}$.

Fig. 5.42 Gyrator

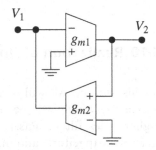

Generally, a floating impedance can be realized with two lossless gyrators embedding a shunt admittance according to Fig. 5.10.

5.9.3.3 Current Conveyor-Based Gyrator

Figure 5.43 shows a circuit that realizes a gyrator and for which the following equations are valid if the current conveyors are ideal.

$$\begin{cases} V_{y2} = -R_2 I_{x2} \\ I_{x2} = I_{z2} \\ V_{y2} = -Z I_{z1} \\ I_{x1} = -I_{z1} \\ V_{y1} = -R_1 I_{x1}. \end{cases}$$

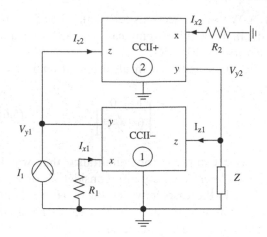

Fig. 5.43 Gyrator

Elimination yields the input impedance

$$Z_{in} = \frac{V_{y1}}{I_1} = \frac{R_1 R_2}{Z}. \tag{5.65}$$

Normally we select $R_1 = R_2$.

5.10 Realization of One-Ports

In this section, we will discuss the realization of one-ports, which will be used for realization of higher-order filters. Passive components such as resistors, capacitors, and inductors are an integral part of every electronic subsystem. In a typical electronic circuit board, 80% of the components are passive, taking up 50% of the PCB area, and requiring 25% of all solder joints. It is therefore of interest due to both technical and economical reasons to be able to integrate analog filters using a standard digital CMOS process. Particularly interesting is to be able to use integrated analog circuits on the same chip as the digital circuits. This allows complete signal processing systems to be integrated in a single chip. The cheap digital CMOS processes are unfortunately not suitable for implementation of analog circuits and often special processes with multiple layers of polycrystalline silicon are therefore used to implement good resistors and capacitors.

5.10.1 Integrated Resistors

Integration of resistors can essentially take place in two main ways, either using long thin conductors of, e.g., polycrystalline silicon, or using a MOSFET, see [54, 56, 65, 92].

5.10.1.1 Polycrystalline Resistors

Polycrystalline-based resistors can only be made with small resistor values because they require a large chip area [65]. The polycrystalline layer is often referred to as simply the polylayer. Figure 5.44 shows an example of a layout of the polylayer for a resistor of $10\,k\Omega$. The layout of the conductor is in meander form in order to obtain a more square shape. Often special layout techniques are used with adjacent dummy resistors to reduce the edge effects.

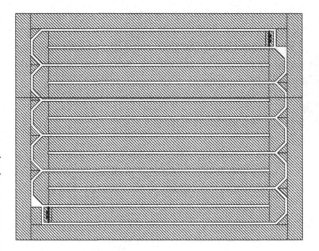

Fig. 5.44 Example of layout of a polycrystalline resistor

5.10.1.2 Active Realization of Integrated Resistors

Figure 5.45 shows how a grounded resistor can be realized with a transconductor. For the circuit we have $I_1 = g_m V_{in}$ and $I = I_1$, which yields

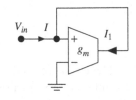

Fig. 5.45 Realization of a grounded resistor

$$R = \frac{V_{in}}{I} = \frac{1}{g_m}.$$

A general floating resistor with $R = 1/(2g_m)$ can be realized with two identical transconductors according to the circuit shown Fig. 5.46. A floating resistor, which is driven by a voltage source, can be realized with a single transconductor as shown in Fig. 5.47. This circuit can be used for simulation of the resistor in series with the signal source in a doubly resistively terminated LC filter.

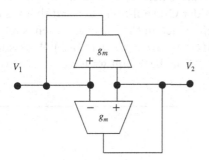

Fig. 5.46 Realization of a floating resistor

Fig. 5.47 Realization of a floating with voltage source

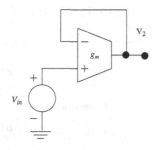

We have for the circuit shown in Fig. 5.48

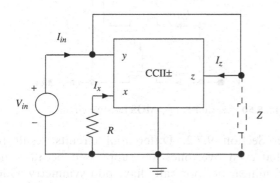

Fig. 5.48 Grounded resistor

$$\begin{cases} V_{in} = V_y = V_x \\ V_x = -RI_x \\ I_z = \pm I_x \\ I_{in} = I_z \end{cases}$$

and after elimination of V_y, V_x, I_x, and I_z, we obtain

$$\frac{V_{in}}{I_{in}} = \mp R. \qquad (5.66)$$

Thus, the circuit realizes a (negative/positive) resistor using a current conveyor.

5.10.1.3 MOSFET Circuits

An important class of circuits, so-called MOSFET circuits [56], utilizes the fact that we can implement good capacitors and MOS transistors instead of resistors. The later is, however, nonlinear, and special circuit techniques must be used to linearize the circuits. MOSFET-C filters [110, 111] are the most common implementation technique for integrated analog filters after transconductor-based filters. MOSFET-C filters often have less distortion than transconductor filters.

Suitable active circuit elements are operational amplifiers or current conveyors. Operational amplifiers, however, require a large gain ($>10^3$) in the feedback loop in order for acceptable function to be obtained. This limits the usable frequency range.

It is in practice necessary to use differential realizations to suppress the distortion due to the nonlinearities. Often is synthesis, analysis, etc., made using a single-ended realization, which then is transformed to a differential realization. A differential realization not only improves the linearity of the filter, but also increases the signal dynamic and suppresses noise on the power supply voltage, which can be large in an integrated digital circuit.

5.10.1.4 MOSFET Resistors

Because it is difficult to effectively implement large resistance values with a polysilicon, it is attractive to instead use a MOSFET, which unfortunately is nonlinear, as resistor. We shall show that this problem can be solved with the help of a differential circuit.

We have for an nMOS transistor according to Fig. 5.49 that operates in the nonsaturated region

Fig. 5.49 MOS transistor

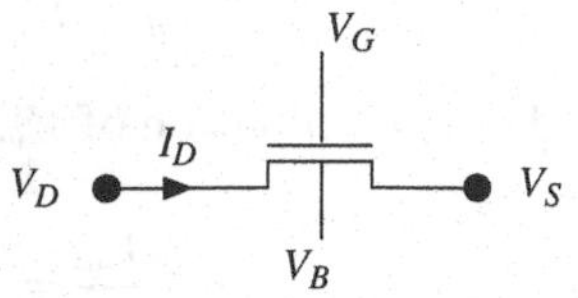

$$I_D = C_{ox}\mu_0 \frac{W}{L}\left((V_{GS} - V_T)V_{DS} - \frac{V_{DS}^2}{2}\right) \quad (5.67)$$

which can be written as

$$I_D = F(V_D, V_G) - F(V_S, V_G) \quad (5.68)$$

$$F(V_X, V_G) = 2K(V_G - V_B - V_{FB} - \Phi_B)$$
$$- K(V_X - V_B)^2 - (4K_\gamma(V_X - V_B + \Phi_B))^{2/3} \quad (5.69)$$

$$K = \gamma C_{ox}\frac{W}{L} \quad (5.70)$$

$$\gamma = \frac{2qN_A\varepsilon_S}{C_{ox}}. \quad (5.71)$$

We assume here that the mobility μ_0 is constant. Consider Czarnul-Song's circuit that is shown in Fig. 5.50, which in the literature often is referred to as the *MOS resistive circuit* (*MRC*). We assume that $V_3 = V_4$. For the circuit we have

$$I_1 - I_2 = (I_{D1} + I_{D3}) - (I_{D2} + I_{D4}) =$$
$$= F(V_1, V_{G1}) - F(V_3, V_{G1}) + F(V_2, V_{G2}) - F(V_3, V_{G2})$$
$$- F(V_1, V_{G2}) + F(V_3, V_{G2}) - F(V_2, V_{G1}) + F(V_3, V_{G1})$$
$$= F(V_1, V_{G1}) - F(V_1, V_{G2}) + F(V_2, V_{G2}) - F(V_2, V_{G1}). \quad (5.72)$$

Insertion of Equation (5.69) yields

$$I_1 - I_2 = 2K(V_{G1} - V_{G2})(V_1 - V_2). \quad (5.73)$$

Hence, $\Delta I = I_1 - I_2$ is proportional to $\Delta V = V_1 - V_2$, i.e., the circuit corresponds to a controllable, linear resistor.

It is not necessary that the transistors operate in the nonsaturated region. The circuit eliminates the nonlinearities even if all transistors operate in the saturated region or if the two cross-coupled transistors operate in the nonsaturated region and the two other in the saturated region.

5.10.2 Differential Miller Integrators

Figure 5.51 shows a differential Miller integrator. A simple approach to derive a differential realization for a single-ended circuit is discussed

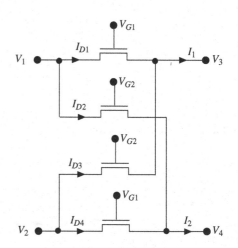

Fig. 5.50 Czarnul-Song's MOS resistive circuit

in Section 9.7.2. Differential circuits result in that even overtones are suppressed because the nonlinearities normally have odd symmetry. Odd overtones are not suppressed. Furthermore,

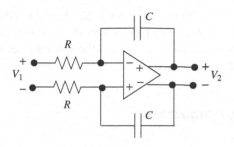

Fig. 5.51 Differential RC integrator

differential realizations have higher CMRR (common mode rejection ratio) and PSRR (power supply rejection ratio).

The resistors in the Miller integrator can be replaced with MOSFETs according to Fig. 5.52, but because they are strongly nonlinear, the integrators can only be used for very small signal swings. Figure 5.53 shows the corresponding Miller integrator, but with resistors replaced with the circuit shown in Fig. 5.50. The circuit has a significantly improved linearity, i.e., the linearity can increase up to 30 dB. Note that the

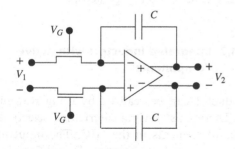

Fig. 5.52 Differential MOSFET-C integrator

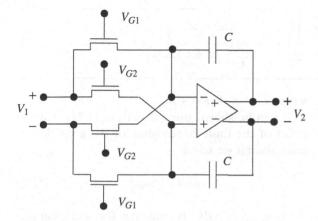

Fig. 5.53 Linearized Miller integrator

nonlinearities influence will be reduced if the integrator is used in a feedback loop, e.g., in a leapfrog filter. A disadvantage of using differential structures is that additional circuitry controlling the common-mode voltage is needed.

5.10.3 Integrated Capacitors

High-quality capacitors are essential components in integrated subsystems, e.g., sample-and-hold, analog-to-digital (ADC), and digital-to-analog (DAC) converters, radio frequency (RF) front-ends, switched-capacitor, and continuous-time filters. Implementations with high capacitance density are desirable, as the capacitors typically requires a significant part of the chip area. Furthermore, additional desired properties are close matching of pairs of capacitors, linearity, small bottom-plate stray capacitor, and accuracy of the capacitance values. In RF circuits, it is also necessary that the self-resonance frequencies are well in excess of the frequency of interest.

Capacitors with good linearity and small losses can be implemented in a CMOS process. The best capacitors are obtained if the capacitor is realized between two polycrystalline plates. Unfortunately, all CMOS processes do not have multiple polylayers.

Implementation of nine unit capacitors using multiple stacked (parallel) plates of metal-to-metal or metal-to-polysilicon have very good linearity and high Q factors, but they suffer from a low capacitance density. This is mainly due to the large vertical spacing between different layers that determines the capacitance. Unfortunately, as the geometries of the process technologies are reduced, the vertical dimension does not shrink as fast as the horizontal dimension. In fact, the width of the wires in an integrated circuit will tend to become smaller whereas the thickness increases in order to keep the cross-area about constant. Thus, parallel plate capacitors will tend to consume a larger fraction of the chip area.

Several innovative approaches have therefore been proposed to improve the density of integrated capacitors [6]. Most of them exploit both lateral and vertical dimensions; for example, interdigital or parallel wires structures, where the small distance

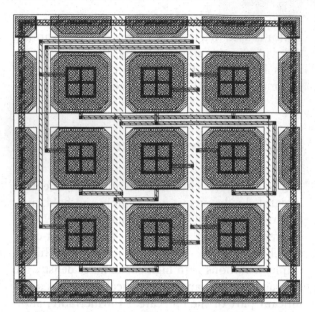

Fig. 5.54 Layout of two matched capacitors consisting of nine unti capacitrs

between two wires in the same layer is used as a capacitor. Many complex structures that exploit both interlayer and intralayer capacitors are possible [6]. In addition, higher capacitance density can be achieved by using insulators with higher dielectric constant.

Some circuits, e.g., SC filter, exploit the fact that the ratio of two capacitors can be very accurate, but their absolute values are of less importance. Figure 5.54 shows a typical layout of two matched capacitors, which consist of nine unit capacitors.

The ratio of the two capacitances can be very accurate (0.1%), while the accuracy of the absolute

values may be of the order 10%. The capacitors in the four corners and in the center constitute one of the capacitors and the others the second one. Unity capacitors are used to reduce the edge effects.

5.10.4 Inductors

5.10.4.1 Integrated Inductors with Passive Components

Inductors can be integrated on various substrates, e.g., thin film and silicon [25]. Magnetic material is rarely used, and the inductances that can be implemented using high-density technologies are relatively small. Furthermore, the Q factors are often in the range 5–20 on silicon substrates, which often is too low for many applications. The effective Q factors can, of course, be increased by combining a passive coil and an active circuit that realizes a negative resistor. For high-performance integrated filters, however, inductorless, or more correctly coil-less solutions are used in practice because integrated coils have too low Q factors.

5.10.4.2 Integrated Inductors with Active Components

An inductor can be realized by using Antoniou's GIC. There exist several alternative selections of the circuit elements in the GIC. The input impedance, with nonideal operational amplifiers, is

$$Z_{in} = \frac{Z_1 Z_3 Z_5 + Z_1 (Z_4 + Z_5)\left(\frac{Z_2}{A_1} + \frac{Z_3}{A_2} + \frac{Z_2 + Z_3 + Z_2 Z_3 Y_s}{A_1 A_2}\right)}{Z_2(Z_4 - Y_s Z_5 Z_3) + (Z_4 + Z_5)\left(\frac{Z_2}{A_1} + \frac{Z_3}{A_2} + \frac{Z_2 + Z_3 + Z_2 Z_3 Y_s}{A_1 A_2}\right)}. \tag{5.74}$$

5.10.4.3 Antoniou's GIC of Type A

There are several ways of choosing the impedances Z_1 through Z_5 so the circuit shown in Fig. 5.41 realizes, for example, an inductor. By choosing $Z_1 = R_1$, $Z_2 = 1/sC_2$, $Z_3 = R_3$, $Z_4 = R_4$, and $Z_5, = R_5$, a GIC of type A is obtained. The input impedance is

$$Z_{in} = Ls$$

where $L = R_1 R_3 R_5 C_2 / R_4$.

It can be shown using Equation (5.74) that the effect of the finite bandwidths of the amplifiers is minimized if we select

$$\omega_{t1} R_5 = \omega_{t2} R_4. \tag{5.75}$$

Antoniou's GIC is sensitive for stray capacitances parallel with R_5. The detrimental effect of a stray capacitance C_p can for type A be compensated with a capacitance $C_s = C_p R_4 / R_3$ between the operational amplifier's inverted input and ground.

5.10.4.4 Antoniou's GIC of Type B

By choosing $Z_1 = R_1$, $Z_2 = R_2$, $Z_3 = R_3$, $Z_4 = 1/sC_4$, and $Z_5 = R_5$, a GIC of type B is obtained with the input impedance

$$Z_{in} = Ls$$

where $L = R_1 R_3 R_5 C_4 / R_2$.

Usually the B type is used, also called type II, because it has better high frequency properties, i.e., is less sensitive for the amplifier's finite gain-bandwidth products, GB [112]. It is advisable to choose [106]

$$Z_2 = Z_3 = R \tag{5.76}$$

to obtain good high-frequency properties and to use matched amplifiers.

The detrimental effect of a stray capacitance C_p at port 2 can for type B be compensated with a resistance $R_s = C_4 R_3 / C_p$ between the operational amplifier's inverted input and ground.

5.10.4.5 Transconductor-Based Inductors

A grounded g_m-C inductor with the inductance $L = C/g_{m1}g_{m2}$ can be realized with a gyrator according to Fig. 5.55.

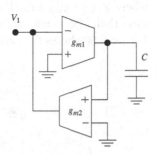

Fig. 5.55 Realization of a grounded inductor

Figure 5.56 shows the corresponding small-signal model where the input and output impedances have been included in the model [18].

We get by summing the currents in the output and input nodes

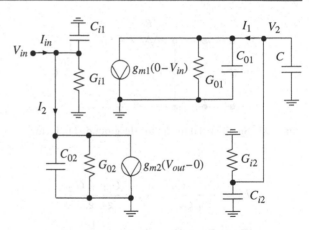

Fig. 5.56 Small-signal model for the grounded inductor

$$\begin{cases} -g_{m1}V_{in} + (Y_{out1} + Y_{in2} + sC)V_2 = 0 \\ I_{in} = Y_{in1}V_{in} + I_2 \\ I_2 = Y_{out2}V_{in} + g_{m2}V_2 \end{cases}$$

where

$$Y_{in1} = G_{i1} + sC_{i1}$$

$$Y_{out1} = G_{01} + sC_{01}$$

$$Y_{in2} = G_{i2} + sC_{i2}$$

$$Y_{out2} = G_{02} + sC_{02}.$$

We get the input admittance

$$Y_{in} = Y_{in1} + Y_{out2} + \frac{g_{m1}g_{m2}}{Y_{in2} + Y_{out1} + sC}$$

and after simplification

$$Y_{in} = G_1 + sC_1 + \frac{g_{m1}g_{m2}}{G_2 + sC_2} \tag{5.77}$$

where

$$G_1 = G_{i1} + G_{01} \quad G_2 = G_{i2} + G_{02}$$
$$C_1 = C_{i1} + C_{02} \quad C_2 = C_{i2} + C_{01} + C.$$

The inductor has the inductance

$$L = \frac{C}{g_{m1}g_{m2}}$$

and the parasitic elements in the model of a grounded inductor that is shown in Fig. 5.57

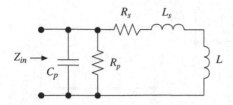

Fig. 5.57 Small-signal model for the grounded inductor

A grounded inductor requires two transconductors whereas a floating inductor requires three.

A floating parallel resonance circuit, which is difficult to realize because it contains a floating capacitor, can thus be realized with two gyrators and a grounded series resonance circuit as shown in Fig. 5.59. The element values in the realized series resonance circuit are

$$L_s = \frac{C_{i1} + C_{02}}{g_{m1}g_{m2}} \quad R_s = \frac{G_{01} + G_{i2}}{g_{m1}g_{m2}}$$

$$C_p = C_{i1} + C_{02} \quad R_p = \frac{1}{G_{i1} + G_{02}}.$$

For two identical gyrators with $G_{i1} = G_{i2}$, $C_{i1} = C_{i2}$, $G_{01} = G_{02}$, $C_{01} = C_{02}$, and $g_{m1} = g_{m2}$, which is the requirement for the gyrators being lossless, we get the Q factor for the inductor

$$Q_L = \frac{\omega L}{R_s} = \omega \frac{C_{i2} + C_{02} + C}{(G_{i2} + G_{02})}.$$

For example, a g_m-C inductor with the inductance $100\ \mu H$ at $\omega = 2\pi\ 10$ MHz obtains a Q factor of more than 3000 if $G_{02} = 0.1\ \mu S$ and $g_m = 220\ \mu S$. This is a high Q factor.

A floating inductor can be realized according to Fig. 5.10 using the gyrator shown in Fig. 5.42. The circuit corresponds to a shunt capacitor between two gyrators. It is advantageous to use grounded capacitors because the voltage over the bottom plate stray capacitance is not changed. This circuit can, however, be simplified according to Fig. 5.58, where $g_{m1} = g_{m3}$ and the series inductor is $L = C/(g_{m2}g_{m3})$.

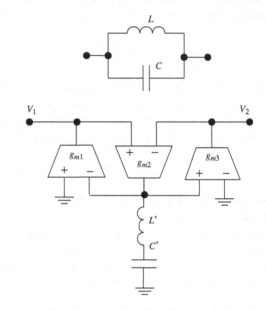

Fig. 5.59 Realization of a floating parallel resonance circuit

$$L' = \frac{C}{g_{m2}g_{m3}} \quad C' = \frac{1}{Lg_{m2}g_{m3}}$$

where $g_{m1} = g_{m3}$ and. The series resonance circuit can then be realized in two different ways. Figure 5.60 shows the case with grounded inductor

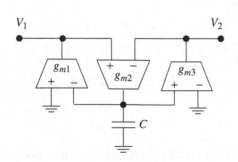

Fig. 5.58 Simplified realization of a floating inductor

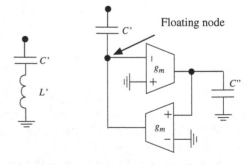

Fig. 5.60 Realization of a series resonance circuit with a floating node

and floating capacitor, which is a poorer realization, as it has a floating node where the stray capacitance of C' will have a large effect.

Figure 5.61 shows the case with a grounded capacitor and floating inductor, which requires three transconductors, but has no floating nodes, which makes it more suitable for implementation.

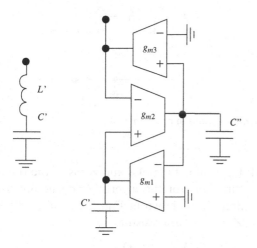

Fig. 5.61 Realization of a series resonance circuit without floating node

5.10.4.6 Current Conveyor-Based Inductors

Integrated inductors can of course also be implemented by using current conveyors [144].

5.10.5 FDNRs

An FDNR can be realized by Antoniou's GIC. Here we are only interested in supercapacitors because superinductors have very poor frequency performance. We have several alternative selections of the elements in the GIC, which mainly differs with respect to the sensitivity to the bandwidth of the operational amplifiers. The best realization of a supercapacitor is obtained by selecting $Z_1 = 1/sC_1$, $Z_2 = R_2$, $Z_3 = R_3$, $Z_4 = R_4$, and $Z_5 = 1/sC_5$. In order to minimize the influence of finite bandwidths of the amplifiers, we should select

$$R_2 = R_3. \tag{5.78}$$

A supercapacitor can also be realized by cascading two lossless integrators that can be realized by, e.g., tranconductors and current conveyors. Figure 5.62 shows a current conveyor circuit that realizes a supercapacitor with the value

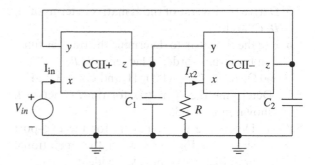

Fig. 5.62 Supercapacitor

$$D = \frac{C_1 C_2 R}{\alpha_1 \alpha_2} \tag{5.79}$$

where α_1 and α_2 are the current gains for CCII+ and CCII−, respectively. An operational amplifier (transconductance) of type OPA2662 typically has the current gains $\alpha_1 = 3.1$ and $\alpha_2 = 2.9$.

5.11 Problems

5.1 a) Show that
$|Z|^2 = Z Z^*$.
b) $2\,Re\{Z\} = Z + Z^*$.
c) If $V = V_r + j V_i$ and $I = I_r + j I_i$, then $P = Re\{VI^*\} = V_r I_r + V_i I_i$.

5.2 Determine which of the following circuit elements are passive and lossless:

a) resistor
b) inductor
c) capacitor
d) negative inductor $(-L)$
e) negative capacitor $(-C)$
f) negative resistor $(-R)$

5.3 Determine the $\mathbb{K}$ matrix for a two-port consisting of

a) a series impedance
b) a shunt admittance

5.4 Derive the $\mathbb{K}$ matrix for

    a) VCVS
    b) VCCS
    c) CCVS
    d) CCCS

5.5 Derive the inverse of the $\mathbb{K}$ matrix in terms of A, B, C, and D.

5.6 Use the $\mathbb{K}$ matrix to determine the transfer function for a third-order T ladder with $R_s = R_L = 600\ \Omega$, $L_1 = L_3 = 600$ mH, and $C_2 = 3.3\ \mu$F.

5.7 Determine the $\mathbb{K}$ matrix for the LC network shown in Fig. 3.5.

5.8 a) Determine the $\mathbb{K}$ matrix for the two-port shown in Fig. 5.63 when the operational amplifier is assumed to be ideal.
    b) Determine the input impedance to the two-port.

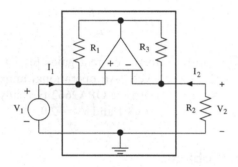

Fig. 5.63 Two-port in problem 5.8

5.9 a) Determine the input impedance to the circuit shown in Fig. 5.64 by using chain matrices. Identify the type of two-port when Z_5 is considered to be the load impedance.
    b) Determine the input impedance when $Z_1 Z_3 = Z_2 Z_4$.

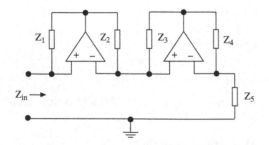

Fig. 5.64 One-port in problem 5.9

5.10 Determine the transconductance, $g_m = -I_2/V_1$, for the circuit shown in Fig. 5.65 and propose how it can be used in a possible application.

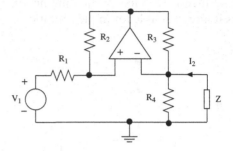

Fig. 5.65 Circuit in problem 5.10

5.11 Determine the poles and zeros of the transfer functions for the inverting and noninverting amplifiers shown in Fig. 5.27 and 5.28 when Z_1 and Z_2 are resistors.

    a) $R_1 = R_2 = 10$ kΩ
    b) $R_1 = 10$ kΩ and $R_2 = 100$ kΩ
    c) $R_1 = \infty$ and $R_2 = r = 10$ kΩ

    The amplifiers are nonideal and are modeled with the simple single-pole model with $A_0 = 2\ 10^5$ and a real pole $s_p = -100$ rad/s. Determine the 3 dB bandwidth of the circuits.

5.12 Determine suitable values for the resistor r in the circuits shown in Figs. 5.28 and 5.29. What is the function of the resistor? How should the resistor, r, be selected if R_2 in the inverting amplifier is changed to a capacitor?

5.13 Estimate A_0, ω_{3dB}, and ω_t for the operational amplifier MAX426x that has a 3 dB bandwidth of 450 MHz when the open-loop gain equals 1 and gain = 10 at 50 MHz.

5.14 Propose a model that includes the frequency response for the inverting amplifier, shown in Fig. 5.27. The model should only contain passive components and an ideal operational amplifier. Hint: Place an extra impedance in parallel with Z_2.

5.15 Derive the transfer function for the Miller integrator shown in Fig. 5.32. Determine the poles and zeros when $R = 10$ kΩ, $C = 10$ pF, $A_0 = 10^5$, and $\omega_{3dB} = 100$ rad/s.

5.16 Derive the transfer function for the Deboo integrator shown in Fig. 5.66. Determine the poles and zeros when all resistors are equal, e.g., $R = 10\text{ k}\Omega$ and $C = 10\text{ pF}$, $A_0 = 10^5$ and $\omega_{3dB} = 100$ rad/s. The integrator is sensitive for resistor ratio mismatch and is not recommended.

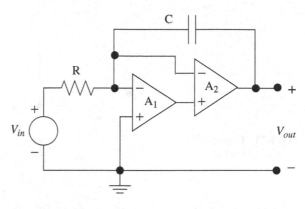

Fig. 5.68 Circuit in problem 5.17

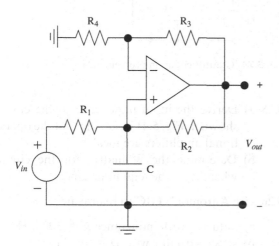

Fig. 5.66 Circuit in problem 5.17

5.17 Determine the transfer functions for the circuits shown in Figs. 5.67, 5.68, and 5.69.
5.18 Derive Equation (5.79).
5.19 Propose a circuit with only one operational amplifier that realizes the function: $V_{out} = 2V_1 - 5V_2$ where V_1 and V_2 are inputs.
5.20 Determine the output signal V_3 as a function of V_1 and V_2 for the circuit shown in Fig. 5.70.
5.21 Determine the output signal V_3 as a function of V_1 and V_2 for the circuit shown in Fig. 5.71.

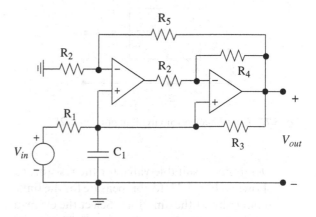

Fig. 5.69 Circuit in problem 5.17

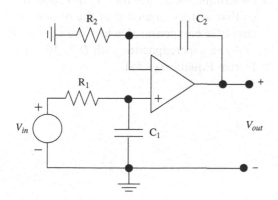

Fig. 5.67 Circuit in problem 5.17

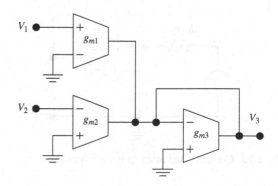

Fig. 5.70 Transconductor circuit in problem 5.20

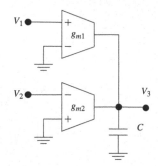

Fig. 5.71 Transconductor circuit in problem 5.21

5.22 Determine the chain matrix and identify the two-port that is realized by the circuit shown in Fig. 5.72.

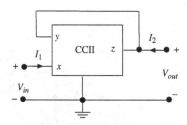

Fig. 5.72 Current conveyor circuit in problem 5.21

5.23 Determine a suitable value for the resistor, r_x, shown in Fig. 5.73, to compensate for the finite bandwidth of the amplifier. Select the element values $R = 22\ \text{k}\Omega$ and $C = 10\ \text{nF}$. The operational amplifier has very low power consumption and has therefore low bandwidth, e.g., $\omega_t = 100\ \text{krad/s}$.

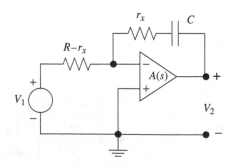

Fig. 5.73 Compensated integrator used in problem 5.23

5.24 Determine the Q factor for the circuit shown in Fig. 5.74 and the resonance frequency when $R = 10\ \text{k}\Omega$, $C_1 = 10\ \text{nF}$, $C_2 = 27\ \text{nF}$, and $r = 500\ \Omega$.

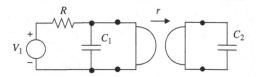

Fig. 5.74 Gyrator circuit in problem 5.24

5.25 a) Derive the input impedance to the circuit shown in Fig. 5.41 assuming that the operational amplifiers are ideal.
   b) Determine the $\mathbb{K}$ matrix for the circuit where Z_5 is the load impedance.

5.26 Use Antoniou's GIC to realize an

   a) inductor with inductance $L = 300\ \text{mH}$
   b) supercapacitor with $D = 10^{-8}\ \text{Fs}$

5.27 Determine the admittance between the operational amplifiers' inverting inputs and ground in Antoniou's GIC so that a stray capacitor at the output of the circuit is compensated when $Z_1 = R_1$, $Z_2 = R_2$, $Z_3 = R_3$, and $Z_4 = 1/sC_4$. Assume that Z_5 consists of R_5 in parallel with $1/sC_{stray}$ and that the operational amplifiers are ideal.

5.28 Determine suitable element values to realize an inductor with inductance $100\ \text{mH}$ when there is a stray capacitor of $10\ \text{pF}$ in parallel with Z_5.

5.29 Determine the admittance Y_s in Problem 5.27 so that a stray capacitor at the output of the circuit is compensated when the circuit shall realize a supercapacitor with $D = 10^{-9}\ \text{Fs}$.

5.30 Derive Equation (5.55).

Chapter 6
First- and Second-Order Sections

6.1 Introduction

Many active filter structures are based on first- and second-order sections. In this chapter, we will first discuss different types of sections with respect to their transfer functions. Next, we will classify, analyze, and compare different realizations that are based on resistors, capacitors, and with one or several operational amplifiers, transconductors, or current conveyors [58, 130]. These circuits can, of course, be implemented with discrete components, but this is of less interest because a large number of active elements are often needed and the cost and the power consumption becomes high. Integrated active elements are considerably cheaper and consume less power.

A problem with integrating analog circuits is that the capacitors require large chip area and that the tolerances in the capacitance values vary significantly due to variations in the manufacturing process. The resistance values are restricted to a relatively small interval and the tolerances in the resistance values vary significantly as well.

In *SC* filters, circuit techniques are used that are based on the fact that the frequency response depends on the ratio between two capacitances and not on the absolute capacitance values. The error in the ratio between two capacitances can be made less than 0.1% even though the capacitance values can vary ±10% or more. The capacitors dominate the chip area required for an integrated analog filter. Typically, the total capacitance, which can currently be economically integrated on a chip, is less than 30 pF.

Circuit techniques that rely on ratios of component values are not directly applicable for integrated active filters. Instead, special circuits for tuning of the frequency response are used because the active elements and capacitance values vary strongly and the filter properties depend on the absolute values. The tuning must also be made while the filter is operating because the element values vary with the temperature and the power supply voltage. A large number of transconductor-*C* filters are manufactured and used in, e.g., the read and write channels in hard drives. These filters are programmable because the bandwidth must be adjusted depending on which track is being accessed.

Transconductor-*C* filters and other methods for integrating analog filters are active research topics due to the trend to integrate whole signal processing systems on a single chip. Integrating a whole system tends to reduce the physical size and power consumption, and cost is therefore suitable for manufacturing cheap consumer products in large volumes.

6.2 First-Order Sections

A first-order section can either be of lowpass (LP), highpass (HP), or allpass (AP) type.

6.2.1 First-Order LP Section

The transfer function for a first-order LP section is

L. Wanhammar, *Analog Filters Using MATLAB*, DOI 10.1007/978-0-387-92767-1_6,
© Springer Science+Business Media, LLC 2009

$$H(s) = \frac{G}{s - \sigma_p} \qquad (6.1)$$

$$H(s) = \frac{Gs}{s - \sigma_p}. \qquad (6.2)$$

with a real pole on the negative real axis and one zero at $s = \infty$. Figure 6.1 shows the magnitude and phase functions for a first-order LP section with a real pole for the two cases $\sigma_{p1} = -1$ rad/s and $\sigma_{p2} = -5$ rad/s. By choosing $G = -\sigma_p$, the magnitude function is normalized to 1 for $\omega = 0$. Note that the magnitude function decays faster the closer the pole is to the $j\omega$-axis. The phase goes from 0 to $-\pi/2$.

Figure 6.2 shows the magnitude and phase functions for a first-order highpass section with a real pole for the two cases $\sigma_{p1} = -1$ rad/s and $\sigma_{p2} = -5$ rad/s and a zero at the origin. The gain at high frequencies has been normalized to 1 by choosing $G = 1$.

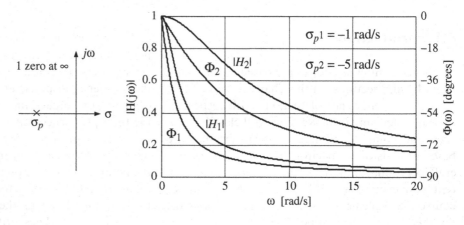

Fig. 6.1 Pole-zero configuration and magnitude and phase functions for a first-order LP section for the two cases $\sigma_{p1} = -1$ rad/s and $\sigma_{p2} = -5$ rad/s

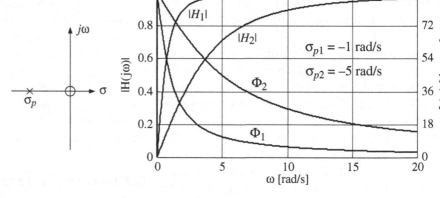

Fig. 6.2 Pole-zero configuration and magnitude and phase functions for a first-order HP section for the two cases $\sigma_{p1} = -1$ rad/s and $\sigma_{p2} = -5$ rad/s

According to Equation (5.27), an operational amplifier can thus be modeled with a first-order transfer function with $G = A_0\omega_{3dB}$ and $\sigma_p = -\omega_{3dB}$.

6.2.2 First-Order HP Section

The transfer function for a first-order HP section is

6.2.3 First-Order AP Section

The transfer function for a first-order AP section is

$$H(s) = G\frac{s - \sigma_z}{s - \sigma_p} \qquad (6.3)$$

where $\sigma_z = -\sigma_p$. The poles and the zeros for an allpass section lie symmetrically around the $j\omega$-axis.

Figure 6.3 shows the magnitude and phase functions for an allpass filter for the two cases $\sigma_{p1} = -1$ rad/s and $\sigma_{p2} = -5$ rad/s where $G = 1$, i.e., $|H(j\omega)| = 1$.

and

$$H(s) = \frac{s}{s + \dfrac{1}{RC}} \qquad (6.5)$$

Fig. 6.3 Pole-zero configuration and magnitude and phase functions for a first-order AP section for the two cases $\sigma_{p1} = -1$ rad/s and $\sigma_{p2} = -5$ rad/s

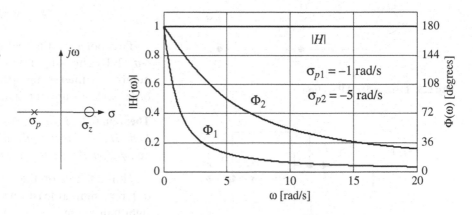

6.3 Realization of First-Order Sections

The first-order sections of LP and HP type can be realized with simple RC sections according to Fig. 6.4. The sections must, however, be driven by a low impedance signal source, e.g., an operational amplifier. Furthermore, if a section has high output impedance, it cannot be loaded by a subse-

respectively. To reduce the influence of the bias currents, a resistor R should be inserted into the feedback loop. Henceforth, we will always explicitly show these resistors in order to simplify the schematics.

Theorem 6.1 *A lowpass filter that is realized by a linear active RC network is converted to a highpass filter if all resistors and capacitors are replaced by capacitors and resistors, respectively.*

Fig. 6.4 Realization of first-order LP and HP sections

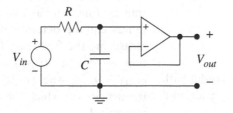 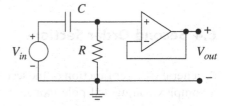

quent circuit. In this case, we must use a buffer amplifier with high impedance input and low impedance output. In some cases, a first-order section is combined with a second-order section into a third-order section to reduce the number of operational amplifiers.

The LP and HP sections have the transfer functions

An amplifier is required in order to realize a first-order AP section. Figure 6.5 shows an example of a first-order AP section with the transfer function

$$H(s) = \frac{s - \dfrac{1}{RC}}{s + \dfrac{1}{RC}} \qquad (6.6)$$

$$H(s) = \frac{1}{RC} \cdot \frac{1}{s + \dfrac{1}{RC}} \qquad (6.4)$$

where we select $R_1 \approx R/2$ in order to minimize the offset voltage due to the bias currents.

Figure 6.6 shows a first-order g_m-C section with two transconductors. We have for the circuit

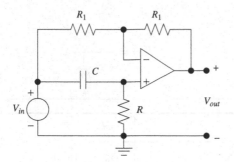

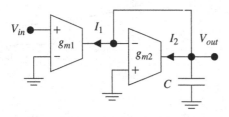

Fig. 6.5 First-order AP section

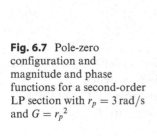

Fig. 6.6 First-order LP section

$$\begin{cases} I_1 = g_{m1} V_{in} \\ I_2 = -g_{m2} V_{out} \\ I_1 + I_2 = -s^{CV_{out}}. \end{cases}$$

After simplification we obtain

$$H(s) = \frac{V_{out}}{V_{in}} = \frac{g_{m1}}{C} \frac{1}{s + \frac{g_{m2}}{C}}.$$

6.4 Second-Order Sections

A general transfer function of the second order with a complex-conjugated pole pair $s_p = \sigma_p \pm j\omega_p$ and arbitrary zeros can be written

$$H(s) = \frac{as^2 + bs + c}{s^2 - 2\sigma_p s + r_p^2} \qquad (6.7)$$

where the *pole radius*, r_p, is

$$r_p = \sqrt{\sigma_p^2 + \omega_p^2}. \qquad (6.8)$$

Two poles on the real axis are obtained if $r_p \leq -\sigma_p/2$. By choosing different values of a, b, and c, a number of interesting special cases of the second-order sections are obtained.

Theorem 6.2 *A necessary condition for a polynomial to be Hurwitz is that all coefficients of s are present, i.e., $\neq 0$, and have the same sign.*

Thus, if any of the terms is missing or has a different sign, at least one root (pole) will lie in the right half plane.

6.4.1 Second-Order LP Section

The transfer function for a second-order LP section is

$$H(s) = \frac{G}{s^2 - 2\sigma_p s + r_p^2}. \qquad (6.9)$$

The corresponding pole-zero configuration, magnitude and phase response is shown in Fig. 6.7. The transfer function has two zeros at $s = \infty$. The pole radius can be determined with high accuracy by measuring the frequency for which the phase function is $-90°$ as $H(jr_p) = G/(-\sigma_p jr_p)$ and $2\sigma_p = \omega_{-45} - \omega_{-135}$, where ω_{-45} and ω_{-135} are the angular frequencies for which the phase is $-45°$ and $-135°$, respectively.

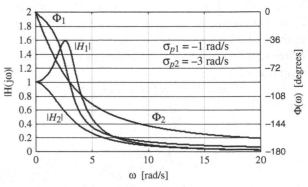

Fig. 6.7 Pole-zero configuration and magnitude and phase functions for a second-order LP section with $r_p = 3$ rad/s and $G = r_p^2$

6.4.1.1 Q Factor for a Pole Pair

The transfer function for an LP section can also be written

$$H(s) = \frac{G}{s^2 + \left(\dfrac{r_p}{Q}\right)s + r_p^2} \qquad (6.10)$$

where the Q factor (*quality factor*) of the poles is defined as

$$Q \triangleq -\frac{r_p}{2\sigma_p}. \qquad (6.11)$$

According to Equation (6.11), the Q factor depends on the angle between the $j\omega$-axis and the vector from origin to the pole. The Q factor for a real pole is 0.5. The Q factor for a resonance circuit is defined according to Definition 3.1 as the ratio between maximum stored energy and dissipated energy during a period. Figure 6.8 shows the

magnitude function for different Q factor when $r_p = 3$ rad/s. The magnitude function at $\omega = 0$ has been normalized to 1 by choosing $G = r_p^2$.

There is a peak in the magnitude function if $Q > 1/\sqrt{2}$ and it occurs at

$$\omega_{peak} = r_p\sqrt{1 - \frac{1}{2Q^2}}. \qquad (6.12)$$

The Q factor is a simple measure of how difficult it is to implement a second-order transfer function. It is therefore customary to classify pole pair in second-order sections with respect to the Q factor for the poles according to

- low Q factors : $Q \leq 2$
- medium high Q factors : $2 < Q \leq 20$
- high Q factors : $Q > 20$

Example 6.1 Determine the pole radius and the Q factor for the pole pair, $s_p = -0.25 \pm j0.95$ krad/s. *The pole radius is* $r_p = \sqrt{\sigma_p^2 + \omega_p^2} = \sqrt{0.25^2 + 0.95^2} = 0.9823441$ krad/s and the Q factor is $Q = \dfrac{-r_p}{2\sigma_p} = \dfrac{-0.9823441}{2 \cdot (-0.25)} = 1.9646883$.

The largest change of the magnitude function due to errors in the Q factor and the pole radius is obtained in the proximity of the resonance frequency.

Figure 6.9 shows the variation in the magnitude function when the Q factor and the pole radius is varied with ±5% where the nominal pole radius is $r_p = 3$ rad/s. From the figure, it is evident that the magnitude function is very sensitive for variations in the pole radius whereas an error in the Q factor has less influence. It can be shown that the variation of the magnitude function is proportional to

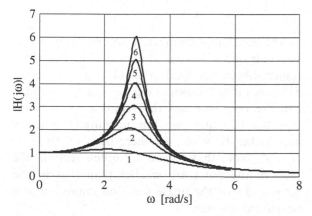

Fig. 6.8 Magnitude function for pole pairs with different Q factors

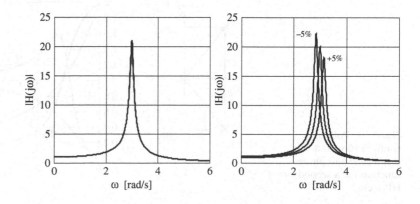

Fig. 6.9 Variation in the magnitude function due to a ±5% variations in the Q factor (*left*) and pole radius (*right*). $Q = 20$

$$\frac{\Delta|H|}{|H|} \propto \frac{\Delta Q}{Q} \qquad (6.13)$$

and

$$\frac{\Delta|H|}{|H|} \propto 2Q\frac{\Delta r_p}{r_p}. \qquad (6.14)$$

The influence on the magnitude function, due to an error in the Q factor, is proportional to the error in Q factor, whereas the influence of an error in the pole radius is much larger for large Q factors.

It can be shown that similar relations as Equations (6.13) and (6.14) are valid for the phase function as well. It is therefore essential to use realizations of the sections in which any deviation from the nominal component values have small influence on the pole radius. In this chapter, we will therefore focus on this issue when we compare different realizations of second-order sections.

6.4.2 Second-Order HP Section

The transfer function for a second-order HP section (*highpass section*) is

$$Hs = \frac{G_s{}^2}{s^2 - 2\sigma_p s + r_p^2} = \frac{G_s{}^2}{s^2 + \left(\frac{r_p}{Q}\right)s + r_p^2}. \qquad (6.15)$$

The transfer function has two zeros at the origin. The pole radius in Fig. 6.10 is $r_p = 3$ rad/s and the gain at high frequencies has been normalized to 1 by choosing $G = 1$. An implementation of a HP section

will of course have finite bandwidth due to parasitic capacitances and the active element's finite bandwidth, but this will occur at such high frequencies that are of no interest.

The phase function at $s = jr_p$ is $+90°$. The peak of the magnitude function occurs at

$$\omega = \frac{r_p}{\sqrt{1 - \frac{1}{2Q^2}}}. \qquad (6.16)$$

6.4.3 Second-Order LP-Notch Section

The transfer function for a second-order LP-notch section (*lowpass-notch section*) is

$$H(s) = \frac{G(s^2 + r_z^2)}{s^2 - 2\sigma_p s + r_p^2} = \frac{G(s^2 + r_z^2)}{s^2 + \left(\frac{r_p}{Q}\right)s + r_p^2}, \quad r_z > r_p. \ (6.17)$$

The transfer function has a complex conjugating zero pair on the $j\omega$-axis at $\omega = r_z$. The frequency response becomes equal to zero for this frequency. The pole radius, shown in Fig. 6.11, is $r_p = 3$ rad/s, and the zero pair has the radius $r_z = 6$ rad/s. For high frequencies, the magnitude function approaches G, which has here been set to $G = (r_p/r_z)^2$. Note that for an LP-notch section, the radii of the poles are smaller than that of the zeros and that the phase function jumps $-\pi$ at a zero at the $j\omega$-axis.

If the pole radius $\approx$ the zero radius, we have a second-order *LP notch filter*, i.e., a filter with

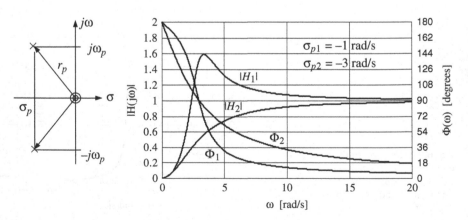

Fig. 6.10 Pole-zero configuration and magnitude and phase function for a second-order HP section

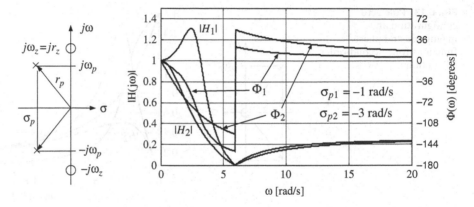

Fig. 6.11 Magnitude and phase functions and corresponding pole-zero configuration for a second-order LP-notch section with $G = (r_p/r_z)^2$

approximately constant magnitude function, except for a notch at the zero. Such filters can be used as simple bandstop filters, i.e., to attenuate disturbances originating from the mains (50 or 60 Hz).

The transfer function has a finite zero pair. The pole radius, shown in Fig. 6.12, is $r_p = 3$ rad/s, and the complex conjugating zero pair has the radius $r_z = 2$ rad/s and $G = 1$. The phase jumps $-\pi$ at the zero on the $j\omega$-axis.

6.4.4 Second-Order HP-Notch Section

An HP-notch section is obtained if instead the pole radius is larger than the zero radius, as shown in Fig. 6.12. The transfer function for a second-order HP-notch section (*highpass-notch section*) is

$$H(s) = \frac{G(s^2 + r_z^2)}{s^2 - 2\sigma_p s + r_p^2} = \frac{G(s^2 + r_z^2)}{s^2 + \left(\dfrac{r_p}{Q}\right)s + r_p^2}, r_z < r_p. \quad (6.18)$$

6.4.5 Second-Order BP Section

The transfer function for a second-order BP section (*bandpass section*) is

$$H(s) = \frac{G_s}{s^2 - 2\sigma_p s + r_p^2} = \frac{G_s}{s^2 + \left(\dfrac{r_p}{Q}\right)s + r_p^2}. \quad (6.19)$$

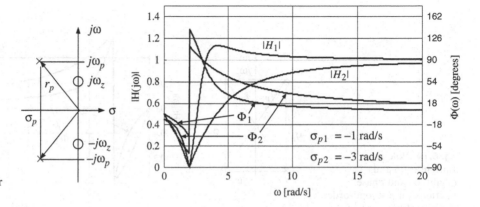

Fig. 6.12 Pole-zero configuration and magnitude and phase functions for a second-order HP-notch section

Fig. 6.13 Pole-zero configuration and magnitude and phase functions for a second-order BP section

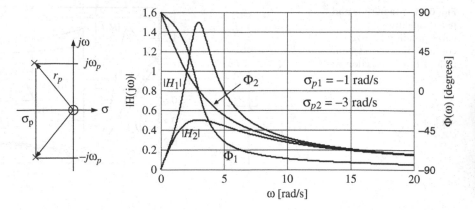

The pole radius, shown in Fig. 6.13, is $r_p = 3$ rad/s and $G = r_p$. The transfer function has a finite zero at $s = 0$ and one at $s = \infty$. The peak of the magnitude function occurs at $\omega = r_p$. The phase function at $s = jr_p$ is $0°$.

6.4.5.1 Second-Order AP Section

The transfer function for a second-order AP section (*allpass section*) is

$$H(s) = G\frac{s^2 - 2\sigma_z s + r_z^2}{s^2 - 2\sigma_p s + r_p^2} \qquad (6.20)$$

where the poles are

$$s_p = \sigma_p \pm j\omega_p$$

and the zeros, which equal the poles mirrored in the $j\omega$-axis, are

$$s_z = \sigma_z \pm j\omega_z = -\sigma_p \pm j\omega_p.$$

The magnitude function is constant, but the section has a frequency-dependent phase function. Allpass sections are often used for correction of the phase function of a filter by placing AP sections in series with the filter so the total phase function becomes approximately linear. The AP section, which has its zeros in the right half plane, is a maximum-phase filter.

6.4.6 Element Sensitivity

As a measure of sensitivity of a function, Y, with respect to a parameter, x, we use the ratio of the normalized derivatives.

Definition 6.1 The element sensitivity is defined as

$$S_x^Y \triangleq \frac{x}{Y}\frac{\partial Y}{\partial x}. \qquad (6.21)$$

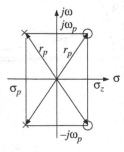

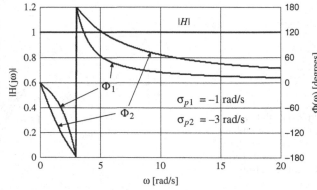

Fig. 6.14 Pole-zero configuration and magnitude and phase functions for a second-order allpass section

For an LP second-order section, we have

$$S_Q^{H(s)} = \frac{r_p}{Q} \cdot \frac{s}{D(s)} \qquad (6.22)$$

and

$$S_{r_p}^{H(s)} = -\frac{r_p}{Q} \cdot \frac{s + 2r_p Q}{D(s)} \qquad (6.23)$$

where $D(s)$ is the denominator of a second-order transfer function. The numerator in Equation (6.22) has a zero at $s = 0$ whereas Equation (6.23) has a zero on the negative real axis, far from the origin. Hence, the sensitivity of the transfer function with respect to Q is much less than the sensitivity with respect to r_p. Thus, we have, if $\omega << 2r_p Q$ and $Q >> 1$,

$$\left| S_Q^{H(j\omega)} \right| << \left| S_{r_p}^{H(j\omega)} \right|. \qquad (6.24)$$

The difference in sensitivity was earlier illustrated in Fig. 6.9. According to Equations (6.13) and (6.14), the influence of errors in the pole radius is about $2Q$ times larger than for corresponding error in the Q factor. Hence, we should therefore only use second-order sections where the errors in the element values have minimal influence on the pole radius and we will henceforth only discuss sections that have minimal *pole radius sensitivity*.

The sensitivity of the pole radius for errors in a resistor and capacitance are

$$S_R^{r_p} \triangleq \frac{R}{r_p} \frac{\partial r_p}{\partial R} \qquad (6.25)$$

and

$$S_C^{r_p} \triangleq \frac{C}{r_p} \frac{\partial r_p}{\partial C}, \qquad (6.26)$$

For a useful second-order section, the sensitivities $S_R^{r_p}$ and $S_C^{r_p}$ for all R and C elements should be small. For all second-order sections, we have

$$\sum_i S_{R_i}^{r_p} = \sum_i S_{C_i}^{r_p} = -1 \qquad (6.27)$$

$$\sum_i S_{R_i}^{Q} = \sum_i S_{C_i}^{Q} = 0. \qquad (6.28)$$

6.4.7 Gain-Sensitivity Product

As a measure of sensitivity of the poles radius and Q factor due to the finite amplifier gain, A_0, we use the *gain-sensitivity product* that is defined for the pole radius [105, 106, 112].

Definition 6.2 The gain-sensitivity product for the pole radius is

$$GS_{A_0}^{r_p} \triangleq A_0 S_{A_0}^{r_p} = \frac{A_0^2}{r_p} \frac{\partial r_p}{\partial A_0} \qquad (6.29)$$

and for the Q factor

$$GS_{A_0}^{Q} \triangleq A_0 S_{A_0}^{Q} = \frac{A_0^2}{Q} \frac{\partial Q}{\partial A_0}. \qquad (6.30)$$

The relative errors in the pole radius and the Q factor can be expressed in terms of the relative errors in resistors, capacitances, and gain-sensitivity products of the operational amplifiers

$$\frac{\Delta r_p}{r_p} \approx \sum_i S_{R_i}^{r_p} \left(\frac{\Delta R_i}{R_i} \right) + \sum_i S_{C_i}^{r_p} \left(\frac{\Delta C_i}{C_i} \right) + \sum_i GS_{A_{0i}}^{r_p} \left(\frac{\Delta A_{0i}}{A_{0i}^2} \right) \qquad (6.31)$$

$$\frac{\Delta Q}{Q} \approx \sum_i S_{R_i}^{Q} \left(\frac{\Delta R_i}{R_i} \right) + \sum_i S_{C_i}^{Q} \left(\frac{\Delta C_i}{C_i} \right) + \sum_i GC_{A_{0i}}^{Q} \left(\frac{\Delta A_{0i}^2}{A_{0i}} \right) \qquad (6.32)$$

where the summation is over all components.

For a usable section, we must have $GS_{A_0}^{r_p} \approx 0$, because the error in the pole radius, according to Equation (6.14), must be small. In the literature there are, however, sections that have higher gain-sensitivity products. These sections are of course of no practical use as the variation in A_0 is very large (typicaly $\pm 50\%$).

6.4.8 Amplifiers with Finite Bandwidth

The pole radius and the Q factor also depend on the bandwidth of the operational amplifier, i.e., $\omega_t \approx 2\pi\, GB$, where GB is the gain-bandwidth product. We use the following two measures of the deviations from nominal values with ideal amplifiers.

The relative error in the pole radius, due to the finite bandwidth of the amplifiers, is

$$\delta \triangleq \frac{\Delta r_p}{r_{pnominal}} = \frac{r_p - r_{pnominal}}{r_{pnominal}}. \tag{6.33}$$

The deviation in the Q factor, due to the finite bandwidth of the amplifiers, is

$$\eta \triangleq \frac{Q}{Q_{nominal}}. \tag{6.34}$$

It can be shown that both δ and η are independent of the RC network and only depend on the gain-sensitivity product with respect to the Q factor, i.e., both δ and η can be expressed in terms of $GS_{A_0}^Q$ [112].

6.4.9 Comparison of Sections

According to Equation (6.14), it is important to use sections and design these so the sensitivities in the pole radius with respect to the passive elements become small. Even more important is to use sections and design these so the gain-sensitivity products are minimized. The sections, which we will discuss henceforth, will have gain-sensitivity products that are almost zero or approximately 1

(low), but there exist sections with gain-sensitivity products that are proportional to Q^2.

While comparing different sections, we first compare the sensitivities $GS_{A_0}^{r_p}$, $S_R^{r_p}$, and $S_C^{r_p}$, which should be minimal, and thereafter the sensitivities $GS_{A_0}^Q$, S_R^Q, and S_C^Q.

Because of sensitivity reasons, normally only sections of the first and second order are used, but in some cases, sections of the third order may also be used [31]. Furthermore, when selecting the type of sections, we may also consider other factors. For example, we may consider:

- cost
- power consumption
- requirement on the amplifier's bandwidths
- spreads of the element values
- low output impedance so the sections can be cascaded
- high input impedance for all frequencies because the previous section may have a limited drive capability (this can cause large distortion and may even cause an oscillation)
- possibility to add several input signals
- realize several simultaneous filter functions (LP, BP, HP, and AP)
- easy to tune, etc.

6.5 Single-Amplifier Sections

Second-order sections can be realized in many different ways, and many circuit solutions have been proposed in the literature. In this section, we will classify the majority of circuit solutions into a few classes, which will have similar sensitivity properties [31, 38, 64, 68, 79, 81, 112, 113]. There are, however, a few realizations that do not fit into these classes, but, in general, they do not provide any additional advantages. Hence, we can neglect these realizations.

Because of cost and their power consumption, sections with only a single operational amplifier are especially interesting. Such second-order sections are called *SAB* (*single amplifier biquad*) as the transfer function consists of the ratio between two second-order polynomials. Note that the power consumption of active RC filters with high dynamic signal range is relatively large in comparison with competitive filter technologies such as *SC* filters and digital filters.

Figure 6.15 shows a general single op-amp section, but the structure is so general that it is difficult to analyze. Hence, we need to specify and specialize the *RC* network into more detail.

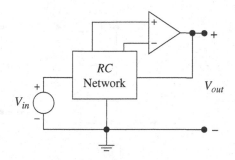

Fig. 6.15 General single-op-amp section

6.5.1 *RC Networks*

We are from a sensitivity point of view interested in how the poles are created. That is, the poles are created by feedback loops. The zeros, which are less of a problem, can be obtained by injecting the input signal into several nodes in the *RC* network.

The poles of an *RC* two-port are single on the negative real axis, excluding $s = 0$ and $s = \infty$. The zeros can be anywhere in the *s*-plane, except for a sector defined by $j\omega \leq \pi\sigma/n$ and $j\omega \geq -\pi\sigma/n$ where *n* is the number of poles. Hence, a second-order *RC* two-port cannot have its zeros on the $j\omega$-axis.

To realize a complex conjugate pole pair, an amplifying element is required. In most second-order sections, either a *bridged-T* or *bridged twin-T RC* network is used for the network N_1. The class of sections, with the network N_1 of the type bridged-*T* network, which was introduced by Deliyannis (1969), is called NF1 sections, and the class of sections with bridged twin-*T* networks is called NF2 sections.

6.5.2 *Gain-Sensitivity Product for SAB*

The transfer function for a SAB can be written

$$H(s) = \pm \frac{N(s)}{D(s) + \dfrac{E(s)}{A}}. \tag{6.35}$$

We get with an ideal operational amplifier, i.e., $A \rightarrow \infty$,

$$H(s) = \pm \frac{N(s)}{D(s)} \tag{6.36}$$

where $E(s)$, which is due to the finite bandwidth of the amplifier, represents an error polynomial [39]. Generally, for a operational amplifier described by Equation (5.27), $E(s)$ is a polynomial of the form

$$E(s) = a\left(s^2 + b\left(\frac{r_p}{Q}\right)s + cr_p^2 \right) \tag{6.37}$$

when $D(s)$ has been normalized so that $D(s) = s^2 + \left(\dfrac{r_p}{Q}\right)s + r_p^2$.

The gain-sensitivity product for SAB sections is

$$GS_{A_0}^Q = a(b - 1). \tag{6.38}$$

6.5.3 *Sections with Negative Feedback*

Figure 6.16 shows the general case for a *negative feedback section*. The networks N_1 and N_2 only contain resistors and capacitors. There are several different ways of choosing the networks, but all have in many respects similar sensitivity properties.

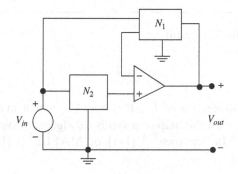

Fig. 6.16 NF section with negative feedback

It can be shown that single op-amp sections with negative feedback have the following properties. The relative error in the pole radius, due to the finite bandwidth of the operational amplifier, is

$$\delta = \frac{\Delta r_p}{r_{pnominal}} \approx -\frac{1}{2Q}\frac{r_p}{\omega_t} GS_{A_0}^Q. \qquad (6.39)$$

The deviation in the Q factor, due to the finite bandwidth of the operational amplifier, is

$$\eta = \frac{Q}{Q_{nominal}} \approx \frac{1}{1 + \left(\dfrac{r_p}{\omega_t}\right)\left(\dfrac{r_p}{\omega_t} - \dfrac{1}{2Q}\right) GS_{A_0}^Q} \qquad (6.40)$$

where $\omega_t = 2\pi\, GB$. The only factor in Equations (6.39) and (6.40) that depends on the network N_1 is $GS_{A_0}^Q$. For NF sections, we have $GS_{A_0}^Q \geq 2Q$. For a given r_p, Q factor, and network, N_1, can we only reduce the right-hand sides of Equations (6.39) and (6.40) by selecting an amplifier with larger ω_t.

The network N_1 is a part of the feedback loop and will therefore determine the poles. The sensitivity of the pole radius is minimal for the sections, which we will discuss, due to the choice of structure for N_1. Therefore any remaining degrees of freedom in selecting the network N_1 and its element values are chosen to minimize the sensitivity of the Q factor, while at the same time the spreads of the element values is kept within reasonable limits.

6.5.3.1 NF1 Sections

Figure 6.17 shows a general NF1 section, which is characterized by a bridged-T network in the feedback loop.

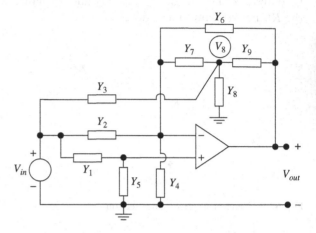

Fig. 6.17 NF1 section with bridged-T network

The transfer function for the NF1 section can be determined by summing the currents into all nodes except for the amplifier's output. We get, for the currents into the nodes V_8, V_-, V_+ and for the output node

$$-Y_8 V_8 + Y_3(V_{in} - V_8) + Y_7(V_- - V_8) + Y_9(V_{out} - V_8) = 0$$

$$Y_2(V_{in} - V_-) + Y_7(V_8 - V_-) + Y_6(V_{out} - V_-) - Y_4 V_- = 0$$

$$Y_1(V_{in} - V_+) - Y_5 V_+ = 0$$

and for the amplifier

$$V_{out} = A(V_+ - V_-).$$

Elimination of V_8, V_+, and V_-, which may be done with the help of a symbolic algebra program, e.g., Mathematica™ [141] or MAPLE™ [120],[1] yields

$$\begin{cases} N(s) = -KN_2 + Y_7 Y_3 + N_1 Y_2 \\ D(s) = Y_7 Y_9 + N_1 Y_6 \\ K \quad = Y_1/(Y_1 + Y_5) \\ N_1 \quad = Y_9 + Y_8 + Y_7 + Y_3 \\ E(s) = N_2 = -Y_7^2 + (Y_7 + Y_6 + Y_4 + Y_2)N_1. \end{cases} \qquad (6.41)$$

Note that $N(s)$, $D(s)$, and $E(s)$ in Equation (6.41) must be normalized so that $D(s)$ is monic, i.e., of the form

$$D(s) = s^2 + \left(\frac{r_p}{Q}\right)s + r_p^2$$

in order to get the desired error polynomial in Equation (6.37).

---

[1] The reader should recognize that solving of these equation systems and simplifying the expressions by hand is a very tedious and error-prone process.

We get with an ideal operational amplifier, i.e., letting $A \to \infty$,

$$H(s) = -\frac{N(s)}{D(s)} = -\frac{-KN_2 + Y_7 Y_3 + N_1 Y_2}{Y_7^2 + N_1 Y_6}. \quad (6.42)$$

Normally, a symmetric T network is used, i.e., $Y_7 = Y_9$. Note that only the admittances Y_6, Y_7, Y_8, and Y_9 in the feedback loop affect the poles while Y_1, Y_2, Y_3, Y_4, and Y_5 only affect the zeros.

It can be shown that NF1 sections have the following sensitivity properties:

- $\left| S_{R,C}^{r_p} \right| \leq 0.5$ for R and C
- $GS_{A_0}^{r_p} \approx 0$
- $\left| S_{R,C}^{Q} \right| \leq 0.5$ for R and C
- $GS_{A_0}^{Q} \propto Q^2$ (μ means proportional to)
- Spreads in passive element values $\propto Q^2$.

The sensitivity to R and C elements are minimal, see Problem 6.9. Because the sensitivity of the pole radius is minimal for all sections, which we will discuss, it only remains to minimize the sensitivity of the Q factor. In practice, a passive sensitivity $\left| S_{R,C}^{Q} \right| \leq 1$ is sufficiently low. $GS_{A_0}^{Q}$ for NF1 sections are, however, large, and NF1 sections are only usable to realize low and medium high Q factors (<10).

There are also practical limitations of how large the spreads in the element values can be. For example, a discrete operational amplifier cannot drive a load resistance that is smaller than a few kΩ while the largest usable resistor should typically have a resistance that is smaller than 400 kΩ due to stray capacitances and leakage currents. Thus, the spreads in element values should be less than 400, i.e., the Q factor must be less than 20. For high Q factors, the spreads in the element values of the NF1 sections becomes large and the tuning of these becomes difficult. The sections gain should not be chosen higher than 10. Examples of NF1 sections are shown in Figs. 6.18, 6.19, 6.20, 6.21 and 6.22.

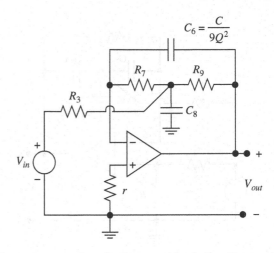

Fig. 6.18 LP section of NF1 type with a bridge-T network

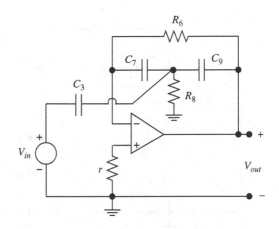

Fig. 6.19 HP section of NF1 type with a bridged-T network

6.5.3.2 NF1 LP Section

The transfer function for the NF1 LP section shown in Fig. 6.18 is

$$H(s) = -\frac{1}{R_3 R_7 C_6 C_8} \frac{1}{s^2 + \left(\frac{1}{R_3} + \frac{1}{R_7} + \frac{1}{R_9}\right)\frac{s}{C_8} + \frac{1}{R_7 R_9 C_6 C_8}} \quad (6.43)$$

We normally select a symmetric T network and $R_7 = R_9 = R_3 = R$, $C_8 = C$, and $C_6 = C_8/(9Q^2)$, which yields

$$r_p = \frac{3Q}{RC}$$

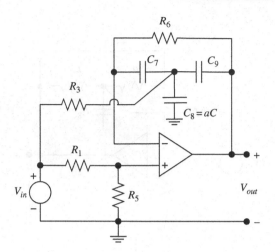

Fig. 6.20 HP-noted section of NF1 type with a bridged-T network

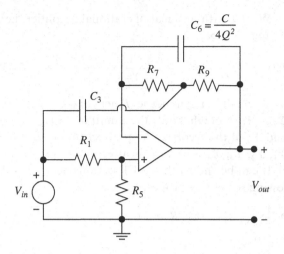

Fig. 6.22 AP section of NF1 type with bridged-T network

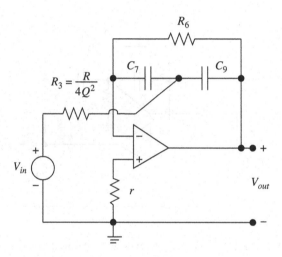

Fig. 6.21 BP section of NF1 type with a bridged-T network

$$E_{LP}(s) = s^2 + (3Q^2 + 1)\left(\frac{r_p}{Q}\right)s + 2r_p^2$$

and from Equation (6.38) we get

$$GS_{A_0}^Q = 3Q^2. \qquad (6.44)$$

R cannot be chosen too low so that the input impedance becomes too low and represents a too high load to the signal source. The output impedance is low. Note that the section gain can be lowered by replacing R_3 with a resistive voltage

divider. The resistor r is selected $r \approx R_3//R_7//R_9$ (the $//$ denote impedances in parallel).

Example 6.2

a) Determine suitable element values for the NF1 section, shown in Fig. 6.18, to realize a pole pair $s_p = -5\pi \pm j50\pi$ krad/s under the assumption that the operational amplifier is ideal.

b) Determine the transfer function and the poles according to Equation (6.35) when the operational amplifier is characterized by the following parameters: $A_0 = 10^5$ and $\omega_t = 20$ Mrad/s.

c) Compare the results with Equations (6.39) and (6.40).

a) The Q factor is $Q_{nominal} = \dfrac{-r_p}{2\sigma_p} \dfrac{-\sqrt{(5\pi)^2 + (50\pi)^2}}{2 \cdot (-5\pi)} = 5.0249378$

We first select $C = 1$ nF, because capacitors contrary to resistors can easily be trimmed. We get $C_6 = C/9Q^2 = 4.400$ pF and

$$R = \frac{3Q}{Cr_p} = \frac{3 \cdot 5.02494}{10^{-9} \cdot 157863.08} = 95.493 \, k\Omega.$$

Note that the spreads in the element values are large, i.e., $C/C_6 = 9Q^2$.

b) We have $Y_1 = Y_2 = Y_4 = 0$, $Y_3 = Y_7 = 1/R$, $Y_6 = sC/9Q^2$, $Y_8 = sC$, and $Y_5 \cong 2/3R$.
   $K = 0$, $N_1 = sC + 3/R$, $N_2 = 1/R^2 - (3/R + sC)(1/R + sC/9Q^2)$, and

$$E(s) = s^2 + (3Q^2 + 1)\left(\frac{r_p}{Q}\right)s + 2r_p^2.$$

The denominator with $A \approx \omega_t/s$ is

$$N(s) + \frac{E(s)}{A} = s^3 + 22411172s^2 + 6.7815977 \cdot 10^{11}s + 4.9841664 \cdot 10^{17}$$

$$= (s + a + jb)(s + a - jb)(s + 22381867)$$

and the poles with a non-ideal operational amplifier are

$$s_p = a \pm jb = -14.652287 \pm j\,148.50586 \,\text{krad/s}$$

and there is a parasitic real pole at $s_p = -22381.868$ krad/s. Because the radius of the real pole is about 141 times larger than the radius for the complex pole pair, it will have an insignificant influence on the frequency response.

The Q factor with a non-ideal operational amplifier is

$$Q = \frac{-\sqrt{(-14652.287)^2 + (148505.86)^2}}{2 \cdot (-14652.292)}$$

$$= 5.0922748.$$

(c) From (b) and Equations (6.39) and (6.40) we get

$$\delta = \frac{|-14652.292 + j148505.86| - |-15707.963 + j157079.63|}{|-15707.963 + j157079.63|} = -0.05470$$

$$\eta = \frac{Q}{Q_{nominal}} = \frac{5.0922748}{5.0249} = 1.0134.$$

The deviations in the pole radius and the Q factor are thus very large, probably too large for the section to be usable. To reduce the deviation in the pole radius and Q factor, we should therefore choose an operational amplifier with larger ω_t.

According to Equation (6.39) through (6.40) we get with $GS_{A_0}^Q = 3Q^2 = 75.75$

$$\delta = -\frac{1}{2Q}\frac{r_p}{\omega_t}GS_{A_0}^Q = -\frac{1}{2(5.0249)}\frac{157863.08}{2 \cdot 10^7}75.75$$

$$= -0.0595$$

$$\eta = \frac{1}{1 + \left(\dfrac{r_p}{\omega_t}\right)\left(\dfrac{r_p}{\omega_t} - \dfrac{1}{2Q}\right)GS_{A_0}^Q} = 1.0579.$$

The error in the estimate of the pole radius is relatively small whereas the deviation in estimated Q factor is larger. The error in the estimates increases if the ratio ω_t/r_p becomes small. Note that the ratio ω_t/r_p should be of the order 50 or higher for the influence of the operational amplifier's finite bandwidth to be insignificant. The availability of symbolic algebra programs decreases the need for these estimates.

6.5.3.3   NF1 HP Section

Figure 6.19 shows the NF1 HP section, which with an ideal amplifier has the transfer function

$$H(s) = -\frac{C_3}{C_9} \cdot \frac{s^2}{s^2 + \dfrac{C_3 + C_7 + C_9}{R_6 C_7 C_9}s + \dfrac{1}{R_6 R_8 C_7 C_9}}. \quad (6.45)$$

We normally choose a symmetric T network, i.e., $C_7 = C_9 = C_3 = C$, $R_6 = R$ and $R_8 = R_6/(9Q^2)$, which gives $r_p = 3Q/RC$ and the error polynomial

$$E_{HP}(s) = 2s^2 + (3Q^2 + 1)\left(\frac{r_p}{Q}\right)s + r_p^2$$

and the gain-sensitivity product is $GS_{A_0}^Q = 6Q^2$, which is slightly larger than for the corresponding LP section. With this choice of element values, the gain becomes equal to 1 for high frequencies. The gain can be reduced by replacing C_3 with a capacitive voltage divider. The resistor r is selected $r \approx R_6$.

We obtain the section shown in Fig. 6.19 if all resistors and capacitors, shown in Fig. 6.18 (except r), are changed to capacitors and resistors, respectively. This procedure, which can be applied to all LP sections, corresponds to a lowpass-to-highpass transformation.

6.5.3.4 NF1 HP-Notch Section

In the NF1 HP-notch section, shown in Fig. 6.20, the input signal is injected into the bridged-T network and through the network, N_2, which in this case is a resistive voltage divider. The finite complex conjugate zero pair is realized by the difference between these two signal paths. This scheme to realize zeros is sensitive to errors in the components.

With the following choices $C_7 = C_9 = C$ and

$$a = \frac{r_p^2}{r_z^2} - 1 > 0 \qquad R_3 = \frac{R_6}{(2+a)^2 Q^2}$$

$$K = \frac{R_5}{R_1 + R_5} = \frac{(2+a)Q^2}{(2+a)Q^2 + 1}$$

we obtain an NF1 HP-notch section. The transfer function and error polynomial of the NF1 HP-notch section are

$$H(s) = (1+a)K \cdot \frac{s^2 + r_z^2}{s^2 + \dfrac{(2+a)}{R_6 C} s + \dfrac{1}{R_3 R_6 C^2}} \qquad (6.46)$$

$$E(s) = (1+a)s^2 + ((2+a)Q^2 + 1)\left(\frac{r_p}{Q}\right)s + r_p^2. \quad (6.47)$$

6.5.3.5 NF1 BP Section

Figure 6.21 shows an NF1 BP section with a simple bridged-T network. The transfer function of the NF1 BP section, with an ideal amplifier, is

$$H(s) = -\frac{1}{R_3 C_9} \cdot \frac{s}{s^2 + \left(\dfrac{C_7 + C_9}{R_6 C_7 C_9}\right)s + \dfrac{1}{R_3 R_6 C_7 C_9}}. \quad (6.48)$$

With $C_7 = C_9 = C$, $R_6 = R = r$, and $R_3 = \frac{R}{4Q^2}$, we obtain

$$H(s) = \frac{-2r_p Q_s}{s^2 + \left(\dfrac{r_p}{Q}\right)s + r_p^2} \qquad (6.49)$$

$$E_{BP}(s) = s^2 + (2Q^2 + 1)\left(\frac{r_p}{Q}\right)s + r_p^2 \qquad (6.50)$$

$$r_p = \frac{2Q}{RC} \qquad GS_{A_0}^Q = 2Q^2.$$

The spreads in the resistors are proportional to $4Q^2$. Also in this case, the gain can be reduced by replacing R_3 with a resistive voltage divider.

6.5.3.6 NF1 AP Section

Figure 6.22 shows an NF1 AP section, which also was proposed by Delyannis. Also in this case, a network, N_2, is used to realize the complex conjugate zero pair. The transfer function of the NF1 AP section, with an ideal amplifier, is given by Equation (6.20).

Selecting $R_7 = R_9 = R$, $C_3 = C$, and $C_6 = C/(4Q^2)$, we get

$$r_p = \frac{2Q}{RC} \qquad G = \frac{R_5}{R_1 + R_5} = \frac{Q^2}{Q^2 + 1}$$

$$E_{AP}(s) = s^2 + (2Q^2 + 1)\left(\frac{r_p}{Q}\right)s + r_p^2. \qquad (6.51)$$

A resistor r in series with C_6 may be used to compensate for the finite bandwidth of the amplifier [116] where

$$r = 2\frac{Q}{\omega_t C_6}$$

and R_7 is changed to $R_7 - r$.

6.5.3.7

NF2 sections, which were introduced by Hamilton and Sedra (1972), have a bridged twin-T network in the feedback loop, as shown in Fig. 6.23. There exists, in principle, two types of bridged twin-T networks, i.e., T networks of notch and bandpass type.

A drawback of the bridged twin-T network is that it realizes a third-order transfer function. Ideally, it should be only a second-order transfer function in the feedback loop. However, the two T networks can be designed so that a real pole and a real zero cancels.

The condition for the pole and zero to be cancelled is

$$Y_3 + Y_7 + Y_8 + Y_9 = k(Y_{10} + Y_{11} + Y_{12} + Y_{13})$$

where k is a positive real constant.

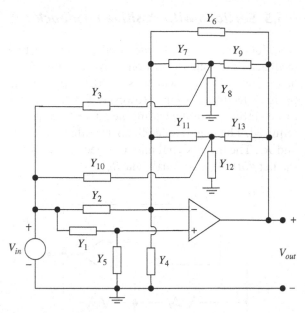

Fig. 6.23 NF2 section with a bridged twin-T network

The generic NF2 section, shown in Fig. 6.23, has the transfer function

$$H(s) = -\frac{N(s)}{D(s) + \dfrac{E(s)}{A}} \quad (6.52)$$

where

$$
\begin{cases}
N(s) = N_1 Y_{11} Y_{10} + (Y_7 Y_3 + N_1 Y_2)N_2 - KE(s) \\
D(s) = N_1 Y_{11} Y_{13} + N_2 Y_7 Y_9 + N_1 N_2 Y_6 \\
N_1 \quad = Y_3 + Y_7 + Y_8 + Y_9 \\
N_2 \quad = Y_{10} + Y_{11} + Y_{12} + Y_{13} \\
N_3 \quad = Y_2 + Y_4 + Y_6 + Y_7 + Y_{11} \\
E(s) = N_1 Y_{11}^2 + N_2 Y_7^2 - N_1 N_2 N_3 \\
K \quad = \dfrac{Y_1}{Y_1 + Y_5}.
\end{cases}
\quad (6.53)
$$

Normally, symmetric T networks are used, i.e., $Y_7 = Y_9$ and $Y_{11} = Y_{13}$.

It can be shown that the NF2 sections have the following sensitivity properties:

- $|S_{R,C}^{r_p}| \le 0.5$ for R and C
- $GS_{A_0}^{r_p} \approx 0$
- $|S_{R,C}^{Q}| \le 0.5$ for R and C if pole and zero cancels
- $|S_{R,C}^{Q}| \propto Q$ for R and C if pole and zero cancels
- $GS_{A_0}^{Q} \propto Q$
- Spreads of passive elements ∞ Q.

$GS_{A_0}^{Q}$ and the spreads of passive elements in the NF2 sections are, thus, much lower than for NF1 sections, but they have more passive elements and $|S_{R,C}^{Q}|$ is also higher if pole and zero do not cancel. The higher passive sensitivities are a relativey small problem as the element values can be trimmed to desired values. Through a suitable choice of components, the temperature coefficients can be chosen so the RC products become almost temperature and frequency independent. The high-frequency properties of the NF2 sections are very good [24]. An essential drawback with NF2 sections is, however, the large number of passive circuit elements. NF2 sections are usable to realize medium to high Q factors. Figs. 6.24 and 6.25 show two examples of NF2 sections.

6.5.3.8 NF2 Generic Section

A generic NF2 section is shown in Fig. 6.24. The transfer function, assuming an ideal operational amplifier, is

$$H(s) = -\frac{fs^2 + \dfrac{g}{RC}s + \dfrac{b}{R^2 C^2}}{s^2 + \dfrac{r_p}{Q}s + r_p^2} \quad (6.54)$$

where

$$r_p = \frac{\sqrt{1 + a(2+b)}}{RC} \qquad Q = \frac{r_p RC}{a(2+f)} \qquad g = b - e + 2$$

and the gain-sensitivity product is

$$GS_{A_0}^{Q} = \frac{(f+1)(f+b+4)}{a(f+2)}.$$

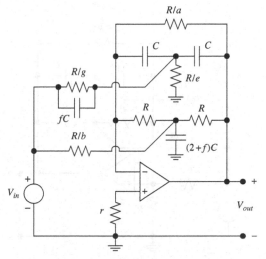

Fig. 6.24 NF2 section with a bridged twin-T network

With these element values, a real pole and a real zero will be cancelled. This section can realize all types of second-order transfer functions, except an allpass section.

Example 6.3 Determine the element values in the NF2 section that are shown above to realize a BP section with $r_p = 5$ krad/s, $Q = 15.473398$, and $G = 323.13523$.

Obviously $f = b = 0$ as the section shall be of BP type. Elimination of the parameter a in the equations above yields

$$RC = \frac{\sqrt{1 + 4Q^2} + 1}{2Qr_p}.$$

We select a fixed capacitance $C = 10$ nF and trim the resistors. We get
$R = 20656.71\,\Omega$

$$a = \frac{RCr_p}{2Q} = 0.03337455 \qquad R/a = 618.93592 \text{ k}\Omega$$

$$g = GRC = 0.066749105 \qquad R/g = 309.46796 \text{ k}\Omega$$

$$e = 2 - g = 1.9332509 \qquad R/e = 10.68496 \text{ k}\Omega.$$

6.5.4 NF2 AP Section

The transfer function of the NF2 AP section, which is shown in Fig. 6.25, is given by Equation 6.20) where $G = 1$. A suitable choice of element values is

$$r_p = \frac{\sqrt{1 + 2a}}{RC} \qquad Q = \frac{\sqrt{1 + 2a}}{2a} \qquad k = \frac{1}{1 + a}$$

and the error polynomial is

$$E(s) = s^2 + \left(\frac{a + 2}{a}\right)\left(\frac{r_p}{Q}\right)s + r_p^2.$$

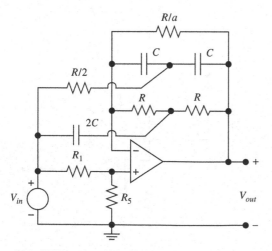

Fig. 6.25 NF2 section with a bridged twin-T network

6.5.5 Sections with Positive Feedback

Another class of second-order sections uses positive feedback. The generic section, shown in Fig. 6.26, has a network N_1, which is of BP type, in the positive feedback loop. However, for the circuit to be stable, a compensating negative feedback is required. The negative feedback is realized with R_1 and R_2. The input signal may be injected partly via the network N_1 and partly via R_3.

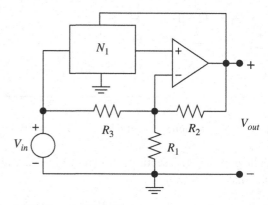

Fig. 6.26 Generic PF section

PF sections with positive feedback can be divided into four subgroups depending on the network structure and the degree of negative feedback.

6.5.5.1 PF1 Sections

N_1 is a RC ladder network in a PF1 section. Figure 6.27 shows a PF1 section with the ladder elements Y_1, Y_2, Y_3, and Y_4 and an amplifier with finite gain $K > 1$. The

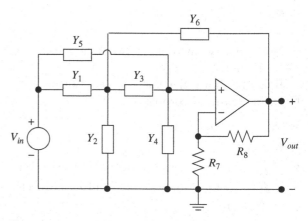

Fig. 6.27 General PF1 section with ladder network

feedback loop is closed by injecting the output voltage, V_{out}, by replacing the first grounded ladder arm with an impedance divider, Y_2 and Y_6.

The PF1 section,[2] shown in Fig. 6.27, has the transfer function

$$H(s) = \frac{N(s)}{D(s) + \dfrac{E(s)}{A}} \qquad (6.55)$$

where

$$
\begin{cases}
N(s) = K(N_1 Y_5 + Y_3 Y_1) \\
D(s) = N_2 + K Y_6 Y_3 \\
N_1 \;\; = Y_6 + Y_3 + Y_2 + Y_1 \\
N_2 \;\; = Y_3^2 - (Y_5 + Y_4 + Y_3) N_1 \\
E(s) = K N_2 \\
K \;\;\; = \dfrac{R_7 + R_8}{R_7}.
\end{cases}
\qquad (6.56)
$$

Figures 6.28, 6.29, and 6.30 show some examples of PF1 sections. It can be shown that PF1 sections have the following sensitivity properties:

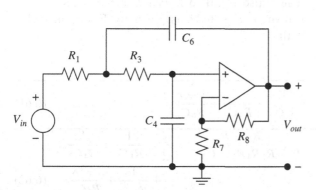

Fig. 6.28 PF1 section of LP type

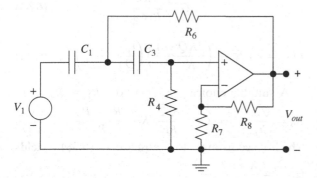

Fig. 6.29 PF1 section of HP type

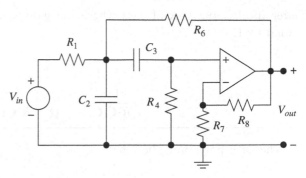

Fig. 6.30 PF1 section of BP type

- $\left| S_{R,C}^{r_p} \right| \le 0.5$ for R and C
- $GS_{A_0}^{r_p} \approx 0\,0$
- $\left| S_{R,C}^{Q} \right| \propto Q$ for R and C
- $GS_{A_0}^{Q} \propto Q$
- Spreads in passive elements are independent of Q.

Note that the spreads in the element values are independent of Q. It can be shown that PF1 sections have the following properties with respect to the gain-bandwidth product, GB.

The relative error in the pole radius, due to the finite bandwidth of the operational amplifier, is

$$\delta \approx -\frac{1}{2Q}\frac{r_p}{\omega_t} GS_{A_0}^{Q}. \qquad (6.57)$$

The deviation in the Q factor, due to the finite bandwidth of the operational amplifier, is

$$\eta \approx \frac{1}{1 + \delta - 2GK\delta\left(\dfrac{r_p}{\omega_t}\right)Q} \qquad (6.58)$$

where $\omega_t = 2\pi\, GB$. PF1 sections have good high-frequency properties, and the spreads in the element values are less than for the NF sections. The sections are usable for Q factors up to 50.

6.5.5.2 PF1 LP Section

Figure 6.28 shows a PF1 LP section, also called *Sallen-Key LP* section, where the amplifier has a

---

[2]Sections of this type, which are based on VCVS, were described by R.P. Sallen and E.L. Key in 1955.

later discuss the special case where the gain is equal to 1.

The transfer function for the PF1 LP section, shown in Fig. 6.28, with an ideal amplifier, is given by Equation (6.10), with

$$r_p^2 = \frac{1}{R_1 R_3 C_4 C_6} \qquad \frac{r_p}{Q} = \frac{1}{R_1 C_6} + \frac{1}{R_3 C_6} + \frac{1-K}{R_3 C_4} \qquad G = K r_p^2. \tag{6.59}$$

The error polynomial becomes

$$E(s) = K\left(s^2 + (KQ+1)\left(\frac{r_p}{Q}\right)s + r_p^2\right). \tag{6.60}$$

By choosing $R_1 = R_3 = R$, $C_4 = C_6 = C$, and $C_6 = \dfrac{1}{r_p R_3}$, we obtain $r_p = \dfrac{1}{RC}$, $G = k = 3 - \dfrac{1}{Q}$, and $GS_{A_0}^Q = K^2 Q$.

The gain, K, must thus be < 3 in order for the filter to be stable. But with this choice of element values, the passive sensitivities with respect to Q become proportional to Q, i.e., the passive sensitivities become large.

W. Saraga proposed the following selection of the element values. First, select C_4 to a suitable value and then $R_3 = 1/(\sqrt{3} r_p C_4)$, $R_1 = \sqrt{3}(R_3/Q)$,

and $C_6 = \sqrt{3} Q C_4$, which gives $K = 4/3$. With this choice, the active sensitivity with respect to Q becomes small, but the passive sensitivities increase and the spreads in the element values become proportional to Q. By choosing $1 \le K \le 4/3$, we can obtain a good compromise between the active and passive sensitivities.

6.5.5.3 PF1 HP Section

The transfer function for the PF1 HP section, which also is called *Sallen-Key HP* section [102], shown in Fig. 6.29, is given by Equation (6.15) with

$$r_p = \frac{1}{\sqrt{R_4 R_6 C_1 C_3}} \qquad \frac{r_p}{Q} = \frac{1}{R_4 C_1} + \frac{1}{R_4 C_3} + \frac{1-K}{R_6 C_1}. \tag{6.61}$$

The error polynomial becomes

$$E(s) = K\left(s^2 + ((2K+3)Q - 1)\left(\frac{r_p}{Q}\right)s + r_p^2\right)$$

and $GS_{A_0}^Q = K((2K+3)Q - 2)$, which is low. A suitable choice of element values is $C_1 = C_3 = C$ and $R_6 = R_4 = R$ where $r_p = 1/RC$, which gives $R_4 = 1/r_p C_6$ and $G = K = 3 - 1/Q$.

Because a PF1 section of HP type can be obtained from the LP section, shown in Fig. 6.28, by changing resistors and capacitors to capacitors and resistors, respectively, except for R_7 and R_8, it is obvious that both sections have the same element sensitivity.

6.5.5.4 PF1 BP Section

The transfer function for the PF1 BP section shown in Fig. 6.30 is given by Equation (6.19) with

$$r_p^2 = \frac{R_1 + R_6}{R_1 R_4 R_6 C_2 C_3} \qquad \frac{r_p}{Q} = \frac{1}{R_1 C_2} + \frac{1}{R_4 C_2}$$
$$+ \frac{1}{R_4 C_3} + \frac{1-K}{R_6 C_2} \tag{6.62}$$

$$G = \frac{K}{R_1 C_2} \tag{6.63}$$

$$E(s) = K\left(S^2 + \left(\frac{3KQ-1}{K-1}\right)\left(\frac{r_p}{Q}\right)s + r_p^2\right). \tag{6.64}$$

A suitable choice is $r_p = 1/RC$, $R_1 = R_4 = R$, $R_6 = \dfrac{\sqrt{2}}{Cr_p}$, $C_2 = C_3 = \dfrac{\sqrt{2}}{Rr_p}$, $K = \dfrac{R_7 + R_8}{R_7} = 4 - \dfrac{\sqrt{2}}{Q}$, which according to Equation (6.38) yields $GS_{A_0}^Q = \dfrac{K^2(3Q-1)}{k-1}$.

positive gain of $K = (1 + R_8/R_7) > 1$. We shall

6.5.5.5 PF2 Sections

In the PF2 sections, as well as in PF1 sections, a ladder network is used, but instead a voltage follower is used as amplifier. By selecting the gain $K = 1$ in a PF1 section, the corresponding PF2 section is obtained, see Fig. 6.27. The PF2 sections are therefore often referred to as *Unity Gain sections*.

PF2 sections have the transfer function

$$H(s) = \frac{N(s)}{D(s) + \dfrac{E(s)}{A}} \qquad (6.65)$$

where

$$\begin{cases} N(s) = -N_1 Y_5 - Y_3 Y_1 \\ D(s) = N_2 + Y_6 Y_3 \\ N_1 = Y_1 + Y_2 + Y_3 + Y_6 \\ E(s) = N_2 = Y_3^2 - (Y_3 + Y_4 + Y_5)N_1. \end{cases} \qquad (6.66)$$

It can be shown that PF2 sections have the following sensitivity properties:

- $\left| S_{R,C}^{r_p} \right| \leq 0.5$ for R and C
- $GS_{A_0}^{r_p} \approx 0$
- $\left| S_{R,C}^{Q} \right| \leq 0.5$ for R and C
- $GS_{A_0}^{Q} \propto Q^2$
- Spreads in passive elements are $\propto Q^2$.

PF2 sections have, thus, the same spreads in the element values as NF1 sections and they are also from a sensitivity point of view similar to NF1 sections, i.e., have large gain-sensitivity products. An advantage with PF2 sections is, however, the low passive sensitivity and they have few circuit elements.

Figures 6.31, 6.32 and 6.33 show some examples of PF2 sections, which also are known as *Sallen-Key unity gain sections* [102].

6.5.5.6 PF2 LP Section

The transfer function for the PF2 LP section shown in Fig. 6.31 is given by Equation (6.10) where

$$r_p = \frac{1}{\sqrt{R_0 R_3 C_4 C_6}} \qquad Q = \frac{R_0 R_3 C_6 r_p}{R_0 + R_3} \qquad (6.67)$$

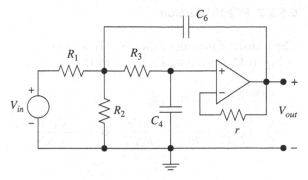

Fig. 6.31 PF2 section of LP type

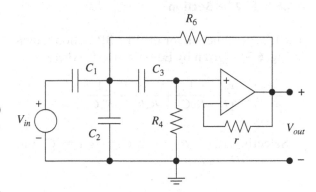

Fig. 6.32 PF2 section of HP type

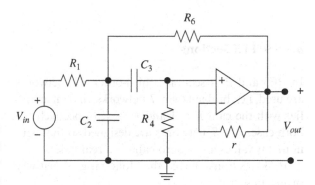

Fig. 6.33 PF2 section BP type

$$G = \frac{R_2}{R_1 + R_2} r_p^2 \qquad R_0 \frac{R_1 R_2}{R_1 + R_2} \qquad (6.68)$$

and the error polynomial

$$E(s) = s^2 + Q^2 \left(\frac{R_0 + R_3}{R_3} \right) \left(\frac{r_p}{Q} \right) s + r_p^2. \qquad (6.69)$$

A small resistor in series with C_4 may be used to compensate for the finite bandwidth of the amplifier [116].

6.5.5.7 PF2 HP Section

The transfer function of the PF2 HP section shown in Fig. 6.32 is given by Equation (6.15) where

$$r_p^2 = \frac{1}{(C_1 + C_2)C_3R_4R_6}$$

$$\frac{r_p}{Q} = \frac{1}{R_4C_1} + \frac{1}{R_4C_3} - \frac{GC_2}{R_4C_1^2} \quad G = \frac{C_1}{C_1 + C_2}.$$

6.5.5.8 PF2 BP Section

The transfer function for the PF2 BP section shown in Fig. 6.33 is given by Equation (6.19) where

$$\frac{r_p}{Q} = \frac{1}{R_1C_2} + \frac{1}{R_4C_2} + \frac{1}{R_4C_3}$$

Selecting $R_1 = R_4 = R$, $C_2 = C_3 = C$, and $R_6 = R/(9Q^2-1)$ yields

$$GS_{A_0}^Q = 3Q^2 - \frac{1}{3}.$$

6.5.5.9 PF3 Sections

In PF3 and PF4 sections, third-order RC networks are used, i.e., bridged twin-T networks, and an amplifier with the gain $K > 1$ and $K = 1$, respectively. In both cases, the T networks are designed so that with nominal values a real zero cancels a real pole.

PF3 sections have the following sensitivity properties:

- $\left|S_{R,C}^{r_p}\right| \leq 0.5$ for R and C
- $GS_{A_0}^{r_p} \approx 0$
- $\left|S_{R,C}^Q\right| \leq 0.5$ for R and C if cancelling occurs
- $\left|S_{R,C}^Q\right| \propto Q$ for R and C without cancelling
- $GS_{A_0}^Q \propto Q$
- Spreads in passive elements are $\propto Q$.

The passive sensitivities for bridged twin-T networks are, as for the NF2 sections, large, i.e., proportional to Q, if cancellation does not occur. Furthermore, PF3 sections do not have better

properties with respect to $GS_{A_0}^Q$ than PF1 sections, which have fewer circuit elements. However, in some cases PF3 sections may be easier to tune, but there are in practice few reason to use PF3 sections.

6.5.5.10 PF4 Sections

Voltage followers are used in PF4 sections as well as in PF2 sections. A PF4 section has the following sensitivity properties:

- $\left|S_{R,C}^{r_p}\right| \leq 0.5$ for R and C
- $GS_{A_0}^{r_p} \approx 0$
- $\left|S_{R,C}^Q\right| \propto Q$ for R and C
- $GS_{A_0}^Q \propto Q$
- Spreads in the passive elements are $\propto Q$.

The gain-sensitivity product has been reduced due to the bridged twin-T network, but the passive sensitivities have increased compared with PF1 sections. PF4 sections have almost the same sensitivity properties as NF2 sections, but they have no special advantage compared with PF1 sections. Figures 6.34, 6.35 and 6.36 show some examples of PF4 sections [112]. In all sections, the element values have been chosen so a real pole and a real zero are cancelled.

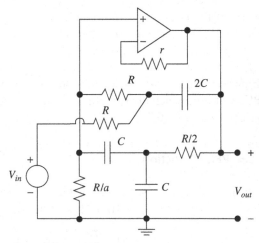

Fig. 6.34 PF4 LP section

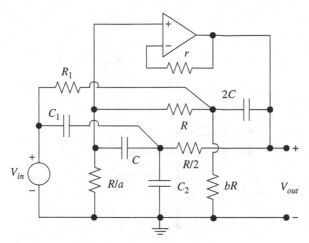

Fig. 6.35 PF4 LP-notch and HP-notch section

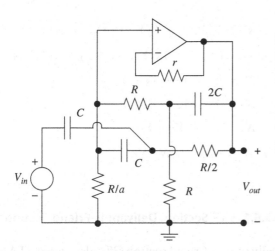

Fig. 6.36 PF4 section of HP type

6.5.5.11 PF4 LP Section

Figure 6.34 shows a PF4 LP section. The transfer function is given by Equation (6.10) where

$$r_p = \frac{\sqrt{1+2a}}{RC} \quad Q = \frac{\sqrt{1+2a}}{2a} \quad G = \frac{1}{4a^2 Q^2}.$$

6.5.5.12 PF4 LP-Notch and HP-Notch Sections

Figure 6.35 shows a PF4-notch section. The transfer function for the section is given by Equations (6.17) and (6.18) where

$$r_p = \frac{\sqrt{1+2a}}{RC} \quad Q = \frac{\sqrt{1+2a}}{2a} \quad G = \frac{C_1}{C}.$$

To obtain an LP-notch section, which has $r_z > r_p$, we may select $R_1 = R$, $b = 0$ and $C_1 = \dfrac{dC}{1+2a}$, $d = (r_z/r_p)^2$ and $C_2 = C_1(1 + 2a - d)/d$.

To obtain a HP-notch section, with $r_z < r_p$ and $G = 1$, we may select $C_1 = C$, $C_2 = 0$, $b = 1$, $R_1 = Rd^2/(1 + 2a)$, and $R_2 - R_1/(1 - d^2)$.

6.5.5.13 PF4 HP Section

Figure 6.36 shows a PF4 HP section. The transfer function is given by Equation (6.15) where $G = 1$ and

$$r_p = \frac{\sqrt{1+2a}}{RC} \quad Q = \frac{\sqrt{1+2a}}{2a}.$$

6.5.5.14 PF4 BP Section

Figure 6.37 shows a PF4 BP section. The transfer function is given by Equation (6.19) where

$$r_p = \frac{\sqrt{1+a}}{RC} \quad Q = \frac{\sqrt{1+a}}{a}.$$

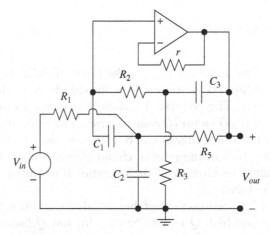

Fig. 6.37 PF4 BP section

6.5.6 ENF Sections

NF and PF sections either have low sensitivities for errors in the passive circuit elements or high sensitivities for variations in the gain-bandwidth product, GB, or vice versa. For example, NF1 sections have low sensitivities for errors in the passive elements and high sensitivities for errors in the GB, whereas the opposite is true for PF1 sections.

By using positive feedback in an NF section, a trade-off can be made between the two types of sensitivities so that the active sensitivities decrease on behalf of an increase in the passive sensitivities and the spreads in the element values decrease as well [112].

Figure 6.38 shows an ENF section (*enhanced negative feedback*). The resistor R_3 can in the general case be an arbitrary RC network. The positive feedback is realized with R_1 and R_2. For $R_2 = \infty$, an NF section is obtained.

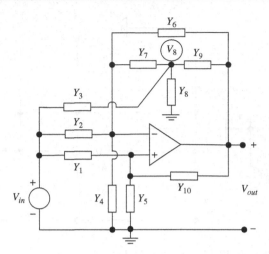

Fig. 6.39 ENF section

where

$$
\begin{cases}
N(s) = Y_7^2 Y_1 + N_3 Y_3 Y_7 + (N_3 Y_2 - N_2 Y_1) N_1 \\
D(s) = (N_1 N_2 - Y_7^2) Y_{10} - (Y_7 Y_9 + N_1 Y_6) N_3 \\
N_1 = Y_3 + Y_7 + Y_8 + Y_9 \\
N_2 = Y_2 + Y_4 + Y_6 + Y_7 \\
N_3 = Y_{10} + Y_5 + Y_1 \\
E(s) = (Y_7^2 - N_1 N_2) N_3.
\end{cases}
\tag{6.70}
$$

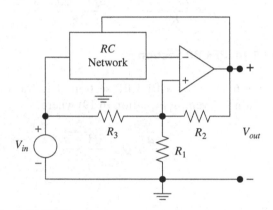

Fig. 6.38 ENF section

If we choose $R_2 < \infty$, the poles of the corresponding NF section will be moved closer to the $j\omega$-axis. The positive feedback thus increases the section's Q factor (*Q enhancement*). The amount of the positive feedback can be chosen so the effect of the errors in the passive circuit elements and the limited bandwidth of the operational amplifier is minimized [38].

ENF sections are suitable to realize sections with medium high Q factors because for low Q factors the NF1 sections are simpler.

The generic ENF section shown in Fig. 6.39 has the transfer function

$$
H(s) = \frac{N(s)}{D(s) + \dfrac{E(s)}{A}}
$$

6.5.6.1 ENF Section: Deliyannis-Friend Section

Deliyannis[3]- Friend section [37], also called STAR, is shown in Fig. 6.40 and can realize transfer functions of type LP-notch, HP, HP-notch, BP, and AP.

6.5.6.2 ENF BP Section

We get a BP section by selecting the element values as shown in Fig. 6.40. A suitable choice of the element values is: $R_1 = R_2 = R_4 = R_8 = \infty$, $R_6 = R$, $R_3 = R/(4a^2 Q^2)$, $C_7 = C_9 = C$, $R_5 = r$, and $R_{10} = (K-1)r$ where $r_p = 1/RC$, $Q = 1/2a$,

---

[3]Teodor Deliyannis, Patras University, Greece.

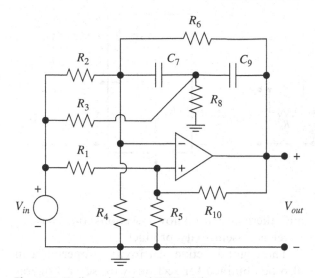

Fig. 6.40 Deliyannis-Friend ENF section

$G = 2(a - 1 - 2a^2Q^2)/RC$, and $K = 1 + 2a^2Q^2/(1 - a)$.

The gain-sensitivity product for the Q factor is

$$GS_{A_0}^Q = \frac{(1 - a + 2a^2Q^2)^2}{2a^3Q^2}. \quad (6.71)$$

If we select $aQ > 3$, we get $GS_{A_0}^Q = 2aQ^2$. If a is selected less than 1, the section becomes less dependent on the operational amplifier's bandwidth, but the sensitivities with respect to the passive components increases [32, 36].

6.5.7 Complementary Sections

Consider the two sections shown in Fig. 6.41 The networks N are the same but the terminals 2 and 3 have

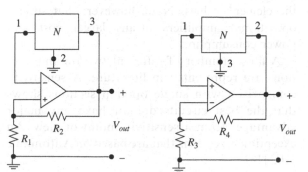

Fig. 6.41 Complementary sections

been interchanged and the inputs to the operational amplifiers have also been interchanged. If we have $R_2 = k R_1$ and $R_3 = k R_4$ where $k > 0$, the sections have the same poles and element sensitivities.

Example 6.4 Fig. 6.42 shows the PF1 BP section, shown in Fig. 6.30, and corresponding complementary section.

6.6 Transconductor-Based Sections

Transconductor-C filters, also called g_m-C filters, are one of the filter techniques that are recommended for filtering of high frequencies. Transconductor-C filters with a passband edge of more than 100 MHz have been manufactured. Bipolar transistors are preferable at these frequencies because they have higher g_m, lower noise, less DC offset, and use less power than MOS transistors. For filtering of signals up to 100 MHz, MOS transistors can also be used. Important applications for such filters are in hard drives.

It is essential to minimize the number of transconductors in a structure because the power consumption, noise, and chip area is directly proportional to the number of transconductors.

First- and second-order sections can be realized in many different ways and with one or more transconductors. A structure for realization of first- and second-order sections with only one transconductor and three admittances is shown in Fig. 6.44.

The transfer function is

$$H(s) = \frac{g_m(Y_1 + Y_2)}{Y_1 Y_2 + Y_1 Y_3 + Y_2 Y_3 + g_m Y_2}. \quad (6.72)$$

By a suitable choice of admittances, which can consist of resistors and capacitors, all types of first- and second-order sections can be realized, see [32]. The resistors can, of course, be realized with transconductors.

A circuit with one transconductor and four admittances is shown in Fig. 6.43. The section's transfer function is

$$H(s) = \frac{Y_1(Y_3 - g_m)}{Y_1 Y_2 + Y_1 Y_4 + Y_2 Y_3 + Y_2 Y_4 + Y_3 Y_4 + g_m Y_3}.$$

Analysis and design of these types of sections are discussed in detail in [32].

Fig. 6.42 PF1 BP section
and corresponding
complementary section

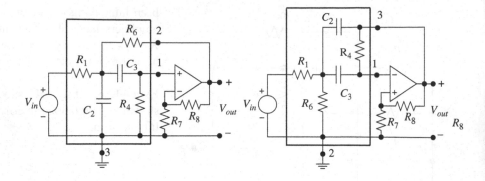

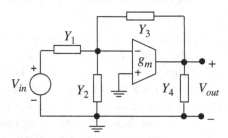

Fig. 6.43 Section with one transconductor and four admittances

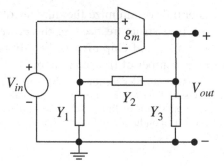

Fig. 6.44 Section with one transconductor and three admittances

6.7 GIC-Based Sections

In the previous chapter, the sensitivity of amplifier sections was discussed. The gain-sensitivity product for the pole radius was very small for all sections. If that is not the case, then the section has no practical use. The gain-sensitivity product for the Q factor was at best proportional to Q

and there are no known sections that have a lower gain-sensitivity product.

These good active sensitivity properties can also be obtained for sections with several operational amplifiers. For the single op-amp sections, the sensitivity for errors of the passive elements is at the lowest proportional to Q. However, sections with low passive sensitivities did not have low active sensitivities and vice versa. A motive to use several operational amplifiers is therefore to obtain low passive sensitivities while at the same time the active sensitivities are small. Of course, the gain-sensitivity product for the pole radius must be very small; otherwise, the section is not useful.

In the literature, there exist second-order sections that have been optimized with respect to the sensitivity of the pole with respect to the amplifier gain [20, 40, 41]. The pole sensitivity is improved for certain types of sections if several operational amplifiers are used.

Another reason to use several operational amplifiers is that the sections may be easier to tune, realize at the same time several types of transfer functions, and have smaller spread in the element values. Note, however, that several operational amplifiers means larger cost and power consumption.

A large number of different two op-amp sections are represented in literature. A sensitivity comparison with single op-amp sections shows that the best circuits do not have any major advantages from a sensitivity point of view. An exception is sections that are based on Antoniou's GIC [18].

Second-order sections can be realized with a GIC that simulates the inductor in an LC circuit. Figures 6.45 6.46, 6.47, 6.48, 6.49, and 6.50 show examples of different GIC sections. GIC sections of this class have the following sensitivity properties:

- $\left| S_{R,C}^{r_p} \right| \leq 0.5$ for R and C
- $GS_{A_{01}}^{r_p} \approx 0$
- $GS_{A_{02}}^{r_p} \approx 0$
- $\left| S_{R,C}^{Q} \right| \leq 0.5$ for R and C
- $GS_{A_{01}}^{Q} \approx Q$ and $GS_{A_{02}}^{Q} \approx Q$
- $\delta = \dfrac{\Delta r_p}{r_{p\,nominal}} \approx -r_p \left(\dfrac{1}{\omega_{t1}} + \dfrac{1}{\omega_{t2}} \right)$
- $\eta = \dfrac{Q}{Q_{nominal}} \approx \dfrac{1}{(1+\delta)\left(1 - 2Q\left(\dfrac{r_p}{\omega_{t1}} \right)\left(1 - \dfrac{\omega_{t2}}{\omega_{t1}} + 2\delta \right) \right)}$
- The spreads in passive elements are ∞Q.

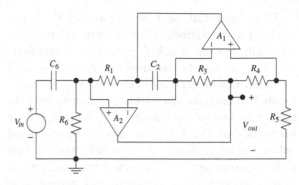

Fig 6.47 HP section

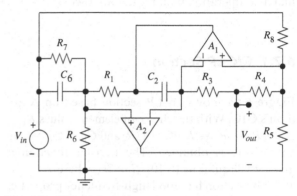

Fig. 6.48 HP-notch section

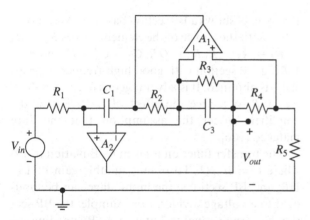

Fig. 6.45 LP section

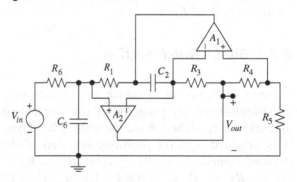

Fig. 6.49 BP section

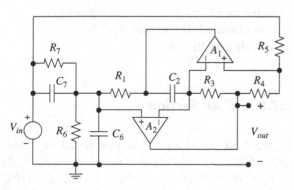

Fig. 6.46 LP-notch section

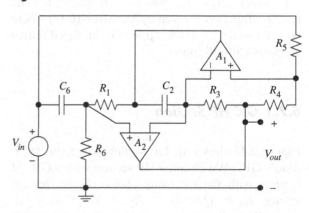

Fig. 6.50 AP section

Thus, these GIC sections are among the better second-order sections. The sensitivity becomes especially low of matched operational amplifiers, i.e., if $\omega_{t1} = \omega_{t2}$ the variations in the Q factor will be small. Operational amplifiers, which are implemented on the same silicon die, are often designed to have matched performances. The high-frequency properties of the GIC sections to be discussed below are equivalent with or better than the best single op-amp sections. The below two op-amp sections have very good high-frequency properties if matched operational amplifiers are used.

6.7.1 GIC LP Section

Figure 6.45 shows an LP section based on Antoniou's GIC. With the choice of element values $R_1 = R_2 = R_4 = R_5 = R, R_3 = QR$, and $C_1 = C_3 = C$ an LP section is obtained. The transfer function is given by Equation (6.10) $G = 2r_p^2$ where $r_p = 1/RC$. The section has good high-frequency properties and its gain can be reduced by using a voltage divider instead of R_1.

6.7.2 GIC LP-Notch Section

Figure 6.46 shows an LP-notch section with very good high-frequency properties if the operational amplifiers are matched. A section with a PIC of type A is obtained with the following selection of the element values: $R_7 = R_6 = 2QR, R_1 = R_3 = R_4 = R_5 = R, C_7 = 0.5(1 + 1/k^2)C, C_2 = C, C_6 = 0.5(1 - 1/k^2)C$, and $k = (r_z/r_p)^2 \geq 1$. The transfer function is given by Equation (6.17) where $G = 1/k$ and $r_p = 1/RC$. Note that the signal source is applied at both ports.

6.7.3 GIC HP Section

Figure 6.47 shows an HP section based on Antoniou's GIC. We obtain a HP section with a GIC of type A with the following choice of the element values: $R_6 = QR, R_1 = R_3 = R_4 = R_5 = R, C_2 = C_6 = C$, and $r_p = 1/RC$.

The transfer function is given by Equation (6.15) where $G = 2$. The section has good frequency properties if the operational amplifiers are matched.

6.7.4 GIC HP-Notch Section

Figure 6.48 shows an HP-notch section with a GIC of type A. The section's transfer function with the choice of element values $R_2 = R_3 = R_8 = R, R_6 = (1 + k^2)QR, R_7 = (1 + 1/k^2)QR, R_4 = 2k^2R/(1 + k^2), R_5 = 2k^2R/(1 - k^2)$, and $C_2 = C_6 = C$ is given by Equation (6.18) where

$$G = 2\frac{2 - k^2}{3 - k^2}, k = (r_z/r_p)^2 \leq 1, \text{and } r_p = 1/RC.$$

6.7.5 GIC BP Section

Figure 6.49 shows a BP section based on Antoniou's GIC. With the choice of the element values $R_1 = R_3 = R_4 = R_5 = R, R_6 = QR, C_2 = C_6 = C$, and $r_p = 1/RC$ a BP section with good high-frequency properties is obtained. It is advantageous to choose $Z_4 = Z_5 = R$ to reduce the effect of finite GB of the amplifiers under the assumption that they are matched [106].

The transfer function is given by Equation (6.19) where $G = 2r_p/Q$. To obtain suitable gain of LP, HP, and BP sections, the input stage can be modified to a voltage divider. For example, the BP section has a gain equal to 2 at $\omega = r_p$. By dividing the R_6 into two equally large resistors, one in series with the voltage source and the other parallel with the capacitance C_6, a halving of the input signal to the GIC is obtained. If the gain shall be larger than 2, we can choose $R_4 > R_5$.

6.7.6 GIC AP Section

Figure 6.50 shows an AP section based on Antoniou's GIC. With the element values $R_1 = R_3 = R_4 = R_5 = R, R_6 = QR, C_2 = C_6 = C$, and $r_p = 1/RC$ we get an AP section with $G = 1$. The sensitivity properties are the same as for the BP section discussed above.

6.8 Two-Integrator Loops

A reason to use several operational amplifiers in a second-order section is that we can simultaneously realize several transfer functions in a single second-order section. That is, the output from different operational amplifiers in the section can have LP, HP, and BP characteristics. Furthermore, the design and tuning procedures often become simpler.

A technique to realize active filter structures with low sensitivities to variations in the bandwidth of the operational amplifiers employs several operational amplifiers to realize a composite amplifier that has very low sensitivities for variations in the gain-bandwidth product. It is also an advantage to use several amplifiers to realize high-performance integrators.

Most three op-amp sections, but not all, realize a so-called *two-integrator loop*. There are several different variations of the loop, and we will here discuss their properties and realization. Note that two-integrator loops are also used to build more advanced filter structures, i.e., leapfrog filters, which will be discussed in Chapter 10.

6.8.1 Two-Integrator Loops with Lossless Integrators

An integrator has the transfer function $H(s) = \pm 1/s$, i.e., the transfer function has a pole at the origin. An integrator with losses has a real pole in the left half of the s-plane, i.e., $H(s) = \pm 1/(s+a)$. In order to distinguish between a true integrator and a lossy integrator, the former is often referred to as a lossless integrator. We will in Section 6.8.4. discuss two types of second-order sections with lossy and/or lossless integrators [32, 107, 115, 131].

For the signal-flow graph shown in Fig. 6.51, which has two lossless integrators, we have

$$V_{out} = V_{in} - \frac{a}{s\tau_1} V_{out} - \frac{b}{s^2\tau_1\tau_2} V_{out} \quad (6.73)$$

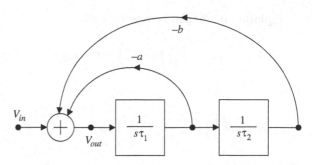

Fig. 6.51 Two-integrator loop with two lossless integrators

and we get

$$H(s) = \frac{V_{out}}{V_{in}} = \frac{s^2\tau_1\tau_2}{s^2\tau_1\tau_2 + as\tau_2 + b}. \quad (6.74)$$

Thus, the signal-flow graph represents a second-order HP filter. The denominator to the two-integrator loop, shown in Fig. 6.51, is

$$D(s) = \tau_1\tau_2 s^2 + a\tau_2 s + b = \tau_1\tau_2 \left(s^2 + \left(\frac{r_p}{Q} \right) s + r_p^2 \right)$$

where

$$r_p = \sqrt{\frac{b}{\tau_1\tau_2}}, \quad Q = \frac{1}{a}\sqrt{\frac{b\tau_1}{\tau_2}}, \quad \frac{r_p}{Q} = \frac{a}{\tau_1}. \quad (6.75)$$

The following sensitivities are obtained directly from Equation (6.75):

$$S_{\tau_1}^{r_p} = S_{\tau_2}^{r_p} = -S_b^{r_p} = -\frac{1}{2} \quad S_a^{r_p} = 0$$

$$S_{\tau_1}^{Q} = -S_{\tau_2}^{Q} = S_b^{Q} = \frac{1}{2} \quad S_a^{Q} = -1.$$

The sensitivities are, thus, very low.

6.8.2 Kerwin-Huelsman-Newcomb Section

The Kerwin-Huelsman-Newcomb section, also known as the state variable realization, is based on the two-integrator loop with two lossless integrators [32, 107, 115, 131]. Consider the HP transfer function

$$H(s) = \frac{Gs^2}{s^2 + \left(\frac{r_p}{Q} \right) s + r_p^2}. \quad (6.76)$$

Equation (6.76) can be rewritten as

$$V_{out}(s)\left(s^2 + \left(\frac{r_p}{Q}\right)s + r_p^2\right) = Gs^2 V_{in}(s)$$

and

$$V_{out} = -\frac{r_p}{Q}\left(\frac{1}{s}\right)V_{out} - r_p^2\left(\frac{1}{s}\right)^2 V_{out} + Gs^2\left(\frac{1}{s}\right)^2 V_{in}$$

and we get

$$V_{out} = -\frac{1}{Q}\left(\frac{r_p}{s}\right)V_{out} - \left(\frac{r_p}{s}\right)^2 V_{out} + GV_{in}. \qquad (6.77)$$

Equation (6.77) corresponds to the signal-flow graph shown in Fig. 6.52, which contains only two lossless integrators and one addition. Because V_{out} has HP characteristic, the outputs $(r_p/s)V_{out}$ and $(r_p/s)^2 V_{out}$ will have BP and LP characteristics, respectively.

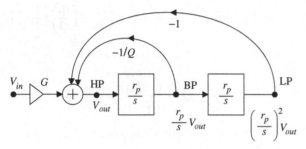

Fig. 6.52   Two-integrator loop

A Miller integrator, with the transfer function $H(s) = -1/sRC$, can be realized with one single operational amplifier.

We modify the signal-flow graph as shown in Fig. 6.52 according to Fig. 6.53 to obtain inverting integrators, i.e., so Miller integrators can be used. Note that the loop gain in each loop in the two signal-flow graphs must be retained after the modification.

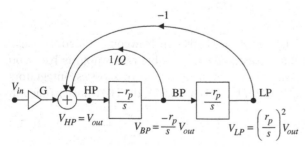

Fig. 6.53   Modified signal-flow graph

The three transfer functions become

$$H_{HP} = \frac{Gs^2}{s^2 + \left(\frac{r_p}{Q}\right)s + r_p^2} \qquad (6.78)$$

$$H_{BP} = \frac{-Gr_p s}{s^2 + \left(\frac{r_p}{Q}\right)s + r_p^2} \qquad (6.79)$$

$$H_{LP} = \frac{Gr_p^2}{s^2 + \left(\frac{r_p}{Q}\right)s + r_p^2}. \qquad (6.80)$$

The relation between the node voltages in the modified signal-flow graph shown in Fig. 6.53 is

$$V_{HP} = \frac{1}{Q}V_{BP} - V_{LP} + GV_{in}. \qquad (6.81)$$

An adder/subtractor circuit that can realize Equation (6.81) is shown in Fig. 6.54.

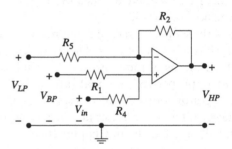

Fig. 6.54   Adder/subtractor

By considering the three input signals, one by one, and superimposing their contributions, we get

$$V_{HP} = \frac{R_4}{R_1 + R_4}\left(\frac{R_2 + R_5}{R_5}\right)V_{BP} - \frac{R_2}{R_5}V_{LP}$$
$$+ \frac{R_1}{R_1 + R_4}\left(\frac{R_2 + R_5}{R_5}\right)V_{in}. \qquad (6.82)$$

Thus, Equation (6.81) can be realized by using the circuit that is shown in Fig. 6.54. We obtain by comparing Equations (6.81) and (6.82):

$$r_p = 1/RC, R_1/R_4 = 2Q - 1, \text{ and } G = 2 - 1/Q$$

where we have selected $R_2 = R_5$, $R_3 = R_6 = R$, and $C_1 = C_2 = C$. The resistor R_4 can be used to select the gain factor G.

Fig. 6.55 KHN section

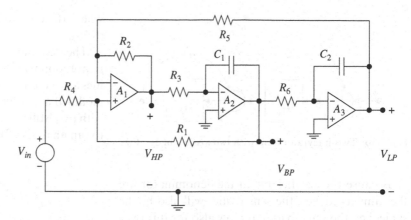

Figure 6.55 shows the complete realization of the section, which is known as the *KHN section* after the originators Kerwin, Huelsman, and Newcomb (1967). Note that the section at the same time realizes an LP, HP, and BP transfer function. Burr Brown manufactures a commercial version of this type of section with the name UAF 42.

The KHN section has, as required, low sensitivity with respect to the pole radius for errors in the passive and active elements. Furthermore, it has low sensitivity with respect to the Q factor for errors in the passive elements and the gain-sensitivity product is proportional to Q. The Q factor is, however, sensitive for the operational amplifier's finite bandwidth, which tends to increase the Q factor and move the poles toward the $j\omega$-axis. If the operational amplifier's bandwidth is too small, the poles move into the right half plane and the section will become unstable. This tendency is referred to as Q enhancement. This effect can partly be alleviated by using a feedback resistor between the output of A_1 and its positive input terminal.

Another problem with this section is that if the input signal contains a large high-frequent component, slew rate-limitation of the output signal may occur in the operational amplifier A_1.

Variations of this circuit can be obtained by instead injecting the input signal via one of the two integrators.

Example 6.5 Determine suitable component values when the KHN section shall realize an HP section with gain = 3 and the poles $s_p = -0.2 \pm j3$ krad/s.

We get $r_p = \sqrt{(-10.2)^2 + 3^2} = 3.00666 \, \text{krad/s}$ and $Q = -r_p/(2\sigma_p) = 7.516648$.

We select $R_2 = R_5$, $R_3 = R_6 = R$, $C_1 = C_2 = C$ and get $r_p = 1/RC$, $G_{HP} = 2 - 1/Q$, and $R_1/R_4 = 2Q - 1$. We select $R_2 = R_3 = R_4 = R_5 = R = 10 \, \text{k}\Omega$, which gives $C = 3.32595$ nF, $G_{HP} = 1.86696$, and $R_1 = 140.333 \, \text{k}\Omega$.

The HP section's maximum gain is obtained at $\omega = \dfrac{r_p}{\sqrt{1 - 1/2Q^2}}$ which gives $|H(j\omega)|_{\text{max}} = 14.0644$. The HP section's gain is, thus, too large and therefore we use a voltage divider on the input. R_4 is therefore replaced with R_{4a} and R_{4b} where $R_{4a} // R_{4b} = R_4$ and $R_{4b}/(R_{4a} + R_{4b}) = 3/14.0644$. We get $R_{4a} = 46.8813 \, \text{k}\Omega$ and $R_{4b} = 12.7114 \, \text{k}\Omega$.

6.8.3 Transposed Two-Integrator Loop

The transposition theorem is useful to generate new signal-flow graphs [108].

Theorem 6.3: Transposition Theorem *If we change the direction on all branches in a signal-flow graph and interchange the input and output, then the new signal-flow graph will have the same transfer function as the original.*

By transposing the signal-flow graph shown in Fig. 6.51, i.e., change the directions on all branches and interchange the input and output, the transposed two-integrator loop, shown in Fig. 6.56, is obtained. The transfer function and its denominator are according to the transposition theorem the same as for the signal-flow graph shown in Fig. 6.51, i.e., the denominator is

$$D(s) = \tau_1\tau_2 s^2 + a\tau_1 s + b = \tau_1\tau_2\left(s^2 + \left(\frac{r_p}{Q}\right)s + r_p^2\right). \tag{6.83}$$

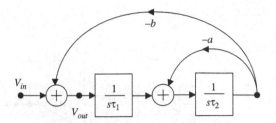

Fig. 6.56 Two-integrator loop with two lossless integrators

Because the coefficients in the denominator are the same as earlier, the sensitivities will also be the same, i.e., the sensitivities are low also for this two-integrator loop.

6.8.4 Two-Integrator Loops with Lossy Integrators

Further variants of two-integrator loops are obtained if one of the integrators is replaced by a lossy integrator. A lossy integrator has a first-order transfer function with a pole on the negative real axis [32, 107, 115, 131].

6.8.5 Tow-Thomas Section

A two-integrator loop with a lossy integrator is obtained if the inner loop is replaced with a first-order LP section [125]. The denominator of the two-integrator loop when one integrator has been replaced with a lossy integrator, according to Fig. 6.57, is

$$D(s) = \tau_1 \tau_2 s^2 + a\tau_2 s + b$$
$$= \tau_1 \tau_2 \left(s^2 + \left(\frac{r_p}{Q} \right) s + r_p^2 \right) \quad (6.84)$$

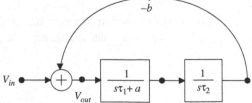

Fig. 6.57 Tow-integrator loop with one lossy integrator

where the coefficients are the same as in Equation (6.75).

The denominator is the same as for the two-integrator loop with lossless integrators and the sensitivities are therefore the same. The feedback loop can be modified according to Fig. 6.58 so it can be realized with two Miller integrators and an inverter. See [107] for an analysis of finite amplifier gain.

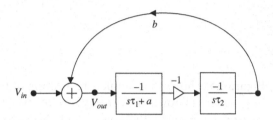

Fig. 6.58 Modified two-integrator loop with one lossy integrator

Figure 6.59 shows the resulting section. This section is named after the originators for the Tow-Thomas section. Here the integrator with losses has been placed to the left but it is of course possible to change the order between the inverter and the integrators or use a positive integrator.

The two transfer functions are

$$H_{LP} = \frac{-R_5}{R_1} \frac{r_p^2}{s^2 + \left(\frac{r_p}{Q} \right)s + r_p^2} \quad (6.85)$$

$$H_{BP} = \frac{-1}{R_1 C_1} \frac{s}{s^2 + \left(\frac{r_p}{Q} \right)s + r_p^2} \quad (6.86)$$

where

$$r_p^2 = \frac{R_4}{R_3 R_5 R_6 C_1 C_2} \quad \frac{r_p}{Q} = \frac{1}{R_2 C_1}. \quad (6.87)$$

Normally, $R_3 = R_4$ and $r_3 \approx R_3//R_4$ are chosen. The gain factor, G, is determined by the resistor R_1. Tow-Thomas section has no HP output and therefore an input signal with high frequency will not cause slew rate-limitation. From a sensitivity point of view, it is equal with the KHN section [32, 107, 115, 131]. With suitable design, both sections have the following sensitivities:

- $\left| S_{R,C}^{r_p} \right| \leq 0.5$ for R and C
- $GS_{A_{01}}^{r_p} \approx GS_{A_{02}}^{r_p} \approx GS_{A_{03}}^{r_p} \approx 1$

Fig. 6.59 Two-Thomas section

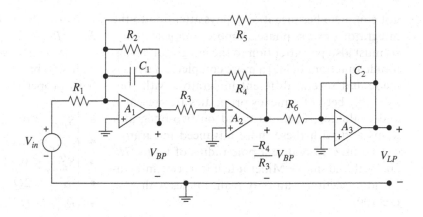

- $\left| S_{R,C}^{Q} \right| \leq 0.5$ for R and C
- $GS_{A_{01}}^{Q} \approx 2Q + 1$
- $GS_{A_{02}}^{Q} \approx 1$
- $GS_{A_{03}}^{Q} \approx Q + 1$
- $\delta \approx -r_p \left(\dfrac{1}{\omega_{t1}} + \dfrac{1}{\omega_{t3}} \right)$ (Relative error in the pole radius)

$$\eta \approx \frac{1}{1 - Q r_p \left(\dfrac{1}{\omega_{t1}} + \dfrac{2}{\omega_{t2}} + \dfrac{3}{\omega_{t3}} \right)} \qquad (6.88)$$

- The spreads in passive elements are ∞ Q.

The passive sensitivities are, thus, minimal and the gain-sensitivity products are proportional to Q, i.e., lowest possible. The error in the pole radius and the increase of the Q factor, due to the limited bandwidth of the operational amplifiers, is with identical operational amplifiers

$$\delta = \frac{-3r_p}{\omega_t} \qquad (6.89)$$

$$\eta = \frac{1}{1 - \dfrac{4Q r_p}{\omega_t}}. \qquad (6.90)$$

For large Q factors and relatively large pole radius, η becomes large and we get $Q = \eta Q_{nominal}$.

Example 6.6 Consider a second-order section with $Q = 25$, $r_p = 2\pi$ 10 krad/s that should be realized with an operational amplifier of the type 741 with $\omega_t = 2\pi$ Mrad/s.

From Equation (6.89) we get $\delta = -\frac{3r_p}{2\omega_t} = \frac{3 \cdot 2\pi \cdot 10^4}{2 \cdot 2\pi \cdot 10^6} = -0.015$. Thus the pole radius is reduced. However, according to Equation (6.90), η becomes $= \infty$. That is, the section becomes unstable. The cause of the increase in the Q factor is due to an excess phase shift in the integrators.

We can generally show that for a two-integrator loop [31],

$$\eta = \frac{1}{1 + Q \left(\dfrac{1}{Q_{I1}} + \dfrac{1}{Q_{I2}} \right)}. \qquad (6.91)$$

Insertion of Equation (5.51), which is valid for the Miller integrator, into Equation (6.91) with $\omega = r_p$ gives Equation (6.88). The real poles cause excess phase in the integrators. This phase function is the cause for the increase of the section's Q factor. The error in the phase function between an ideal integrator and a Miller integrator is

$$\Delta Q = \arctan \left(\frac{R}{Z} \right) = \arctan \left(\frac{1}{Q} \right) \approx \frac{1}{Q_I}. \qquad (6.92)$$

This extra phase shift can be compensated with a capacitor parallel to R_4 shown in Fig. 6.59.

$$C_c = \frac{1}{R_4} \left(\frac{1}{Q_{I1}} + \frac{1}{Q_{I2}} \right) = \frac{1}{R_4} \left(\frac{1}{\omega_{t1}} + \frac{2}{\omega_{t2}} + \frac{1}{\omega_{t3}} \right). \qquad (6.93)$$

Insertion of Equations (5.55) and (5.51), with matched operational amplifiers, into Equation (6.91) yields $\eta \approx 1$. The error in Q factor has, thus, in practice been eliminated. However, as discussed in Section 5.9.2, the passive compensation is not working as well as the active compensation using the phase-lead integrator.

6.8.5.1 Integrators with Several Operational Amplifiers

The above analysis of integrators, which is based on the quality factor of the integrator, is, however, not complete because it only pays attention to the integrator's excess phase. A more complete analysis must also pay attention to the integrator's magnitude function. In [42] a more complete analysis is made and several different integrators with considerably better frequency properties are presented and evaluated. Sections based on two-integrator loops and with these more advanced integrators can be used to realize a pole radius of up to $GB/20$. With the simple Miller integrator, two-integrator loop sections can only realize poles with $r_p < GB/100$.

6.8.6 Åkerberg-Mossberg Section

Figure 6.60 shows a section that is named after the inventors, the Åkerberg-Mossberg section[4] [147]. The section is based on a two-integrator loop with one lossy integrator and with an active compensated integrator (phase-lead integrator).

Normally, we choose $R_1 = R/G$, $R_2 = QR$, $R_3 = R_4 = R_5 = R_6 = R$, and $C_1 = C_2 = C$.

The Åkerberg-Mossberg section is characterized with suitable element values of the following sensitivity properties:

- $\left|S_{R,C}^{r_p}\right| \leq 0.5$ for R and C
- $GS_{A_{01}}^{r_p} \approx GS_{A_{02}}^{r_p} \approx GS_{A_{02}}^{r_p} \approx 1$
- $\left|S_{R,C}^{Q}\right| \leq 0.5$ for R and C
- $GS_{A_{01}}^{Q} \approx 2Q$
- $GS_{A_{02}}^{Q} \approx Q$
- $GS_{A_{03}}^{Q} \approx$ low
- $\delta \approx -r_p\left(\dfrac{1}{\omega_{t1}}\dfrac{1}{2\omega_{t3}}\right)$
- $\eta \approx \dfrac{1}{(1+\delta)\left(1 + Qr_p\left(\dfrac{2}{\omega_{t3}} - (1+2\delta)\left(\dfrac{1}{\omega_{t1}}+\dfrac{1}{\omega_{t3}}\right)\right)\right)}$
- The spreads in passive elements are $\propto Q$.

The sensitivity of the Q factor due to the limited bandwidth of the operational amplifiers is consider-

Fig. 6.60 Åkerberg-Mossberg section

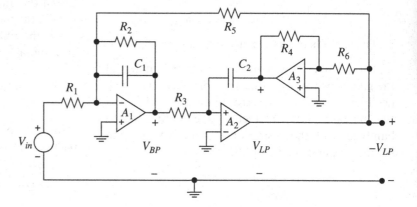

The transfer function is for the V_{LP} output

$$H_{LP}(s) = \frac{G}{s^2 + \dfrac{r_p}{Q}s + r_p^2} \qquad (6.94)$$

where

$$r_p^2 = \frac{R_6}{R_3 R_4 R_5 C_1 C_2} \qquad \frac{r_p}{Q} = \frac{1}{R_2 C_1} \qquad G = \frac{R_5}{R_1}r_p^2.$$

ably less than for KHN and Tow-Thomas sections and equal with the best single- and two-op-amp sections. The reason is that the phase errors in the Miller integrator and the phase-lead integrator are the same, but of different sign, if the amplifiers are matched. For matched operational amplifiers, we obtain

$$\delta \approx -\frac{3r_p}{2\omega_t} \qquad (6.95)$$

$$\eta \approx \frac{1}{(1+\delta)\left(1 - 4\delta Q\dfrac{r_p}{\omega_t}\right)}. \qquad (6.96)$$

---

[4]Dag Åkerberg and Kåre Mossberg, Royal Institute of Technology, Sweden.

Example 6.7 For the Åkerberg-Mossberg section with $Q = 25$, $r_p = 2\pi \, 10$ krad/s and matched operational amplifiers of the type 741 with $\omega_t = 2\pi$ Mrad/s, we obtain

$$\delta = -\frac{3r_p}{2\omega_t} = \frac{3 \cdot 2\pi \cdot 10^4}{2 \cdot 2\pi \cdot 10^6} = -0.015. \text{ The pole radius is reduced}$$

$$\eta = \frac{1}{(1+\delta)\left(1 - \dfrac{4\delta Q r_p}{\omega_t}\right)} = \frac{1}{(1 - 0.015)\left(1 + \dfrac{4 \cdot 0 \cdot 015 \cdot 25 \cdot 2\pi \cdot 10^4}{2\pi \cdot 10^6}\right)}$$

$$= \frac{1}{0.985 \cdot 1.015} = 1.000225 \Rightarrow Q = 1.000225 \, Q_{nominal} \approx 25.0056.$$

The effect of the finite bandwidth of the operational amplifiers on the Q factor is, thus, very small and the effect is somewhat smaller on the pole radius compared with Tow-Thomas section.

6.9 Amplifiers with Low *GB* Sensitivity

Active filters are in general sensitive for variations of the gain-bandwidth product of the operational amplifiers, and typically the ratio r_p/ω_t must be less than about 0.01; otherwise the pole radius and Q factor will differ too much from their desired values. The effect is especially large for large Q factors.

By using several operational amplifiers to realize a composite amplifier, we can make it less sensitive for the finite gain-bandwidth products of the individual operational amplifiers.

Figure 6.61 shows an example of such a BP section, which uses a composite amplifier with low sensitivity with respect to the finite gain-bandwidth products of the operational amplifiers. The section can be used for poles with $r_p/\omega_t < 0.05$, i.e., the

requirement on the gain-bandwidth products of the operational amplifiers is small [41].

The section is actually a single op-amp section with a bridged-T network where the operational amplifier has been replaced by three operational amplifiers. Normally, a symmetric T network is used, i.e., $C_1 = C_2$. The passive sensitivities are, thus, the same as for the corresponding single op-amp section. For the section, we have

$$r_p^2 = \frac{1}{R_3 R_4 C_1 C_2} \qquad Q = \frac{1}{2}\sqrt{\frac{R_2}{R_1}}.$$

To avoid oscillations when the power supply voltage is applied, or for large disturbances via the input signal, a reverse biased diode should be placed between the ($+$)-input on A_3 and ground. The oscillations originate from the operational amplifiers in this case working in a nonlinear mode.

Further examples of several sections with low requirements on the bandwidth of the operational amplifiers can be found in [41].

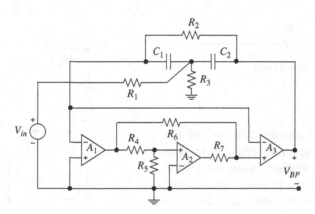

Fig. 6.61 BP section with low sensitivity for variations in *GB*

6.9.1 *Differential Two-Integrator Loops*

The above integrators are suitable to use in a two-integrator loop. Figure 6.62 shows a Tow-Thomas section. Because it is necessary to use differential realizations in integrated circuits, we mirror the structure shown in Fig. 6.62 in the ground plane and obtain a corresponding differential structure as shown in Fig. 6.63. Note that the inverter is realized by interchanging the outputs of the last integrator. Inverters are, thus, without cost in differential realizations.

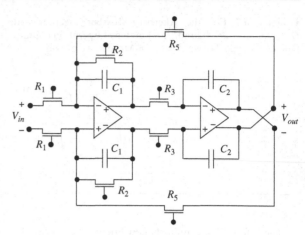

Fig. 6.64 Tow-Thomas section with MOSFETs

6.9.2 *Transconductor Based on Two-Integrator Loops*

In the same way as for active *RC* sections, there are a large number of realizations of first- and second-order g_m-*C* sections. Below are some examples shown of second-order sections, which are based on a two-integrator loop [1, 32]. All these sections have

$$|S^{r_p}_{C_i}| = |S^{r_p}_{g_m}| = |S^Q_{C_i}| = |S^Q_{g_m}| = \frac{1}{2}$$

i.e., the sensitivities are low.

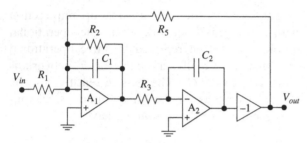

Fig. 6.62 Tow-Thomas section

6.9.2.1 LP Section

Figure 6.65 shows an LP section with the transfer function given by Equation (6.10) where

$$\frac{r_p}{Q} = \frac{g_{m2}}{C_2} \qquad r_p = \sqrt{\frac{g_{m1}}{C_1}\frac{g_{m2}}{C_2}} \qquad G = -\frac{g_{m2}}{C_2}\frac{g_{m3}}{C_1}.$$

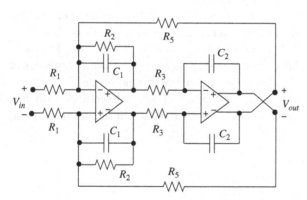

Fig. 6.63 Balanced Tow-Thomas section

By replacing the integrators with any of the earlier discussed integrators, a realization that is suitable for implementation in an integrated circuit is obtained. Figure 6.64 shows the resulting balanced Tow-Thomas section with MOSFETs.

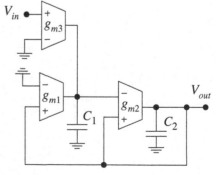

Fig. 6.65 LP section

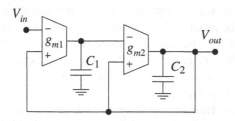

Fig. 6.66 Simplified LP section

If we select $G = r_p^2$, the section can be simplified according to Fig. 6.66, so only two transconductors are required. Note that both the capacitors are grounded, which is advantageous for implementation in an integrated circuit.

6.9.2.2 HP Section

A HP section based on a two-integrator loop is shown in Fig. 6.67. The HP section has the transfer function given by Equation (6.15) where

$$r_p = \sqrt{\frac{g_{m1}}{C_1}\frac{g_{m2}}{C_2}\frac{g_{m3}}{g_{m4}}} \qquad \frac{r_p}{Q} = \frac{g_{m2}}{C_2}\frac{g_{m3}}{g_{m4}} \qquad G = -\frac{g_{m3}}{g_{m4}}.$$

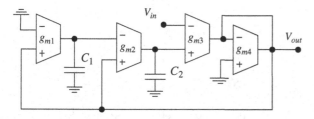

Fig. 6.67 HP section

Most of the previously discussed structures based on one, two, or three op-amps have corresponding realizations based on transconductors. Numerous alternative realizations of second-order section can be found in [32].

6.9.3 *Current Conveyors-Based Sections*

In this section, we will discuss two examples of second-order sections that are based on current conveyors of type II. Additional realizations of

second-order section can be found in [27, 32, 118, 148, 149].

Note that in integrated circuits, it is advantageous to only use grounded passive elements because parasite capacitances then have less effect and the tuning becomes simpler.

6.9.3.1 LP Section

Figure 6.68 shows a second-order section of low-pass type. The active element is here used as an inverting current amplifier as the *y*-port is grounded.

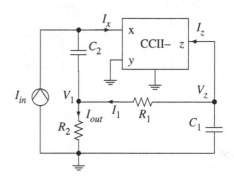

Fig. 6.68 Section-order LP section

We have

$$V_z - V_1 = R_1 I_1$$

$$I_z = -I_x$$

$$V_z = -\frac{I_1 + I_z}{sC_1}$$

$$I_{out} = I_{in} - I_x + I_1$$

$$V_1 = R_2 I_{out}$$

$$R_2 I_{out} = -\frac{I_{in} - I_x}{sC_2}.$$

Elimination yields the transfer function of the section

$$H(s) = \frac{I_{out}}{I_{in}} = \frac{1}{C_1 C_2 R_1 R_2 s^2 + (R_1 + R_2)C_1 s + 1}. \quad (6.97)$$

6.9.3.2 General Section

A general second-order section that operates in voltage mode is shown in Fig. 6.69 and can realize LP, HP, BP, notch, and allpass sections [148].

The section has only three current conveyors and five passive elements, i.e., a minimal number. It has a low impedance output and can therefore easily be cascaded. Furthermore, the last conveyor operates as current amplifier and can in practice be realized easier than a complete CCII+. The component spread is proportional to the Q factor, i.e., the spread is small. This means that the total capacitance for the section also becomes small.

The transfer function for the section shown in Fig. 6.69 is obtained by selecting suitable input signals and from the relation

$$V_{out} = \frac{R_2 V_{in1} - sC_1 R_1 R_2 V_{in2} + (s^2 C_1 C_2 R_1 R_2 R_3 + sC_1 R_1 R_3) V_{in3}}{s^2 C_1 C_2 R_1 R_2 R_3 + sC_1 R_2 R_3 + R_2}.$$

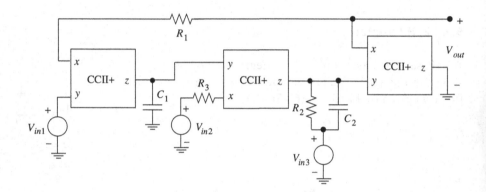

Fig. 6.69 General second-order section

By selecting the input signals and components, we get the transfer functions shown below.

Section type	$V_{in}1$	$V_{in}2$	$V_{in}3$	
LP	V_{in}	0	0	
HP	0	V_{in}	0	$R_2 = R_3$
BP	0	V_{in}	0	
Notch	V_{in}	V_{in}	V_{in}	$R_2 = R_3$
AP	V_{in}	V_{in}	V_{in}	$R_2 = 2R_3$

The pole radius and Q factor are

$$r_p^2 = \frac{1}{C_1 C_2 R_1 R_3} \qquad Q = R_2 \sqrt{\frac{C_2}{C_1 R_1 R_3}} \qquad (6.98)$$

The passive sensitivities, which are low, are

$$S_{C_1}^{r_p} = S_{C_2}^{r_p} = S_{R_1}^{r_p} = S_{R_3}^{r_p} = -\frac{1}{2}$$

$$S_{C_1}^{Q} = S_{R_1}^{Q} = S_{R_3}^{Q} = -S_{C_2}^{Q} = -\frac{1}{2}.$$

See [77] for second-order sections that can be realized with controllable conveyors, i.e., CCCII±.

6.10 Sections with Finite Zeros

The KHN section realizes transfer functions of LP, BP, and HP type whereas Tow-Thomas and Åkerberg-Mossberg sections only realize transfer functions of BP and LP type. To realize a, section with arbitrary finite zeros, two different methods are used.

6.10.1 Summing of Node Signals

Figure 6.70 shows one of two common schemes to generate finite zeros, which are formed by a weighted sum of the signals in three independent nodes. The transfer function is

$$H(s) = \frac{fs^2 + \dfrac{gs}{\tau_1} + \dfrac{e}{\tau_1 \tau_2}}{s^2 + \dfrac{as}{\tau_1} + \dfrac{b}{\tau_1 \tau_2}}. \qquad (6.99)$$

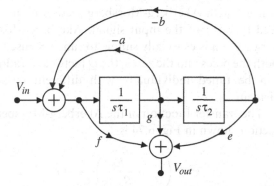

Fig. 6.70 Two-integrator loop with finite zeros

A drawback with this method is that it requires one extra operational amplifier to perform the sum of the node voltages.

Fig. 6.71 shows a typical example of a second-order section with finite zeros. The transfer function is obtained from Fig. 6.71 and from Section 6.8.2. We get

$$H(s) = \left(\frac{-R_{10}}{R_7}\right)H_{HP} + \left(\frac{-R_{10}}{R_8}\right)H_{BP} + \left(\frac{-R_{10}}{R_9}\right)H_{LP}$$

and

$$H(s) = \frac{-R_{10}G\left(\dfrac{s^2}{R_7} + \dfrac{r_p s}{R_8} + \dfrac{r_p^2}{R_9}\right)}{s^2 + \left(\dfrac{r_p}{Q}\right)s + r_p^2}. \qquad (6.100)$$

The section can only realize zeros on the $j\omega$-axis or in the left half plane because only positive coefficients can be realized in the numerator polynomial. Negative coefficients can be realized by using the positive input of the amplifier, as was done in the circuit shown in Fig. 6.54.

6.10.2 Injection of the Input Signal

Another and better method is to inject the input signal into several nodes in the section as shown in Fig. 6.72. This requires no extra operational amplifier, which reduces both the cost and power consumption. Note that the structure in Fig. 6.72 is the transpose of the structure shown in Fig. 6.70 and it

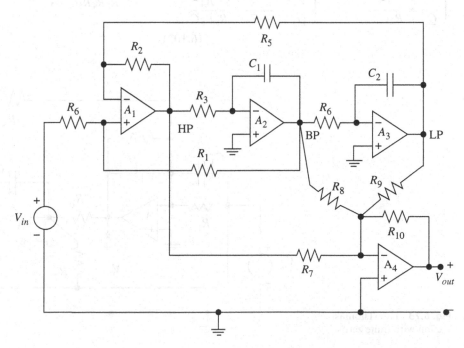

Fig. 6.71 KHN section with zeros realized by summing node voltages

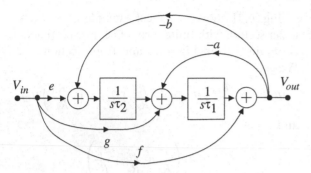

Fig. 6.72 Two-integrator loop with finite zeros

has the same transfer function. Figure 6.73 shows how arbitrary zeros can be realized with the Tow-Thomas section by feed forward of the input signal. In Fig. 6.73, the order of the inverter and the integrator has been changed due to the transposition.

The transfer function to the Tow-Thomas section with injection of the input signal into two nodes becomes

$$H(s) = G \cdot \frac{s^2 + \dfrac{r_z}{Q_z}s + r_z^2}{s^2 + \dfrac{r_p}{Q_p}s + r_p^2} \qquad (6.101)$$

where $G = -\dfrac{C_3}{C_1}$ and

$$\begin{cases} r_p^2 = \dfrac{R_4}{R_3 R_5 R_6 C_1 C_2} \\[3mm] \dfrac{r_p}{Q_p} = \dfrac{1}{R_2 C_1} \end{cases} \begin{cases} r_z^2 = \dfrac{R_4}{R_3 R_5 R_7 C_2 C_3} \\[3mm] \dfrac{r_z}{Q_z} = \dfrac{1}{R_1 C_3} - \dfrac{R_4}{R_8 R_5 C_3}. \end{cases} \qquad (6.102)$$

This section can realize zeros on both sides of the $j\omega$-axis. It is only necessary to use either R_1 or R_4, depending on if the zeros should be in the left- or right- hand side of the s-plane. For zeros on the $j\omega$-axis, we may select $R_1 = R_8 = \infty$.

The element values for all sections should be chosen so the sensitivities become low and that the signal levels at the operational amplifier's output become optimal. An advantage is that the poles and zeros can be trimmed independently.

Figure 6.74 shows how arbitrary zeros can be realized with Åkerberg-Mossberg section by using feed forward of the input signal. Åkerberg-Mossberg section is especially simple to tune because for both the poles and the zeros, the Q factor and radius can be tuned individually with different circuit elements.

The transfer function of the Åkerberg-Mossberg section shown in Fig. 6.74 is

$$H(s) = \frac{-C_3}{C_1} \cdot \frac{s^2 + \left(\dfrac{1}{R_1} - \dfrac{R_6}{R_5 R_8}\right)\dfrac{s}{C_3} + r_z^2}{s^2 + \dfrac{r_p}{Q}s + r_p^2} \qquad (6.103)$$

where

$$r_p^2 = \frac{R_6}{R_3 R_4 R_5 C_1 C_2} \quad Q = R_2 C_1 r_p \quad r_z^2 = \frac{R_6}{R_4 R_5 R_7 C_2 C_3}.$$

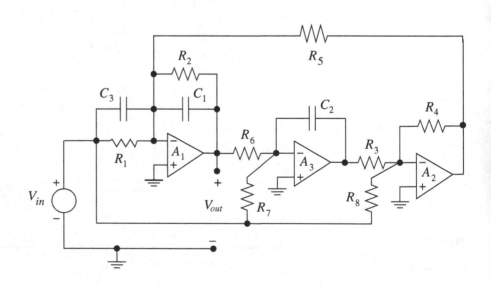

Fig. 6.73 Tow-Thomas section with finite zeros

Fig. 6.74 Åkerberg-Mossberg section with arbitrary zeros

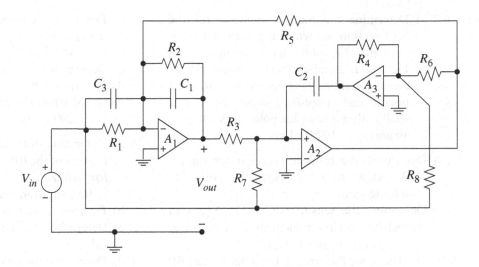

Example 6.8 Select the component values in an Åkerberg-Mossberg section so that

a) A section with the poles corresponding to $Q = 10$ and $r_p = 20\,\mathrm{krad/s}$ and zeros on the $j\omega$-axis at $r_z = 25\,\mathrm{krad/s}$ is realized.
b) An allpass section with the same poles is realized.
a) We select all capacitors equal, i.e., $C_1 = C_2 = C_3 = 10\,\mathrm{nF}$, which makes the gain equal with -1. The condition on the Q factor gives $R_2 = Q/C_1 r_p = 50\,\mathrm{k\Omega}$. From the ratio of the pole and zero radii, we obtain $\left(\dfrac{r_p}{r_z}\right)^2 = \dfrac{R_6 R_4 R_5 R_7 C_2 C_3}{R_3 R_4 R_5 C_1 C_2 R_6} = \dfrac{R_7 C_3}{R_3 C_1}$ and we select $R_7 = 10\,\mathrm{k\Omega}$, which gives $R_3 = 1.5625\,\mathrm{k\Omega}$.

Furthermore, we select $R_5 = R_6 = 10\,\mathrm{k\Omega}$, which yields the condition for the pole radius that $R_4 = 1.6\,\mathrm{k\Omega}$. Insertion of the selected values in the expression for the radius of the zero gives the correct value. The real part of the zero should be 0, hence we select $R_1 = 10\,\mathrm{k\Omega}$, which gives $R_8 = 10\,\mathrm{k\Omega}$.

b) For an allpass filter, we must have $r_p = r_z$, which is why we select $R_3 = R_7 = 10\,\mathrm{k\Omega}$. However, the pole radius is not changed. Further, we have that the real parts of the poles and zeros shall be equal, but with different signs. Note that R_1 and R_8 only affect the zero's real part. We can thus select $R_1 = \infty$ and $R_8 = R_2 = 50\,\mathrm{k\Omega}$. The other components are not affected.

6.10.2.1 Signal Injection Through Grounded Elements

A common method to inject input signals into a network is to place the input voltage source in series with a grounded impedance. An example of this method is the structure shown in Fig. 6.21. Often, only a fraction of the grounded impedance is "lifted from ground" and the remaining part is kept at ground.

6.11 Problems

6.1 Determine the maximum of the magnitude function for a second-order LP section. Assume that $Q > 1/\sqrt{2}$.

6.2 Determine the angular frequency for the peak of the magnitude function for a section with the poles and zeros shown in Fig. 6.3.

6.3 Determine the maximum of the magnitude function for a second-order HP section. Assume that $Q > 1/\sqrt{2}$.

6.4 Determine the coefficients a_0, a_1, and a_2 so that the transfer function given by Equation (6.7) represents a

a) lowpass
b) highpass
c) bandpass
d) allpass
e) lowpass-notch
f) highpass-notch section.

6.5 Determine the Q factors for the poles of a Butterworth, Chebyshev I, and a Cauer filter with a passband ripple corresponding to $\rho = 5\%$ and $\omega_c = 10\,\mathrm{krad/s}$.

6.6 a) Determine the transfer function of the circuit in Fig. 6.5.
   b) Determine the poles and zeros and DC gain.
   c) Determine the group delay.
   d) Determine suitable element values so that $\tau_g(0) = 200\,\mu\mathrm{s}$.

6.7 a) Determine the transfer function for the NF1 section, shown in Fig. 6.19, when the operational amplifier has finite gain.

b) Determine suitable element values and the error function $E(s)$ assuming an ideal operational amplifier when the circuit shall realize a complex pole pair with $Q = 10$ and $r_p = 10\,\text{Mrad/s}$.

6.8 Determine the transfer function for an LP NF1 section with an operational amplifier with finite gain.

6.9 Determine the sensitivities for Y, S_x^Y, with respect to a circuit element x, i.e., when $Y = x^k$ for $k = \pm 0.5$ and ± 1.

6.10 a) Determine the error in the poles for the BP section shown in Fig. 6.21 when
$$C_7 = C_9 = C, R_6 = R, R_3 = \frac{R}{4Q^2}, r_p = 2\frac{Q}{RC}.$$

b) Determine suitable element values for realization of a complex pole pair with radius $4\frac{1}{4}$ Mrad/s and $Q = 5$.

c) Determine the peak gain.

d) Determine the spreads in the element values.

e) Determine the change in the pole radius and Q factor when the operational amplifier has $\omega_t = 200\ \frac{1}{4}\,\text{Mrad/s}$ when $GS_{A_0}^Q = 2Q^2$.

6.11 Determine $S_{R_3}^{r_p}, S_{R_7}^{r_p}, S_{R_9}^{r_p}, S_{C_6}^{r_p}$ och $S_{C_8}^{r_p}$ and $S_{R_3}^{Q}, S_{R_7}^{Q}, S_{R_9}^{Q}, S_{C_6}^{Q}$ och $S_{C_8}^{Q}$ with respect to the passive circuit elements for the LP NF1 section shown in Fig. 6.18 with the following choice of element values $R_3 = R_7 = R_9 = R$, $C_8/C_6 = 9Q^2$, and $C_8 = C$.

6.12 Show that $S_x^{|H(j\omega)|} = Re\left\{S_x^{H(j\omega)}\right\}$ and $S_x^{\Phi} = \frac{1}{\Phi(j\omega)} Im\left\{S_x^{H(j\omega)}\right\}$.

6.13 Determine the gain-sensitivity product for a section that realizes the transfer function

$$H(s) = \frac{G}{as^2 + bs + c + \dfrac{ds^2 + es + f}{A}}$$

6.14 Determine the gain-sensitivity product for an NF1 LP section.

6.15 a) Determine the transfer function of the NF2 section shown in Fig. 6.25.

b) Determine suitable element values for realizing a complex pole pair with $Q = 20$ and $r_p = 4\pi\,\text{Mrad/s}$.

c) Determine the spreads in element values.

d) Determine the errors in pole radius and Q factor when the operational amplifier has $\omega_t = 200\ \pi\,\text{Mrad/s}$ when $GS_{A_0}^Q = 2/a$.

6.16 a) Determine suitable element values for realization of the BP section shown in Fig. 6.24 for realizing a complex pole pair with radius 4π Mrad/s and $Q = 5$.

b) Determine the peak gain.

c) Determine the spreads in the element values.

d) Determine the errors in pole radius and Q factor when the operational amplifier has $\omega_t = 200\ \pi\,\text{Mrad/s}$ when $GS_{A_0}^Q = 2/a$.

6.17 Estimate and compare the relative errors in pole radius and Q factor for a BP realized with NF1 and NF2 sections when the pole radius is $r_p = 40\pi\,\text{Krad/s}$ and $Q = 10$ and the op-amp has $\omega_t = 2\pi\ 10\,\text{Mrad/s}$.

6.18 Determine the transfer function of PF1 section shown in Fig. 6.27 when the operational amplifier has finite gain.

6.19 a) Determine the transfer function of the section shown in Fig. 6.28 when the operational amplifier has finite bandwidth.

b) Determine suitable element values for realization of a complex pole pair with pole radius 4π Mrad/s and $Q = 5$.

c) Determine the peak gain.

d) Determine the spreads in the element values.

e) Determine the change in pole radius and Q factor when the operational amplifier has $\omega_t = 200\ \pi\,\text{Mrad/s}$ when $GS_{A_0}^Q = 2Q$.

f) Determine and compare the gain-sensitivity products for the two choices of element values in Problem 6.19 a).

6.20 Determine suitable element values for the Sallen-Key LP section shown in Fig. 6.28 (PF1) to realize a pole pair with $Q = 5$, $r_p = 2\pi$ krad/s, and DC gain $= 2$. Select $R_1 = R_3$ and $C_4 = C_6$.

6.21 a) Determine suitable element values in a PF1 BP section to realize pole pair with radius 4π Mrad/s and $Q = 5$.

b) Determine the peak gain.

c) Determine the spreads in the element values.

6.22 a) Determine the transfer function of the PF1 section shown in Fig. 6.28.

b) Determine the sensitivity of r_p for variation in the passive elements.

c) Determine the sensitivity of σ_p for variation in the passive elements.

d) Determine the sensitivity of σ_p to the amplifier DC gain.

e) Determine the passive sensitivities when the section realizes a pole pair with $Q = 10$ and $r_p = 4\pi$ Mrad/s.

6.23 a) Determine the transfer function of the branching filter shown in Fig. 6.75.

b) Determine suitable element values so that the crossover frequency becomes 1.7 kHz.

6.27 Derive the transfer function for the HP section in Fig. 6.47 and identify the corresponding *RLC* network.

6.28 Determine suitable element values for a HP GIC second-order section to realize a complex pole pair with $Q = 1/\sqrt{2}$ and $r_p = 20$ krad/s.

6.29 Determine the transfer function for the section shown in Fig. 6.76.

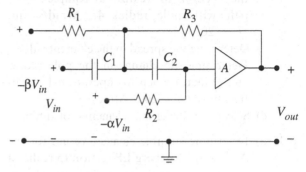

Fig. 6.76 Circuit in problem 6.29

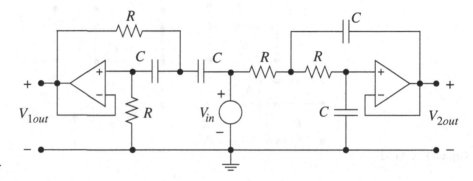

Fig. 6.75 Branching filter

6.24 a) Determine suitable element values for realization of a PF1 BP section with a complex pole pair with the pole radius 4π Mrad/s and $Q = 5$.

b) Determine the peak gain.

c) Determine the spreads in the element values.

6.25 Determine the sensitivity of Q and r_p for variations in C_6 in the PF2 section shown in Fig. 6.31.

6.26 a) Derive the transfer function for the LP section in Fig. 6.45 when the operational amplifiers have very large bandwidths.

b) Determine suitable element values to realize a complex pole pair with $Q = 1/\sqrt{2}$ and $r_p = 20$ krad/s.

6.30 a) Determine suitable element values for a KHN LP section to realize a complex pole pair with pole radius 4π Mrad/s and $Q = 20$

b) Determine the spread in the element values.

c) Determine the change in the pole radius and Q factor when the operational amplifiers have $\omega_t = 200\,\pi$ Mrad/s.

d) Select suitable circuit elements for tuning.

6.31 Determine the transfer function of the Tow-Thomas section when $R_1 = R_2 = R_7 = R_8 = 2R$ and $R_3 = R_4 = R_5 = R_6 = R$ and $C_1 = C_2 = C_3 = C$.

6.32 Derive Equation (10.9).

6.33 Estimate the errors in Q and r_p when the Tow-Thomas section is implemented with 2% resistors and 5% capacitors.

6.34 a) Determine the transfer function for the Tow-Thomas BP section, also called a DIG section (DIG; distributed infinite gain), shown in Fig. 6.73.
b) Determine suitable element values with the section to realize a complex pole pair with pole radius 4π Mrad/s and $Q = 20$.
c) Determine the spread in the element values.
d) Determine the change in the pole radius and Q factor when the operational amplifiers have $\omega_t = 200\pi$ Mrad/s.
e) Select suitable circuit elements for tuning.

6.35 a) Determine suitable element values for the Åkerberg-Mossberg LP section to realize a

complex pole pair with pole radius 4π Mrad/s and $Q = 20$.
b) Determine the spread in the element values.
c) Determine the change in the pole radius and Q factor when the operational amplifiers have $\omega_t = 200 \pi$ Mrad/s. Compare with Problem 6.30.
d) Select suitable circuit elements for tuning.

6.36 Burr-Brown manufactures an integrated circuit (UAF42) with the structure shown in Fig. 6.77. Determine the types of sections that can be realized when $C = 1$ nF $\pm 0.5\%$ and $R = 50$ k$\Omega \pm 0.5\%$.

6.37 a) Determine the transfer functions V_2/V_1 and V_3/V_1 and crossover frequency when the switches are open for the audio system shown in Fig. 6.78.
b) Determine the transfer functions V_2/V_1, V_3/V_1, and V_4/V_1 and crossover frequency when

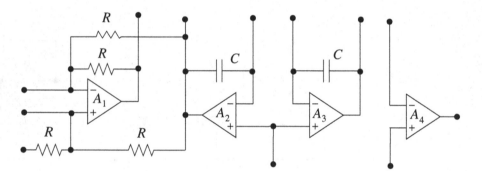

Fig. 6.77 UAF42

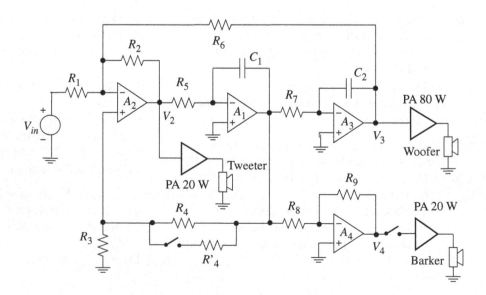

Fig. 6.78 Audio system

the switches are closed. $R_1 = R_2 = R_4 = R_6 = 22$ kΩ, $R_3 = R_5 = R_7 = R_8 = R_9 = 11$ kΩ, $R'_4 = 29$ kΩ, $C_1 = C_2 = 4.7$ nF. TL084 is a suitable choice for the amplifiers.

6.38 Determine suitable element values for the a three op-amp section that realizes the poles and zeros in a second-order notch section with $Q = 5$ and the notch frequency of 50 Hz. Use the KHN section.

6.39 Determine suitable element values for a three op-amp section that realizes the poles and zeros in a second-order notch section with $Q = 5$ and the notch frequency of 50 Hz. Use the Åkerberg-Mossberg section.

6.40 a) Estimate the errors in Q and r_p when the Åkerberg-Mossberg section is implemented with 2% resistors and 5% capacitors and ideal operational amplifiers.

b) Choose one resistor to tune r_p and determine its required range in order to tune r_p to its desired value.

c) Choose another resistor to tune Q and determine its required range in order to tune Q to its desired value.

6.41 Repeat Problem 6.40 for Tow-Thomas section.

6.42 Estimate the upper limit on r_p for the Tow-Thomas section when the error in r_p caused by the operational amplifiers must be less than 0.6%. The operational amplifiers are matched and have a bandwidth of 3 MHz.

6.43 Repeat Problem 6.42 for a KHN section.

6.44 Repeat Problem 6.42 for Åkerberg-Mossberg section.

6.45 An active RC filter has the following normalized element values: $R_1 = 1.34$, $R_2 = 0.713$, $C_1 = 1.0$, and $C_2 = 0.1$. Denormalize the element values so that the passband edge occurs at 2.2 kHz and $R_2 = 12$ kΩ.

6.46 Derive the sensitivities for r_p and Q for a lossless two-integrator loop.

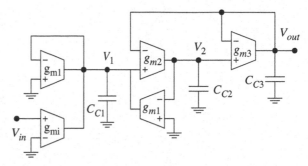

Fig. 6.79 Transconductor-based filter in problem 6.47

6.47 Determine the transfer function of the circuit shown in Fig. 6.79.

6.48 Determine the transfer function of the circuit shown in Fig. 6.80.

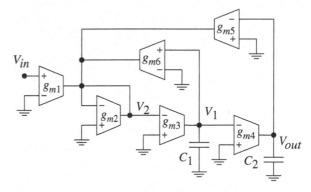

Fig. 6.80 Transconductor-based filter in problem 6.48

6.49 Determine the transfer function of the circuit shown in Fig. 6.81.

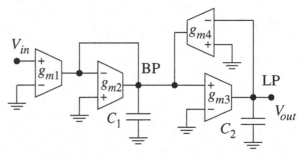

Fig. 6.81 Transconductor-based filter in problem 6.49

Chapter 7
Coupled Forms

7.1 Introduction

The first publication on the active *RC* filter (*ARC*) appeared in 1938 (Scott). J.G. Linvill (1954) is one of the first pioneers of modern active filter theory. At first, electron tubes were used as amplifying elements. They were expensive and had very large power consumption. Filter circuits with only one amplifying element were therefore preferred, but they turned out to be sensitive for variations in resistance and capacitance values and particularly sensitive to the gain of the electron tubes. In practice, it as not possible to design usable active *RC* filters of higher order because of the very high sensitivities. This has led to extensive research to find active filter structures that are less sensitive to the component errors.

High-order filters have inherently much higher element sensitivity compared with the first- and second-order sections discussed in Chapter 6. This can be explained by considering the denominator of a transfer function when the poles lie very close.

Consider the denominator $D(s)$ of a high-order analog filter

$$D(s) = \sum_{k=0}^{N} d_k s^k = \prod_{j=1}^{N} (s + s_{pj}).$$

Taking the derivative of both sides

$$\frac{\partial D}{\partial d_k} = \frac{\partial D \partial s_{pn}}{\partial s_{pn} \partial d_k}$$

yields

$$s^k = \prod_{\substack{j=1 \\ j \neq n}}^{N} (s + s_{pj}) \frac{\partial s_{pn}}{\partial d_k}$$

and

$$\frac{\partial s_{pn}}{\partial d_k} = \frac{s}{\prod_{\substack{j=1 \\ j \neq n}}^{N} \left(1 + \frac{s_{pj}}{s}\right)} \quad \text{for } s = -s_{pn}. \quad (7.1)$$

If the poles are clustered as for narrow-band filters, the factors in the denominator of Equation (7.1) will be small and the derivative will be large. Furthermore, if the degree N of the filter is large, the denominator will contain many small factors, which yields high sensitivity with respect to the coefficient errors.

Example 7.1 Consider a transfer function with the following poles

$$s_{p1,2} = -1 \pm 0.1 i \text{ and } s_{p3,4} = -0.99 \pm 0.1 i.$$

The corresponding denominator is

$$D(s) = s^4 + 3.98 s^3 + 5.9601 s^2 + 3.98 s + 1.000001.$$

Now, assume that errors occur in the circuit and the denominator is changed to

$$D(s) = s^4 + 3.98 s^3 + 5.9601 s^2 + 3.98 s + 1.000000.$$

The new poles are

$$s_{p1,2} = -0.99499985746603237 \pm 0.0998763403064299973 i$$
$$s_{p3,4} = -0.99500014253396762 \pm 0.09987350322800613 i.$$

Hence, an error in the sixth decimal in the denominator results in errors in the poles in the third decimal. The sensitivity is indeed very large. In addition, it was shown above that the sensitivity increases for poles that lie clustered, i.e, very close, and with the number of poles. This phenomenon is well known from numerical analysis; it is difficult to compute accurately the roots of a high-order polynomial, especially if the roots are clustered.

L. Wanhammar, *Analog Filters Using MATLAB*, DOI 10.1007/978-0-387-92767-1_7,
© Springer Science+Business Media, LLC 2009

7.2 Taxonomy for Analog Filters

Many different filter structures have been proposed in order to overcome the sensitivity problem. Fig. 7.1 shows a taxonomy that includes the main analog filter structures and their relations.

In the literature, contradictory statements about the element sensitivities of coupled forms are commonly made. Some authors claim that the sensitivities are low whereas others claim that they are high. The coupled forms use feedback in order to reduce the sensitivities, but it is not clear how the feedback

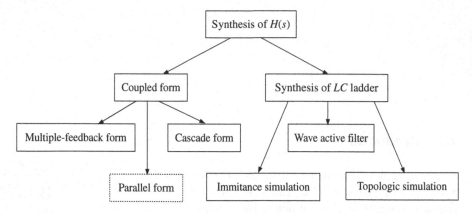

Fig. 7.1 Taxonomy for analog filter structures

An early approach to reduce the sensitivity problem was to realize a high-order transfer function by a cascade of lower-order filters. This approach is now known as the cascade form. The element sensitivity for the whole filter can be reduced by reducing the number of factors, i.e., N in Equation (7.1), and selecting poles that are far apart. In practice, only sections of the first and second order are used.

The cascade form is, however, not very good from a sensitivity point of view and it should only be used in simple applications with relatively low filter requirements. For filters of higher order and for high requirements on the frequency selectivity, more advanced filter structures are recommended. The parallel form is only of theoretical interest and it is not recommended because of its poorer stopband sensitivity.

The search for active filter structures with low element sensitivity, with respect to both passive and active circuit elements, has mainly followed along two development tracks.

7.2.1 Coupled Forms

The first approach is to start from a cascade of the first- and second-order sections and introduce feedback between these sections, in order to reduce the element sensitivity. This class of filter structures, which is referred to as *coupled forms*, will be described in detail in Section 7.5.

shall be introduced and optimized to reduce the sensitivities. Hence, the sensitivity properties may largely depend on the skill of the designer.

Cascade and parallel forms are special cases of coupled forms. The parallel form, however, should not be used, and the cascade form should only be used for simpler applications, i.e., filters of relative low-order and with low to medium high Q factors. However, the cascade form is often used in industry because it is erroneously perceived to be simple to design.

7.2.2 Simulation of Ladder Structures

The second approach starts from an LC network that has minimal element sensitivity, i.e., a doubly resistively terminated LC ladder network. This network can be simulated with a variety of methods using active components. If the simulation technique retains the passivity and maximum power transfer properties of the doubly terminated LC network, then the sensitivity properties will also be retained in the active counterpart. These methods, which are recommended if the requirements are strict, will be discussed in Chapters 8, 9 and 10.

With active filters, we cannot realize any new types of transfer functions compared with passive LC filters, except for unstable filters, which of course only are usable as oscillators. It is only the technologies for the implementations that are different. The filtering becomes the same, though.

7.3 Cascade Form

In this and the following sections, we discuss different methods to realize higher-order active RC filters. The cascade form is one of the simplest and most common realization forms. Because of its relatively large element sensitivity, it is, however, only recommended for simple applications. Typically, it should only be used for filters with low or medium high Q factors and orders less than six to eight. The element sensitivity is relatively high in the passband, but it is low in the stopband.

To realize a filter in cascade form, the numerator and denominator are factorized into first- and second-order polynomials. One pole (zero) on the real axis in the s-plane corresponds to a first-order polynomial. A complex-conjugate pole (zero) pair corresponds to a second-order polynomial. A general transfer function can thus be factorized according to Equation (7.2)

$$H(s) = \frac{(b_0 s + c_0)(a_1 s^2 + b_1 s + c_1)\ldots(a_M s^2 + b_M s + c_M)}{(s - \sigma_0)(s^2 - 2\sigma_{p1}s + r_{p1}^2)\ldots(s^2 - 2\sigma_{pM}s + r_{pM}^2)}. \tag{7.2}$$

By collecting complex-conjugated poles and zeros, the transfer function can be written as a product of the first- and second-order transfer functions

$$H(s) = H_0(s)H_1(s)\ldots H_M(s) \tag{7.3}$$

where

$$H_0(s) = \frac{b_0 s + c_0}{s - \sigma_0} \tag{7.4}$$

and

$$H_i(s) = \frac{a_i s^2 + b_i s + c_i}{s^2 - 2\sigma_{pi}s + r_{pi}^2} \quad i = 1, 2, \ldots, M. \tag{7.5}$$

A transfer function that is factorized into a product of partial transfer functions can be realized by a cascade connection of the corresponding filter sections, as shown in Fig. 7.2. Normally, only second-order sections are used, except for filters of odd-order where one first-order section is required.

The input impedance to each section should be sufficiently large, and the output from each section is taken from an amplifier with low output impedance. This allows the sections to be tuned independently, as the sections do not interact. The trimming is therefore much easier than for filter structures where all components interact.

The design of an active RC filter in cascade form begins by factorizing the transfer function according to Equation (7.3) and then realizing the sections according to the previous chapters. The order in which the sections are cascaded is in principle arbitrary, but in practice their ordering is very important because the order determines the signal levels in and between the sections. We will discuss these issues below.

The cascade form is the most commonly used filter structure, due to its perceived simplicity, in applications with low requirements, i.e., for relatively broad banded filters with low or medium high Q factors.

An advantage, which is shared with many other structures, is the possibility to use a small or minimal number of operational amplifiers. The element sensitivity for the cascade form is essentially equal to the sum of the sensitivities of the individual sections. This means that the sensitivities of the individual sections should be minimized while the number of operational amplifiers should be small. Therefore, sections with only one operational amplifier are used if the Q factors are low. At medium high Q factors, two- or three-op-amp sections may have to

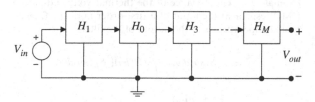

Fig. 7.2 Cascade form

be used. The cascade form should, however, not be used if the filter contains sections with large Q factors. Another important factor at the selection of filter structures is how easy it is to tune. The cascade form is easy to tune because we can tune the poles and the zeros in each section independently.

7.3.1 Optimization of Dynamic Range

Useful signals in active RC filters must be larger than the noise floor and smaller than the power supply voltage. The dynamic signal range is the ratio of the maximum amplitude (rms) of a sinusoidal signal in the passband that does not saturate the amplifiers[1] and the noise voltage (rms). This ratio is called SNR (signal-to-noise-ratio). It is not important to have high SNR in the stopband because these frequencies will be removed. An active RC filter for audio typically has a dynamic signal range of only 80–90 dB. It should be recognized that it is expensive in terms of power consumption to achieve large dynamic range, as well as large bandwidths.

In active RC filters, the output voltage of each amplifier will vary for different frequencies and typically have a peak at some frequency. The variation is determined by the transfer function from the filter input to the output of the amplifier. To maximize the dynamic range, the maximal output voltage of each amplifier should be made equal and correspond to the desired gain of the filter. The design of any filter structure should therefore ensure that the peaks are the same for all amplifiers.

7.3.2 Thermal Noise

All resistors in a circuit generate thermal noise. Thermal noise, also called Johnson noise, appears as a noise voltage in series with the resistor as shown in Fig. 7.3. It has a Gaussian power density function, *pdf,* and a constant power density spectrum, i.e., a white spectrum. The polarity of the noise

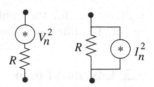

Fig. 7.3 Noise model for a resistor

source is not important because its value is squared anyway. In some cases, a current source in parallel with the resistor may be useful where $\overline{I_n^2} = \overline{V_n^2/R^2}$.

Traditionally, the power density spectrum of the noise for a resistor is given in terms of frequency, i.e., $S(jf) = 4kTR$ [V^2/Hz], where $k = 1.3806504$ 10^{-23} [J/K] is Boltzmann's constant, T is the absolute temperature, and R is the resistance of the resistor.

We, however, prefer to use angular frequency, i.e., the power density spectrum with zero mean is

$$S_R(j\omega) = \frac{2kTR}{\pi} [\text{V}^2/\text{rad}]. \qquad (7.6)$$

The power spectrum density of a 1 MΩ resistor at room temperature is

$$S_R(j\omega) = 2.6368 \cdot 10^{-15} [\text{V}^2/\text{rad}].$$

The power spectrum at the output of a noiseless filter, with the transfer function, $H(s)$, is

$$S(j\omega) = |H(j\omega)|^2 S_R(j\omega). \qquad (7.7)$$

If there is more than one resistor, their individual power spectrum densities at the output can be added, as the thermal noise sources are uncorrected.

The variance of the thermal noise voltage is

$$\overline{v^2} = \int_0^\infty S(j\omega)d\omega. \qquad (7.8)$$

Example 7.2 The variance of the thermal voltage due to a 1 MΩ resistor at the input of a first-order noiseless filter at room temperature is at the output of the filter

$$\overline{v^2} = \int_0^\infty S(j\omega)d\omega = \int_0^\infty |H(j\omega)|^2 S_R(j\omega)d\omega$$

$$= \int_0^\infty \frac{2kTRa^2}{\pi(\omega^2 + a^2)} d\omega = 2.6025 \cdot 10^{-8} \text{ V}^2$$

---
[1] For higher frequencies, slew rate-limitation of the outputs of the amplifiers may occur.

and $\bar{v} = 161$ nV when $H(s) = \dfrac{a}{s+a}$, $a = 2\pi$ Mrad/s, $T = 300$ K.

If the maximal output voltage of the filter is 1.7 V, the dynamic range (signal-to-noise ratio) becomes $10\log\left(\dfrac{1.7^2}{2.6025 \cdot 10^{-8}}\right) = 80.46$ dB. Hence, minimization of the noise is important in most cases.

Thermal noise is a lesser problem in active RC filters that are implemented with discrete components because the capacitances can be chosen relatively large, which reduces the resistances, as the frequency responses are proportional to $1/RC$. In integrated filters, the capacitances cannot be selected too large because this will be expensive in terms of chip area. Typically, the total capacitance is limited for economic reasons to a few hundred pF.

It can be shown that the noise in analog filters is

$$v^2 \propto \frac{1}{C}. \qquad (7.9)$$

Hence, making the capacitors larger will reduce the noise, but the resistors must be reduced with the same factor in order to not change the frequency response. A lower impedance level results in higher power consumption. Moreover, an increase of the required SNR with 3 dB results in a doubling of the power consumption. Techniques to reduce the power consumption using nonlinear circuits have been proposed [98, 129].

7.3.2.1 Flicker Noise

There will also be a $1/f$ (or $1/\omega$) noise component, so-called flicker noise, in the power density spectrum if a DC current flows through the resistor whose magnitude is proportional to the power dissipated in the resistor. Hence, the power density spectrum has the form

$$S_R(j\omega) = \frac{2kTR}{\pi} + c\frac{I^2}{\omega} \qquad (7.10)$$

where c depends on the material and physical design of the resistor [63].

Minimizing the resistance values will reduce the thermal noise due to the passive components, but a typical operational amplifier cannot drive an impedance that is smaller than a few kΩ. Note that the resistance value used above is large.

7.3.3 Noise in Amplifiers

An amplifier also generates noise that is filtered by the circuit [89, 103]. This noise has typically a variance in the range 10–100 μV and it is often larger than the noise generated by the resistors. Hence, the power spectrum for the noise at the filters output depends on circuit structure and the different noise sources. Typically, the noise spectrum will be largest at, or slightly outside, the passband edge.

The noisy operational amplifier is generally represented as a noiseless amplifier, with noise voltage and current sources at the input as shown in Fig. 7.4

Fig. 7.4 Noise model for an operational amplifier

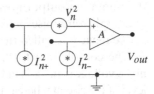

The operational amplifier is assumed ideal in all other respects, i.e., has infinite input impedance and zero output impedance. In addition, we assume that the noise sources are assumed uncorrelated and that the statistics of the noise sources are time-invariant. The magnitude of the noise sources depend on the technology used and differ depending on if MOSFET or bipolar transistors are used [63]. The interested reader is therefore directed to the vendors for further information.

The noise voltage (V_n) and noise current (I_n) power spectral densities are of the forms

$$S_V(j\omega) = S_{V0}\left(1 + \frac{\omega_V}{\omega}\right)$$

$$S_I(j\omega) = S_{I0}\left(1 + \frac{\omega_I}{\omega}\right)$$

where S_{V0}, S_{I0}, ω_V, and ω_I are constants. That is, the noise voltage and noise current power spectral densities have a $1/\omega$ component. Typical orders of magnitude for a CMOS operational amplifier are $S_{V0} \approx 2 \, 10^{-15}$, $S_{I0} \approx 10^{-31}$, $\omega_V \approx 10^3$, and $\omega_I \approx 100$.

Then, if the transfer function from the ith noise voltage source to the output is $H_i(j\omega)$, and the transfer impedance from the kth noise current source to the output is $Z_k(j\omega)$, the total output noise voltage spectral density is given by

$$S(j\omega) = \sum_i |H_i(j\omega)|^2 S_{Vi}(j\omega) + \sum_k |Z_k(j\omega)|^2 S_{Ik}(j\omega).$$
(7.11)

The mean-square total output noise voltage is given by Equation (7.8). The computation of the output noise power spectrum becomes tedious because many transfer functions and impedances need to be computed in a filter with many amplifiers [66, 85, 128].

The total output noise can be divided into two components. The component due to the resistors is the minimum output noise possible and is referred to as "inherent noise" and the component due to the active devices is called the "amplifier noise."

The contribution to the output noise voltage by noise voltage sources is independent of the impedance level of the circuit whereas the contributions due to the resistors and the noise current sources are proportional to the impedance level and to the square of the impedance level, respectively. It therefore follows that if a choice of impedance level is possible, the output noise will be at minimum when the contribution due to noise voltage source is dominant.

7.3.4 Noise in Passive and Active Filters

It can be shown that the thermal noise at the output of *LC* filters and the corresponding active counterparts is proportional to the group delay [44]. In fact, it is proportional to the electrical and magnetic energy stored in the capacitors and inductors. The same is also valid for the sensitivity in the passband, as discussed in Section 3.3.9. Hence, Chebyshev II and Cauer filters generate less thermal noise than Butterworth and Chebyshev I filters.

7.3.5 Distortion

Here we have assumed the idealized case that the circuits realizing the transfer function are linear. This is certainly a reasonable assumption for small signal levels, but for larger signal levels the circuits become weakly nonlinear [91, 103, 136]. This causes nonlinear distortion, which appears as a widening of the signal spectrum. For example, a sinusoidal input signal with frequency ω_0 will give rise to a sinusoidal output signal of the same frequency, but also a number of sinusoidal output signals with frequencies $n\omega_0$, where $n = 2, 3,...$ The magnitude of these so-called "spuriouses" are often larger than the noise discussed above. Hence, the dynamic range estimate discussed in the two previous sections is usually too optimistic.

7.3.6 Pairing of Poles and Zeros

In an Nth-order filter with $N/2$ second-order sections, the poles and zeros can be allocated to the sections in several different ways. Note that an analog filter always has N poles and N zeros.

The first pole pair can be combined with one of $N/2$ different finite zero pairs, or zeros at $s = 0$ and $s = \infty$. The next pole pair can be combined with $(N/2 - 1)$ zero pairs, and so on. There are, thus, $(N/2)!$ ways of combining the pole pairs with the zero pairs.

A simple scheme is

- Assign the pole pair, which has the highest Q factor and its closest zero pair, to section #1, as illustrated in Fig. 7.5.
- Assign to section #2 the pole pair that among the remaining poles has the highest Q factor and the closest zero pair among the remaining zeros, and so on.

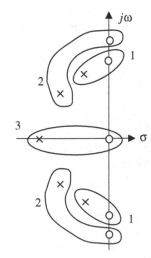

Fig. 7.5 Pairing of poles and zeros

Using this scheme, the variation in the signal level for different frequencies becomes small, but not necessarily minimal, for each section.

Note that, for example, a BP filter may be realized by a cascade of LP, HP, BP, and notch sections depending on how the zeros are allocated to the second-order sections. In some cases, however, it may be favorable to pair poles and zeros into two BP sections instead of a combination of LP and HP sections because the BP sections are easier to tune.

7.3.7 Ordering of Sections

The signal dynamic also depends on the section ordering. After the poles and zeros have been combined into sections, there are $(N/2)!$ orders in which the sections can be cascaded. In all, there are $(N/2)!^2$ different orders and pairings.

A simple rule of thumb, but not necessarily optimal, is

- Place the section with the lowest Q factor first and thereafter increasing Q factors.

A real operational amplifier is not perfect, e.g., there is an offset voltage, i.e., the output voltage has a DC component, which does not originate from the input signal, but comes from the operational amplifier itself. To remove the offset voltage originating from preceding sections, it is often suitable to place a section with a zero at $s = 0$, i.e., an HP or BP section, as the last section.

It may also be suitable to place an LP or BP section as the first section. The reason is to attenuate possible large input signals in the stopband so the following sections will not be driven into saturation or slew rate-limitation. In some cases, there are reasons not to use the optimum order, but instead use a close to optimum order. It is, however, necessary to perform the design to optimize the dynamic signal range. It is advisable to write a computer program to evaluate different alternatives.

Example 7.3 Consider a BP filter meeting the specification $\omega_{c1} = 9.5$ krad/s, $\omega_{c2} = 10.5$ krad/s, $\omega_{s1} = 7.864$ krad/s, $\omega_{s2} = 12.684$ krad/s, $A_{max} = 1.2494$ dB, and $A_{min} = 60$ dB.
The relative bandwidth is $1/10.274 \approx 9.7\%$, i.e., a relatively narrow band filter. The specification is met by a sixth-order BP filter. Modifying the specification in the

MATLAB routine used in Example 2.8, we get for the corresponding Cauer LP filter

$N = 3$
$S_{p0} = -0.4592611$
$S_{p1,2} = -0.2224865$
$\quad\quad \pm j\,0.9533972$
$G = 0.0142881$

$S_{z0} = \infty$
$S_{z1,2} = \pm j\,5.550478$

and for the BP filter

$N = 6$
$s_{z1,2} = \pm j\,13.140996$ krad/s
$s_{z3,4} = \pm j\,7.5905185$ krad/s
$s_{z5} = 0$
$s_{z6} = \infty$
$s_{p1,2} = -0.10593929$
$\quad\quad \pm j\,9.52139476$ krad/s
$s_{p3,4} = -0.11654721$
$\quad\quad \pm j\,10.474792$ krad/s
$s_{p5,6} = -0.22963057$
$\quad\quad \pm j\,9.9847006$ krad/s
$G = -14.288145.$

$Q_1 = 44.940759$

$Q_2 = 44.940759$

$Q_3 = 21.746540$

We select a negative gain factor because we intend to realize the transfer function with three cascaded second-order sections that have negative gain factors.

The Q factors are very high and the filter is indeed challenging to implement. Note that a complex conjugate LP pole pair yields two BP pole pairs with the same Q factors. This fact may be used to validate the design.

The bandpass transfer function is

$$H(s) = \frac{-14.288145s}{s^2 + 211.87858s + 90688181}$$
$$\cdot \frac{(s^2 + 172685790)}{s^2 + 233.0944s + 109734800} \qquad (7.12)$$
$$\cdot \frac{(s^2 + 57615971)}{s^2 + 459.26114s + 99746976}.$$

The attenuation for the bandpass filter is shown in Fig. 7.6. According to the pairing scheme discussed above, we assign to

section   #1: $s_{z3,4}$ and $s_{p1,2}$:
$$H_1(s) = \frac{-G_1(s^2 + 57615971)}{s^2 + 211.87858s + 90688181}$$
section   #2: $s_{z1,2}$ and $s_{p3,4}$:
$$H_2(s) = \frac{-G_2(s^2 + 172685790)}{s^2 + 233.0944s + 109734800}$$
section   #3: s_{z5}, s_{z6},   and $s_{p5,6}$:
$$H_3(s) = \frac{-G_3 s}{s^2 + 459.26114s + 99746976}.$$

Section #1 is a HP-notch, section #2 is an LP-notch, and section #3 is a BP section. The gain factor has been selected negative, i.e., $G = (-G_1)(-G_2)(-G_3)$, because we intend to realize the sections with inverting second-order sections.

We select an ordering of the section with increasing Q factors, according to the rule of thumb, first section #3, then section #2, and finally section #1.

The last two sections have the same Q factors, and we select the LP-notch as the second section in order to remove

Fig 7.6 Attenuation of BP
filter

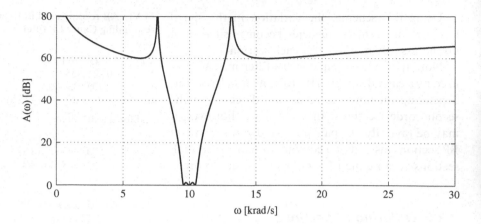

any large high-frequency signals that may cause slew rate-
limitation in the HP-notch section.

7.3.8 Optimizing the Section Gain

The gain of the individual sections should be
selected in order that the filter passband gain
meets the specification and so that the maxima of
the magnitude functions, from the input to the sections outputs, are equal. We demonstrate the optimization of the sections gains by an example.

Example 7.4 Assume that the required gain of the BP filter in
Example 7.3 is 12.

The optimization is done as follows, starting with the first
section, i.e., section #3.

First, we scale section #3 by choosing the gain factor G_3 so
that $|H_3|_{max} = 12$. The maximum of the magnitude function,

Equation (6.19), occurs at $\omega = r_p$. We get $-G_3(Q/r_p) = 12$,
i.e., $G_3 = -5511.1$. The magnitude function is shown in
Fig. 7.7.

Next, choose G_2 so that $|H_3H_2|_{max} = 12$, where H_3 is
the previously optimized section. We get $G_2 = -0.086705$.
The magnitude function from the input to the output of
section #2 (bold) and to the output of section #3 (thin)
are shown in Fig. 7.8.

Note that for $\omega \approx 10.4$ krad/s when $|H_3H_2|$ has its peak,
the $|H_3|$ is only about 5.1 whereas at $\omega \approx 10$ krad/s the
opposite is true, i.e., $|H_3|$ has its peak and $|H_2|$ is about 7.5.

In the last step, the gain factor in section #1 is optimized.
We get $G_1 = -0.3588$. The magnitude function from the input
to the output of sections #3 and #2 (thin) and to the output of
section #1 (bold) are shown in Fig. 7.9.

For $\omega \approx 9.6$ krad/s when $|H| = |H_3H_2H_1|$ has its peak, then
$|H_3|$ is only about 5.4 whereas $|H_3H_2|$ is about 1.8. Hence, for
some frequencies one section has a large output signal whereas
other sections have small output signal and vice versa. That
large and small output signals occur at different frequencies for
the sections is characteristic of the cascade form, and it becomes
more severe for filters with high Q factors.

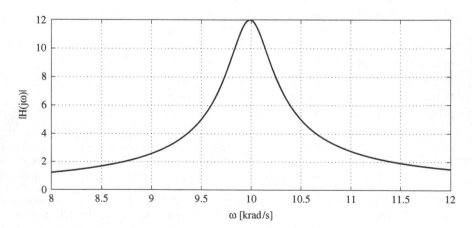

Fig 7.7 Magnitude function
of the scaled section #3

Fig 7.8 Scaled magnitude functions for section #3 and sections #2 and #3

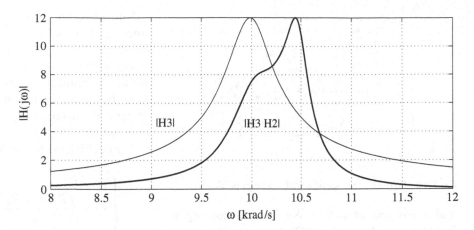

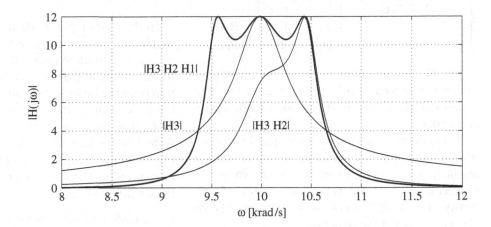

Fig 7.9 Scaled filter

7.3.9 Scaling of Internal Nodes in Sections

The signal levels inside a frequency selective filter structure typically vary significantly. That is, for a sinusoidal input signal, the amplitude in a particular internal node may become very large and at the same time very small in some other node. Moreover, the amplitudes may be of opposite sizes for another frequency. This effect is particularly pronounced for filters with high Q factors. In this section, we will discuss the scaling, i.e., optimization, of the signal levels inside a second-order section with several amplifiers. The outputs of these amplifiers are often referred to as critical scaling nodes.

Consider the two-integrator loop shown in Fig. 6.72 where we assume that the output node already is properly scaled.

We introduce scaling factors, k, into the section. These scaling factors must be introduced in such a way that the poles and zeros are not changed and only the gain is changed. The signal level at the node X is increased with a factor k, and it is possible to select an optimal signal level. Because, the poles and zeros must not be affected, we introduce a second scaling factor $1/k$ that cancels the effect of the first factor. Hence, only the signal level in node X is affected.

7.3.9.1 General Tow-Thomas Section

Consider the Tow-Thomas second-order section shown in Fig. 7.10. In the first design step, we select suitable element values so that the desired poles

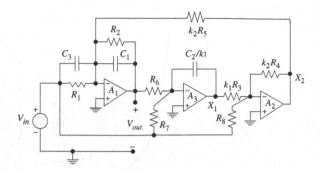

Fig 7.10 Two-Thomas section with scaling coefficients

and zeros are obtained. We have, according to Equations (6.101) and (6.102), $G = -C_3/C_1$ and

$$\begin{cases} r_p^2 = \dfrac{R_4}{R_3 R_5 R_6 C_1 C_2} \\ \dfrac{r_p}{Q_p} = \dfrac{1}{R_2 C_1} \end{cases} \quad \begin{cases} r_z^2 = \dfrac{R_4}{R_3 R_5 R_7 C_2 C_3} \\ \dfrac{r_z}{Q_z} = \dfrac{1}{R_1 C_3} - \dfrac{R_4}{R_5 R_8 C_3}. \end{cases}$$

In addition, we select the ratio of C_1 and C_3 to obtain the desired gain factor for the section. This means that the output of the section becomes properly scaled.

There are two internal nodes X_1 and X_2 that also need to be scaled. We therefore introduce the scaling factors, k_1 and k_2, in order to have two degrees of freedom to scale the two critical nodes independently. The scale factors are used to modify the element values in the structure.

We start from the node closest to the input, i.e., X_1 and divide C_2 by k_1 and multiply R_3 with k_1, then the $1/RC$ constant of the integrator is increased with a factor k_1 and the gain of the subsequent inverter is reduced with the same factor. Hence, neither the poles nor the zeros and their Q factors are changed as the loop gain is the same. The transfer function to the internal node X_i is

$$H_{X_i}(s) = -\frac{a_i s^2 + b_i s + c_i}{s^2 + \dfrac{r_p}{Q} s + r_p^2}. \tag{7.13}$$

The coefficients for node X_1 are

$$\begin{cases} a_1 = 0 \\ b_1 = \left(\dfrac{1}{R_7 C_2} - \dfrac{C_3}{R_6 C_1 C_2} \right) k_1 \\ c_1 = \left(\dfrac{R_1 R_6 - R_2 R_7}{R_1 R_2 R_6 R_7 C_1 C_2} \right) k_1. \end{cases} \tag{7.14}$$

Note that only the gain factor of the transfer function from the input to node X_1 has changed. We determine k_1 so that the maximal magnitude function becomes equal to the maximum of the overall magnitude function. This can easily be done by plotting the magnitude function and numerically determining a suitable value for the scale factor.

The coefficients for node X_2 are

$$\begin{cases} a_2 = 0 \\ b_2 = \left(\dfrac{R_7 C_3 - R_6 C_1}{R_7} \right) R_5 r_p^2 k_2 \\ c_2 = \left(\dfrac{R_2 R_7 - R_1 R_6}{R_1 R_2 R_7} \right) R_5 r_p^2 k_2. \end{cases} \tag{7.15}$$

In the next step, we continue with the modified element values to scale node X_2. This can be done by multiplying both R_4 and R_3 with k_2. Again, we determine k_2 so that the maximal magnitude function from the input to the node X_2 becomes equal to the maximum of the overall magnitude function. Note that the scale factors only appear as gain factors in the numerators of Equations (7.14) and (7.15). In general, the scaling factors can often be introduced in several different ways depending on the circuit topology.

Example 7.5 Consider section #3 in Example 7.4. The BP section has the transfer function

$$H_3(s) = \frac{-G_3 s}{s^2 + 459.26114s + 99746976}.$$

Here, we select reasonable element values, but for the sake of simplicity, we do not select standardized values. Next, we select $C_3 = 0$ and $R_7 = \infty$ to remove the constant and the s^2 term in the numerator. We notice that C_1 affects both the gain factor and the Q factor. We select $C_1 = 20$ nF and we get for the Q factor, i.e., $R_2 C_1 = Q/r_p = 1/459.26114 = 0.00217741$ and $R_2 = 108.871$ kΩ.

The coefficient in the numerator yields $-G_3 = \frac{-1}{R_1 C_1} = -5511.1$ and we get $R_1 = 9.07260$ kΩ. By selecting $R_3 = R_5 = R_6 = 10$ kΩ and $C_2 = 20$ nF, we get from the pole radius $R_4 = 9.07260$ kΩ.

The transfer functions to nodes X_1 and X_2 are given in Equations (7.16) and (7.17), respectively. We get with these values two LP transfer functions

$$H_{X_1}(s) = \frac{-27555500 k_1}{s^2 + 459.26114s + 99746976} \tag{7.16}$$

$$H_{X_2}(s) = \frac{-109.9431 \cdot 10^6 k_2}{s^2 + 459.26114s + 99746976}. \tag{7.17}$$

The maximum of the magnitude function for a BP section is GQ/r_p. Alternatively, we may plot the corresponding magnitude functions to find the maximum. Here we find that $k_1 = 1.99748$ yields a maximum of 12 for the node X_1, and a maximum of 12 is obtained for the node X_2 with $k_2 = 0.500637$. Hence, we change the originally selected element values to $C_2/k_1 = 10.013$ nF, $k_2R_5 = 5.00637$ kΩ, $k_2R_4 = 76.681$ kΩ, and $k_1R_3 = 19.975$ kΩ.

7.3.9.2 General Åkerberg-Mossberg Section

The transfer function of the Åkerberg-Mossberg section is

$$H(s) = -G \cdot \frac{s^2 + \frac{r_z}{Q_z}s + r_z^2}{s^2 + \frac{r_p}{Q}s + r_p^2} \qquad (7.18)$$

where

$$\begin{cases} r_p^2 = \dfrac{R_6}{R_3R_4R_5C_1C_2} \\ \dfrac{r_p}{Q} = \dfrac{1}{R_2C_1} \end{cases} \begin{cases} r_z^2 = \dfrac{R_6}{R_4R_5R_7C_2C_3} \\ \dfrac{r_z}{Q_z} = \dfrac{1}{R_1C_3} - \dfrac{R_6}{R_5R_8C_3} \end{cases} \qquad (7.19)$$

and $G = C_3/C_1$. For example, in the case of a lowpass section, which lacks the s^2 and s terms, we take the limit of the numerator when $C_3 \to 0$.

To scale the Åkerberg-Mossberg section, we proceed in the same manner as a for the Tow-Thomas section. The nodes to be scaled, X_1 and X_2, are

indicated in Fig. 7.11. We scale the node X_1 by multiplying both R_5 and R_6 by k_1, and to scale node X_2, we divide C_2 by k_2 and multiply R_4 with k_2.

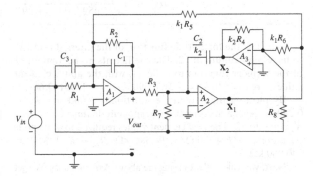

Fig 7.11 Åkerberg-Mossberg section with arbitrary zeros and scaling coefficients

The coefficients in the numerator of Equation (7.13) for the internal node X_1 is

$$\begin{cases} a_1 = \left(\dfrac{1}{R_1} - \dfrac{G}{R_2} + \left(\dfrac{r_p}{Q_p} - \dfrac{r_z}{Q_z}\right)C_3\right)R_5k_1 \\ b_1 = \left(\dfrac{r_p}{Q_pR_1} - \dfrac{Gr_z}{Q_zR_2} + (r_p^2 - r_z^2)C_3\right)R_5k_1 \\ c_1 = \left(\dfrac{r_p^2}{R_1} - \dfrac{r_z^2}{R_2}\right)R_5k_1 \end{cases} \qquad (7.20)$$

and for the node X_2

$$\begin{cases} a_2 = \left(\dfrac{R_2}{R_8} + \left(\dfrac{G}{R_2} - \dfrac{1}{R_1} + \left(\dfrac{r_z}{Q_z} - \dfrac{r_p}{Q_p}\right)C_3\right)\dfrac{R_4R_5}{R_6}\right)k_2 \\ b_2 = \left(\left(\dfrac{R_4}{R_8} - \dfrac{R_4R_5}{R_1R_6}\right)\dfrac{r_p}{Q_p} + \left(\dfrac{Gr_z}{Q_zR_2} + (r_z^2 - r_p^2)C_3\right)\dfrac{R_4R_5}{R_6}\right)k_2 \\ c_2 = \left(\left(\dfrac{R_4}{R_8} - \dfrac{R_4R_5}{R_1R_6}\right)r_p^2 + \dfrac{GR_4R_5r_z^2}{Q_zR_2R_6}\right)k_2. \end{cases} \qquad (7.21)$$

The gain from the input of the filter to the nodes X_1 and X_2 should be made equal to the desired filter gain. It is easily verified from Equations (7.18) and (7.19) that this scaling affects neither the poles nor the zeros, as k_1 and k_2 do not appear in the transfer function.

Example 7.6 Consider section #2 in Example 7.3. The LP-notch section has the transfer function

$$H_2(s) = -0.08671\frac{s^2 + 172685790}{s^2 + 233.0944s + 109734800}.$$

Note that the sign of the gain factor G_2 has to be chosen to fit the selected section. For the sake of simplicity, we select reasonable element values, but not necessarily standardized values. We select $C_1 = 400$ pF, $C_2 = 80$ pF $=> C_3 = G_2C_1 = 34.684$ pF. Next, to obtain the required Q factor, we select $R_2 = Q/C_1r_p = 10.7253$ MΩ. To obtain the required pole radius, we select $R_4 = R_5 = 1$ MΩ, $R_6 = 400$ kΩ, and $R_3 = R_6/(R_4R_5C_1C_2r_p^2) = 113.9109$ kΩ.

For the zero radius, we select $R_7 = R_6/(R_4R_5C_2C_3r_z^2) = 834.803$ kΩ. Finally, the zeros should be on the $j\omega$-axis, i.e., from Equation (7.16) we select $R_8 = R_1 = \infty$. These values yield an LP-notch section with the desired transfer function. The maximum magnitude is in this case $|H_2|_{max} = 2.38245$.

In order to scale the internal nodes, we also must consider the influence of section #3. Hence, we shall scale the internal nodes in section #2 when the transfer function is H_3H_2. We have for node X_1

$$H_{X_1}(s) = \frac{-5511.1s}{s^2 + 459.26114s + 99746976} \cdot \frac{-(a_1s^2 + b_1s + c_1)}{s^2 + 233.0944s + 109734800}.$$

We scale the magnitude function of the internal node X_1 to have a maximum of 12. The coefficients in the numerator of the transfer function to node X_1, with the above element values, are $a_1 = 0$, $b_1 = -2183.3903k_1$, $c_1 = 1396103.6k_1$.

By plotting the magnitude function, we find that $k_1 = 0.24096$ yields a maximum of the magnitude function to node X_1 equal to 12. The new element values required to scale node X_1 are $R_6 = k_1 400$ kΩ $= 96.384$ kΩ, $R_5 = k_1 \cdot 1$ MΩ $= 240.960$ kΩ.

Next, we scale node X_2 using the above element values. We get $a_2 = 0$, $b_2 = 1694.6383k_2$, $c_2 = 1083585.8k_2$.

By plotting the magnitude function, we find that $k_2 = 0.31046$ yields a maximum of the magnitude function to node X_2 equal to 12. The element values required to scale node X_2 are $R_4 = k_2 \cdot 1$ MΩ $= 310.46$ kΩ and $C_2 = 80/k_2$ pF $= 257.68$ pF

Because, the Q factors are very high, it may be advisable to consider a better integrator than the Miller integrator. The passive components should be selected so that the temperature coefficients of the resistors and capacitors are the same, but with different signs.

7.3.10 LTC1562 and LTC1560

LTC1562[TM] is an integrated circuit from Linear Technology that contains four second-order sections of LP, HP, and BP type whose r_p, Q, and G are determined with three external resistors, but the HP section requires an external capacitor as well.

Figure 7.12 shows a chip photo of half of the chip, i.e., two second-order sections. In the center of the photo, different capacitors are seen and on each side of the chip three operational amplifiers. The whole chip is approximately 3.0×4.5 mm including pads.

The usable frequency range is 10–150 kHz, and the error in the pole radius is less than 0.3% and the signal noise ratio, SNR, with $Q = 1$, is 97 dB with $V_{dd} = 5$ V and 107 dB with $V_{dd} = \pm 5$ V. The four sections can be cascaded into two (matched) filters of the fourth order or a single filter of the order eight. The circuit can be used for many different types of applications, i.e., anti-aliasing filter for 14-bit A/D converters, equalizers during data communication, and as two matched (nearly identical) LP filters in I and Q channels in a transceiver.

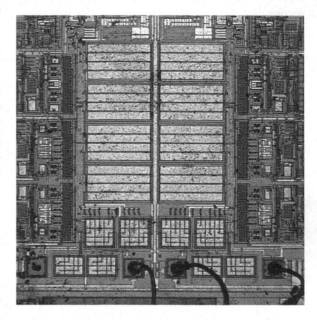

Fig 7.12 Chip photograph of half of the chip LTC 1562

Figure 7.13 shows a photo of half of the chip LTC1560-1. The whole chip is approximately 2.2×3.4 mm including pads and contains a fifth-order LP filter of Cauer type with two selectable passband frequencies: 500 kHz and 1 MHz. An array of unit-size capacitors are seen in the center of the photo. The circuit requires no external components except for two decoupling capacitors. The passband ripple is less than ±0.2 dB up to $0.55\omega_c$ and less than ±0.3 dB up to $0.9\omega_c$, and the stopband attenuation is larger than 63 dB at $2.43\omega_c$. The filter has low noise and low distortion. The signal noise ratio, SNR, is 75 dB with $V_{dd} = \pm 5$ V. The circuit can be used for many different types of applications, i.e., anti-aliasing filter for A/D converters.

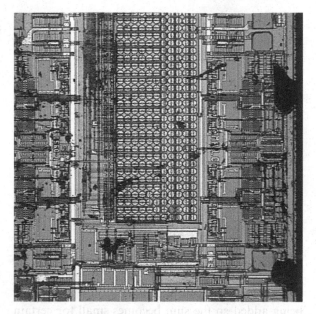

Fig 7.13 Chip photograph of half of the chip LTC 1560-1

7.4 Parallel Form

The transfer function can be rewritten into a partial fraction expansion so it consists of a sum of first- and second-order transfer functions. This is the equivalent to a filter in parallel form that is shown in Fig. 7.14.

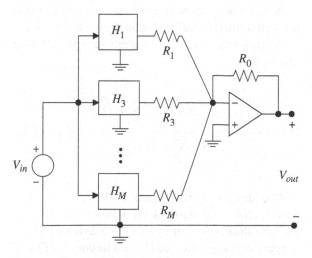

Fig. 7.14 Parallel form

The parallel form is characterized by very high element sensitivity in the stopband and is therefore not practically usable. The cause for the large sensitivity in the stopband is that the input signal goes through several different paths from the input and is summed at the output. The sum of relative large signals should be zero or small in the stopband. Small relative errors in any of the signal paths will therefore deteriorate the stopband. The parallel form or any other structure with several signal paths are not recommended.

7.5 Multiple-Feedback Forms

The sensitivity of the cascade form was improved by partitioning a higher-order filter into several cascaded filter sections of the first and second order. Because the sections do not interact, they can be tuned separately. The sensitivity for the whole filter is, however, relatively large in the passband. An error in an element in one section only affects the corresponding pole or zero pair, whereas in an LC filter an error in an element affects all poles in such a way that the resulting change in the magnitude function becomes small. Active filter structures, which simulate an LC filter, will be discussed in Chapters 8, 9 and 10. The sensitivity in the stopband is, however, low for the cascade form, as each section contributes to the attenuation.

Coupled forms [61, 62] are a more general class of cascaded structures, see Fig. 7.15, where the sensitivity is reduced by using negative feedback between the sections. An error in an element value will affect the poles, and the resulting element sensitivity in the passband becomes less than for the cascade form. The sensitivity in the stopband, however, becomes poorer. A drawback with feedback between all parts of the filter is that the tuning becomes more difficult.

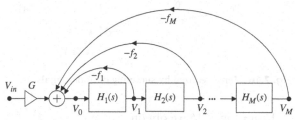

Fig. 7.15 Coupled from of type FLF

It is suitable to use first-order sections for lowpass and highpass filters and second-order sections for bandpass and stopband filters. Coupled forms also known as multiple-loop-feedback forms, which were developed by Perry, Laker, Ghausi, and Schaumann, are mainly used for geometric symmetric bandpass filters.

7.5.1 Follow-the-Leader-Feedback Form(FLF)

Follow-the-leader-feedback structures are described by the signal-flow graph shown in Fig. 7.15. The term *follow-the-leader-feedback* alludes to the children's game "Simon says."

By selecting the sections and the feedback coefficients in different ways, different structure variations are achieved. If the integrators, $H(s) = 1/s$, for the sections, we obtain the *companion form*.

The passband sensitivity is better for this class of structures compared with the cascade form, but the stopband sensitivity is worse. The group delay sensitivity is also higher. This makes this filter structure less interesting. The feedback and feed-forward coefficients in the structure shown in Fig. 7.16 determine the coefficients in the transfer function's denominator and numerator polynomials, respectively.

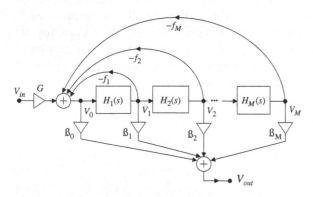

Fig. 7.16 Zeros realized by summing of node signals

From Fig. 7.15 we obtain

$$V_k = \prod_{j=1}^{k} H_j(s) V_0 \qquad (7.22)$$

and

$$V_0 = GV_{in} - (f_1 H_1 + f_2 H_1 H_2 - \ldots + f_M H_1 H_2 \ldots H_M) V_0$$
$$= GV_{in} - \sum_{k=1}^{M} f_k \left(\prod_{j=1}^{k} H_j(s) \right) V_0. \qquad (7.23)$$

The transfer function of the filter shown in Fig. 7.15 is

$$H(s) = \frac{V_M}{V_{in}} = \frac{G \prod_{j=1}^{M} H_j(s)}{1 + \sum_{k=1}^{M} f_k \left(\prod_{j=1}^{k} H_j(s) \right)}. \qquad (7.24)$$

The zeros can be realized by multiple signals being added so the sum becomes small for certain frequencies, see Fig. 7.16. This requires an extra operational amplifier.

The transfer function to the filter shown in Fig. 7.16 is

$$H(s) = \frac{G \sum_{k=0}^{M} \beta_k \prod_{j=1}^{k} H_j(s)}{1 + \sum_{k=1}^{M} f_k \left(\prod_{j=1}^{k} H_j(s) \right)}. \qquad (7.25)$$

A better way to generate finite zeros is to inject the signal into several nodes as shown in Fig. 7.17.

A straightforward derivation yields the transfer function

$$H(s) = \frac{\sum_{m=0}^{M} \beta_m \prod_{j=m+1}^{M} H_j(s) \left(1 + \sum_{k=1}^{m} f_k \prod_{j=1}^{k} H_j(s) \right)}{1 + \sum_{k=1}^{M} f_k \prod_{j=1}^{k} H_j(s)}. \qquad (7.26)$$

The design of these two FLF variants, with finite zeros, and general choices of the sections and simultaneously optimizing the dynamic signal range is very complicated [112]. The choice of the Q factors can be made so the sensitivity is minimized [61, 62].

These two ways of generating zeros are sensitive for errors in the element values especially for transfer functions with clustered zeros, i.e., when the zeros lie close to each other. The sensitivity for

Fig. 7.17 Finite zeros realized by injection of the input signal

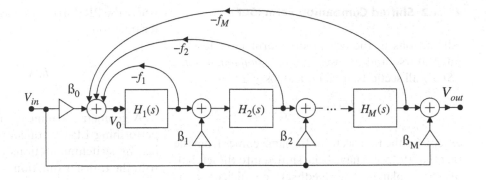

element errors also increases with the number of zeros.

The best way of realizing zeros is as in a ladder network where the input signal is reflected back to the source by an open-circuit series arm or by a short-circuit in a shunt arm. A series resonance circuit and a parallel resonance circuit behave at resonance as a short-circuit and open-circuit, respectively. Furthermore, it is difficult to tune the zeros in the FLF structure because the feedback and feed-forward coefficients affect all poles and zeros.

A large number of different filter structures is obtained by selecting the sections in special ways. The design of these structures is commonly done using a computer program.

7.5.1.1 Companion Form

The *companion form*, which also is referred to as the *state variable form*, is obtained by choosing all sections as identical integrators. This structure is especially usable for geometric symmetric BP filters, but it does not exhibit better sensitivity properties than the cascade form. We demonstrate the design principle with an example.

Example 7.7 Design an FLF filter with integrators as sections, i.e., the companion form, starting from the normalized transfer function for a third-order Cauer filter with $A_{max} = 0.1$ dB, $A_{min} = 24$ dB, $\omega_c = 1$ rad/s, and $\omega_s = 2$ rad/s. The passband edge shall be 200 krad/s.

The transfer function is

$$H(s) = \frac{0.3534086(s^2 + 5.153209)}{(s + 1.117384)(s^2 + 0.763975s + 1.629869)}. \tag{7.27}$$

First the numerator and the denominator are expanded. Note that this actually is not recommended because the

accuracy in the poles and zeros decreases, but here it is difficult to perform the design in terms of poles and zeros.

$$H(s) = \frac{0.3534086s^2 + 1.821188}{s^3 + 1.881359s^2 + 2.483522s + 1.821188}. \tag{7.28}$$

We get from Equation (7.25)

$$H(s) = \frac{G \sum\limits_{k=0}^{M} \beta_k \left(\prod\limits_{j=1}^{k} H_j(s) \right)}{1 + \sum\limits_{k=1}^{M} f_k \left(\prod\limits_{j=1}^{k} H_j(s) \right)} = \frac{G \sum\limits_{k=0}^{M} \beta_k \left(\frac{a}{s} \right)^k}{1 + \sum\limits_{k=1}^{M} f_k \left(\frac{a}{s} \right)^k}$$

where $H_j(s) = \dfrac{a}{s}$. We get with $M = 3$.

$$H(s) = \frac{G \left(\beta_3 \left(\frac{a}{s} \right)^3 + \beta \left(\frac{a}{s} \right)^2 + \beta_1 \left(\frac{a}{s} \right) + \beta_0 \right)}{f_3 \left(\frac{a}{s} \right)^3 + f_2 \left(\frac{a}{s} \right)^2 + f_1 \left(\frac{a}{s} \right) + 1}$$

$$= \frac{G\beta_0 s^3 + G\beta_1 a s^2 + G\beta_2 a^2 s + G\beta_3 a^3}{s^3 + f_1 a s^2 + f_2 a^2 s + f_3 a^3}.$$

We get from identification with Equation (7.28).

$$G\beta_0 = G\beta_2 a^2 = 0 \quad G\beta_1 a = 0.3534086 \quad G\beta_3 a^3 = 1.821188$$

$$f_1 a = 1.881359 \quad f_2 a^2 = 2.483522 \quad f_3 a^3 = 1.821188.$$

We have more free parameters than equations, i.e., there are therefore many possible ways to select the parameters. Assume that the desired gain here is $H(0) = 4$, i.e., we select $K = 4$ and $a = 1$, which gives

$$\beta_3 = \frac{1.821188}{4} \quad \beta_1 = \frac{0.3534086}{4} \quad \beta_0 = \beta_2 = 0$$

$$f_1 = 1.881359 \quad f_2 = 2.483522 \quad f_3 = 1.821188.$$

The denormalization so the passband edge becomes 200 krad/s is done by selecting $a = \omega_c = 2 \, 10^5$. Note that the denormalization does not affect the coefficients f_k and β_k. The coefficients can then be modified so the signal dynamic is optimized, but this will not be discussed here [112].

7.5.1.2 Shifted Companion Form (SCF)

All sections in the companion form have lossless integrators, and in the *shifted companion form (SCF)* all sections are identical lossy integrators,

$$H_i(s) = \frac{a}{s+a}$$

except for the first section. Its name comes from the fact that the poles have been shifted into the left half of the s-plane. The feedback coefficient f_1 is included in the first section, making it different from the others.

7.5.1.3 Primary Resonator Block Structures

In *primary resonator block* structures *(PRBs)* all sections are identical, i.e., they have the same r_p and Q, but the feedback coefficient $f_1 = 0$. This makes the filter structure modular and it may reduce its implementation cost.

The transfer function is obtained directly from Equation (7.24)

$$H(s) = \frac{G \prod_{j=1}^{M} H_j(s)}{1 + \sum_{k=2}^{M} f_k \left(\prod_{j=1}^{k} H_j(s) \right)} = \frac{GH_0^M(s)}{1 + \sum_{k=2}^{M} f_k H_0^k(s)} \quad (7.29)$$

where the PRB structure has sections with the transfer function

$$H_0(s) = \frac{a}{s+a}. \quad (7.30)$$

A geometric symmetric BP filter is obtained by performing LP-BP transformation of the sections, i.e., by switching sections to identical BP sections with the transfer function

$$H_0(s) = \frac{a}{S+a} = \frac{a}{\dfrac{s^2 + \omega_I^2}{s} + a} = \frac{as}{s^2 + as + \omega_I^2}. \quad (7.31)$$

This approach, to use frequency transform structures, can be used in most LP structures to realize HP and geometrically symmetric BP and BS filters.

Example 7.8 Design a geometric symmetric BP filter with $A_{max} = 1$ dB, geometric center frequency $\omega_I = 5$ krad/s, and the bandwidth $\omega_b = 500$ rad/s. The filters gain shall be equal to 1. Start from a PRB structure, which realizes a third-order Chebyshev I LP filter. The LP filter shall according to the BP transformation in Chapter 3 have $\Omega_c = 500$ rad/s.

The normalized poles are $S_{p0} = -0.4941706$ and $S_{p1,2} = -0.2470853 \pm j0.9659987$.

The transfer function for the LP filter is

$$H(S) = \frac{6.141334 \cdot 10^7}{(S + 247.0853)(S + 123.5427 + j482.9993)(S + 123.5427 - j482.9993)}.$$

The Q factors are $Q = 0.5$ and $Q = 2.01772$. We write, for the sake of simplicity, the numerator and the denominator as polynomials[2]

$$H(S) = \frac{6.141334 \cdot 10^7}{S^3 + 494.1706 S^2 + 309602.3 S + 6.141334 \cdot 10^7}.$$

In this case, Equation (7.29) and Equation (7.30) yield

$$H(S) = \frac{GH_0^3}{H_0^3 f_3 + H_0^2 f_2 + 1}$$

$$= \frac{Ga^3}{S^3 + 3aS^2 + a^2(f_2 + 3)S + a^3(f_3 + f_2 + 1)}.$$

Through identification with the transfer function we obtain

$$3a = 494.1706$$

$$a^2(f_2 + 3) = 309602.3$$

$$a^3(f_3 + f_2 + 1) = 6.141334 \cdot 10^7$$

$$Ga^3 = 6.141334 \cdot 10^7$$

---

[2] Note that this should be avoided because we lose accuracy in the poles and zeros.

which gives

$$a = 164.7235 \qquad f_2 = 8.41019$$
$$f_3 = 4.330096 \qquad G = 13.74029.$$

By performing a BP transformation, i.e., changing S to s according to Equation (7.31) with $\omega_I = 5$ krad/s, the BP transfer function is obtained with the poles

$$s_{p1,2} = 123.54266 \pm j4998.473 \qquad Q = 20.23593$$
$$s_{p3,4} = -58.79102 \pm j4763.9489 \qquad Q = 40.51904$$
$$s_{p5,6} = -64.75163 \pm j5246.948 \qquad Q = 40.51904.$$

The zeros are $s_z = 0$ and $s_z = \infty$. These are very high Q factors. The feedback coefficients, f_k, are not affected by the transformation. These computations are easily performed using the MATLAB function **PRB** [51].

According to Equation (7.31), the sections' Q factors becomes only

$$Q_{\text{section}} = \frac{\omega_I}{a} = \frac{5000}{164.7235} = 30.35389.$$

Figure 7.18 shows the realization of the BP filter. All sections are thus equal and have $r_p = 5000$ rad/s and $Q = 30.35389$. The element values are the same as computed in Example 6.3.

Because the Q factors are large, we here select NF2 BP sections. Because the sections and the adder on the input is inverting, one extra inverter is used for the coefficient f_3 to obtain the correct sign. The later inverter can, however, be eliminated if instead the feedback signal is applied to the positive input of the operational amplifier (A_1). With ideal operational amplifiers and components, we get the realization of the filter shown in Fig. 7.18.

PRB and SCF are special cases of FLF. The sensitivity properties of FLF filters are better than for the cascade form,

but not as good as for leapfrog filters, when the transfer function has finite zeros. Leapfrog filters will be discussed in Chapter 10. For transfer functions without finite zeros, the sensitivity properties for FLF and leapfrog filters are similar and both are better than the cascade form. FLF filters are suitable for BP filters without finite zeros.

Coupled forms are sensitive to excess phase in the freedback paths. Typically, the feedback coefficients, f_i, are realized using resistors as shown in Fig. 7.18. The excess phase can be compensated for by adding suitable, small capacitors across the the feedback resistors [104].

7.5.2 Inverse Follow-the-Leader-Feedback Form

In *inverse follow-the-leader-feedback form (IFLF)* the direction of all branches in the signal flow graph have been changed, according to Fig. 7.19. This operation is called transposition and it does not change the transfer function. The IFLF structure has similar sensitivity as FLF, but it often has better signal dynamic. This structure was originally introduced by D.J. Perry (1975), and it is an analog counterpart to the direct form I^t in the digital domain.

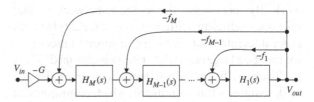

Fig. 7.19 Signal-flow graph for IFLF

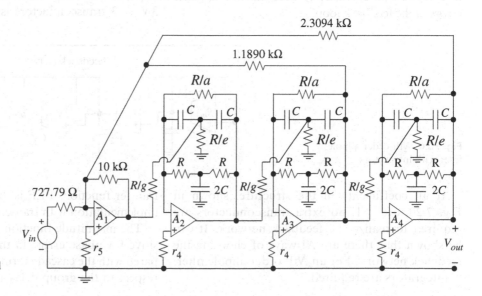

Fig. 7.18 Geometrically symmetric BP filter realized with the PRB structure

7.5.3 Minimum Sensitivity Form

Minimum sensitivity form (MSF) is the most general form of coupled structures as it has a general feedback network. MSF is designed with the help of computer software, which minimizes one sensitivity measure for the filter. The design is relatively complicated.

7.6 Transconductor-Based Coupled Forms

Higher-order filters based on transconductors can be realized using any of the coupled forms. The structures that were used for active *RC* filters are however not directly suitable because every feedback coefficient requires one transconductor, which leads to high power consumption, large chip area, etc. Instead, we can use special feedback networks that only allow the coefficients $+1$, i.e., direct feedback, which can be realized without transconductors.

Consider first an allpole filter according to Fig. 7.20 with integrators and a general feedback network. By specializing the feedback network so that only one coefficient, with the value $+1$, exists between every integrator output to one or more transconductor inputs, a large class of structures is obtained [32]. A feedback coefficient, which does not assume values ±1 or 0, requires a transconductor, which contributes an extra phase shift in the feedback loop. This is unsuitable because the coupled forms are sensitive for phase errors in the feedback loop.

Figure 7.21 shows a first- and a second-order allpole section. Figure 7.22 shows two examples of third-order allpole sections. It can be shown that there are $3! = 6$ different sections, but four of these are of less interest because they can only realize real poles or cannot realize simple LP filters [32].

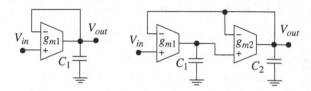

Fig. 7.21 First- and second-order allpole sections

For a fourth-order section, there are $4! = 24$ different structures, but only 10 of these are generally usable. Some of these structures coincide with IFLF and leapfrog filters, which will be discussed in Chapter 10.

7.6.1 Inverse Follow-the-Leader-Feedback Form

Inverse follow-the-leader-feedback form can be realized using g_m-C integrators as sections. Figure 7.23 shows the structure for an Nth-order IFLF structure of allpole type. Note that the upper structure shown in Fig. 7.22 is a third-order IFLF structure.

In follow-the-leader feedback form, a summing of all the integrators' output voltages occur and this requires one transconductor per integrator. Totally $3N + 3$ transconductors is required for a general

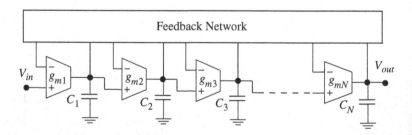

Fig. 7.20 Nth-order allpole filter with integrators

If all coefficients in the structure, shown in Fig. 7.20, are $+1$, no extra transconductors are required to realize the feedback network. It can be shown that there are $N!$ ways of choosing the feedback network. For an Nth-order allpole filter, N integrators are required.

transfer function. FLF is, thus, not well suited for implementation with transconductor-C techniques.

The magnitude function of the IFLF is less sensitive for phase errors in the feedback loops compared with the cascade form, but the sensitivity with respect to the group delay is larger. Moreover, the

Fig. 7.22 Third-order allpole sections

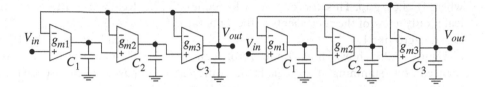

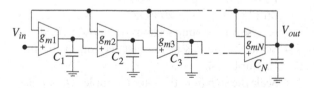

Fig. 7.23 IFLF

7.6.2 Finite Transmission Zeros

large global wiring introduces parasitic poles that significantly degenerate the group delay of multi-feedback structures and they do not appear to be practical for applications where the group delay performance is important.

Finite transmission zeros can in principle be realized either by injection of the input signal in suitable nodes (*feed forward*) or by summing the signals in suitable nodes. Figure 7.24 shows how the input signal is injected in the integrators and how a floating resistor is used on the filter's output to add the output signal from the allpole network and the input signal. The numerator to the transfer function is

$$N(s) = \tau_1\tau_2\tau_3\ldots\tau_N s^N + \tau_1\tau_2\tau_3\ldots\tau_{N-1}s^{N-1}\ldots\tau_1\tau_2 s^2 + \tau_1 s + 1 \tag{7.32}$$

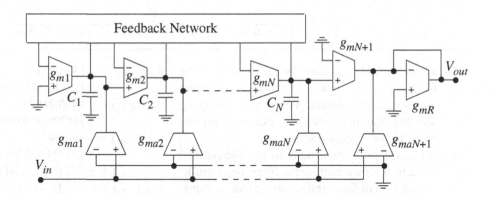

Fig. 7.24 Finite zeros realized by injection of the input signal

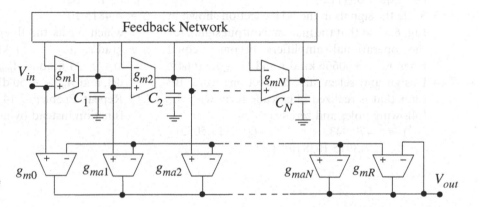

Fig. 7.25 Finite zeors realized by summing of node voltages

where $\tau_i = C_i/g_{mai}$. This approach may be economical if only a few of the coefficients in the numerator are not zero or ± 1.

Figure 7.25 shows how finite zeros can be realized by summing the signals in different nodes. The transconductor g_{mR} corresponds to a resistor, and the transconductor g_{mi} corresponds to the weights of the node signals. In Section 7.1 we showed that the sensitivity for errors in the transconductor g_{mai} becomes high for higher-order filters and if the zeros lie close to each other. Hence, these two structures for realizing the zeros are not recommended.

7.7 Problems

7.1 Determine suitable order and pairing of the poles and zeros for the filter in Example 2.4.

7.2 Realize a Butterworth filter with $A_{max} = 3$ dB, $f_c = 3.5$ kHz, $A_{min} = 25$ dB, and $f_s = 10$ kHz in cascade form with PF2 sections.

7.3 Design an active RC filter with maximally flat passband in cascade form. The specification for the filter is
Passband: $0 \leq f \leq 7.0$ kHz $A_{max} \leq 0.1$ dB
Stopband: $f \geq 20$ kHz $A_{min} \geq 25$ dB
     Use one-op-amp sections with positive feedback.

7.4 Scale the signals in the last section in the filter Example 7.4.

7.5 Scale the signals in the filter Example 7.4, but place the section in the following order: #1, #2, and #3.

7.6 Suggest a suitable realization for the HP filter in Problem 2.31.

7.7 Suggest a suitable realization for the filter in Example 2.8.

7.8 Suggest a suitable realization for the filter in Example 2.9.

7.9 Determine suitable element values for a high-pass filter in cascade form that realizes the transfer function

$$H(s) = \frac{s^5}{(s^2 + 400s + 3.5 \cdot 10^6)(s^2 + 2300s + 7.6 \cdot 10^6)(s + 500)}.$$

7.10 Select suitable sections for realization of the BP filter in Example 3.3 in cascade form and scale the signal levels. Verify the scaling by plotting the magnitude functions from the input to the critical nodes.

7.11 Why is it not a good idea to use the parallel form to realize a high-order filter, i.e., parallel connection of first- and second-order sections?

7.12 Derive the transfer function of the FLF filter, i.e., Equation (11.5).

7.13 Scale the signals in the KHN section shown in Fig. 6.55 so that the maximal output signal for the operational amplifiers becomes equal when $r_p = 3.00666$ krad/s and $Q = 7.516648$.

7.14 Design and select suitable sections for a BP filter that is realized in cascade form with the following poles and zeros
$s_{p1}, (s_{p1}^*) = -479.9438$ $Q_1 = 14.505217$
          $\pm j13915.1038$ rad/s

$s_{p2}, (v_2^*) = -1327.3315$ $Q_2 = 5.8722883$
          $\pm j15532.3355$ rad/s
$s_{p3}, (s_{p3}^*) = -1440.76978$ $Q_3 = 6.3382487$
          $\pm j18206.9763$ rad/s
$s_{p4}, (s_{p4}^*) = -566.7471$ $Q_4 = 17.735021$
          $\pm j20094.5532$ rad/s
$s_{z1}, (s_{z1}^*) = \pm j7521.0955$ rad/s
$s_{z2}, (s_{z2}^*) = \pm j25823.6$ rad/s
$s_{z3}, (s_{z3}^*) = 0$
$s_{z4}, (s_{z4}^*) = \infty$
$G = 4.4815 \cdot 10^6$
     which meets the filter specification $\omega_{c1} = 14$ krad/s, $\omega_{c2} = 20$ krad/s, $\omega_{s1} = 8$ krad/s, $\omega_{s2} = 25$ krad/s, $A_{max} = 0.4$ dB, $A_{min1} = 70$ dB, and $A_{min2} = 50$ dB.

7.15 Repeat Problem 7.14 but realize the transfer function instead by using FLF.

Chapter 8
Immitance Simulation

8.1 Introduction

A doubly resistively terminated LC filter that has been designed using the insertion loss method, i.e., for maximum power transfer in the passband, has minimal element sensitivity. Hence, we can expect that an active RC filter, which simulates the energy relationships in the LC filter, will also have low element sensitivity. In this chapter, several methods are discussed that use active components to realize some undesired circuit elements in an LC network [4, 18, 32, 74, 105, 112, 115]. These methods are called immitance simulation methods as immitances are simulated, i.e., **imp**edances and ad**mittances**.

An inductor can be realized with a generalized immitance converter GIC, which has the input impedance

$$Z_{in}(s) = K(s)Z_L^{\pm 1}. \qquad (8.1)$$

By selecting suitable $K(s)$ and exponent in Equation (8.1), we obtain according to Section 5.4.4 different special cases of converters and inverters, e.g., *PIC* (*positive impedance converter*) with $K(s) = ks$ and $(+)$-sign and gyrator *PII* (*positive impedance inverter*) with $K(s) = k$ and $(-)$-sign [18].

8.2 PIC-Based Simulation

An inductor can be simulated with a *PIC* loaded with a resistor. The most common and best realizations are based on Antoniou's GIC. In practice, however, it is difficult to realize floating inductors,

i.e., inductors that are not grounded at one end. Suitable LC ladder structures that only have grounded inductors are HP and BP filters, which do not have finite zeros in the upper stopband and at most one zero at s = ∞. We demonstrate the method with an example.

Example 8.1 Realize a Cauer HP filter with the structure shown in Fig. 3.69 using the impedance simulating method with a PIC. The specification is $A_{max} = 0.01087$ dB, $A_{min} = 60$ dB, $f_c = 16$ kHz, $f_s = 6$ kHz, and $R_s = R_L = 50\,\Omega$

First, we map the requirement to a corresponding LP requirement.

$$\Omega_c = \omega_I^2/\omega_c \text{ where } \omega_c = 2\pi 16 \text{ krad/s.}$$
$$\Omega_s = \omega_I^2/\omega_s \text{ where } \omega_s = 2\pi 6 \text{ krad/s.}$$

We select ω_I^2 so that $\Omega_c = 1$, in order to get the requirement for a normalized LP filter, i.e., $=> \omega_I^2 = \omega_c$. We get $\Omega_s = \omega_I^2/\omega_s = \omega_c/\omega_s = 16/6 = 2.666$.

We select the LP filter C051523 and a T ladder with only two capacitors. Hence, the corresponding HP filter will have only two inductors that need to be simulated. The normalized element values are $L_1 = 0.732110$, $L_2 = 0.044113$, $C_2 = 1.261137$, $L_3 = 1.496225$, $L_4 = 0.121060$, $C_4 = 1.149490$, $L_5 = 0.662813$, and $R_s = R_L = 1$.

Next, we transform the LP elements to elements in the HP filter and at the same time change the impedance level to R_0 using the mappings:

$$l_{HP} = R_0/(\omega_I^2 C_{LP}) = R_0/(\omega_c C_{LP}) \text{ and}$$
$$c_{HP} = 1/(R_0\omega_I^2 L_{LP}) = 1/(R_0\omega_c L_{LP})$$

where $R_0 = 50\,\Omega$.

We get

$c_1 = 271.7401$ nF	
$c_2 = 4.509865\ \mu$F	$l_2 = 0.3943736$ mH
$c_3 = 132.9637$ nF	
$c_4 = 1.643348\ \mu$F	$l_4 = 0.4326781$ mH
$c_5 = 300.1505$ nF	

L. Wanhammar, *Analog Filters Using MATLAB*, DOI 10.1007/978-0-387-92767-1_8,
© Springer Science+Business Media, LLC 2009

Each grounded inductor is replaced with a PIC that is loaded with a resistor, according to Fig. 8.1, where $l_2 = kr_2$ and $l_4 = kr_4$ where the conversion function is $K(s) = ks$.

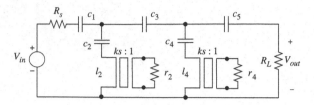

Fig. 8.1 Active HP filter of type impedance simulation using PIC

Figure 8.2 shows the corresponding realization with Antoniou's GIC. For sake of simplicity, the symbol, shown to the right in Fig. 8.2, is used for Antoniou's GIC where $R_2 = R_3$. It is important to use matched operational amplifiers.

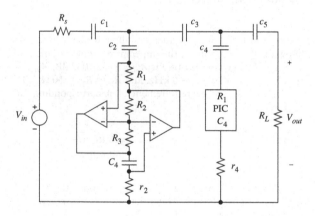

Fig. 8.2 Active HP filter of type inductance simulation using Antoniou's GIC

The input impedance according to Equation (5.63) with ideal operational amplifiers is

$$Z_{in} = \frac{R_1 R_3}{R_2 (1/sC_4)} r.$$

A suitable design, which minimizes the effect of the finite bandwidth of the operational amplifiers, is obtained if we choose $\omega_c C_4 r_2 = 1$, $\omega_c C_4 r_4 = 1$, and $R_2 = R_3$ [112].

We select $C_4 = 1$ nF, which gives $r_2 = r_4 = 1/\omega_c C_4 = 9.94719$ kΩ, and from Equation (5.63) with $R_2 = R_3 = 10$ kΩ we obtain $R_1 = l_2/(r_2 C_4) = 39.6468$ Ω and $R_1 = l_4/(r_4 C_4) = 43.4975$ Ω for the resistor R_1 in the first and second PIC, respectively.

8.3 Gyrator-Based Simulation

An inductor can also be realized with a gyrator loaded with a capacitor. This method is also best suited for LC structures with grounded inductors.

A gyrator is a nonreciprocal two-port, which was defined by Equation (5.16) [18]

$$\begin{cases} V_1 = -r_1 I_2 \\ I_1 = \dfrac{V_2}{r_2} \end{cases} \qquad (8.2)$$

where r_1 and r_2 are real constants > 0. The gyrator is a lossless circuit element if $r_1 = r_2$. According to Section 5.4.4, the gyrator is a PII (*positive impedance inverter*) with the input impedance for port 1

$$Z_{in1} = \frac{B(s)1}{C(s)Z_2} = \frac{r_1 r_2}{Z_2}. \qquad (8.3)$$

The K matrix for a lossless gyrator is obtained directly from Equation (8.2)

$$K_{Gyrator} = \begin{bmatrix} 0 & r \\ \dfrac{1}{r} & 0 \end{bmatrix}$$

where $r = r_1 = r_2$. For the circuit shown in Fig. 8.3, we have the input impedance

$$Z_{in}(s) = r^2 Cs. \qquad (8.4)$$

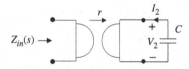

Fig. 8.3 Gyrator loaded with a capacitor

Thus, an inductor can be simulated with a gyrator loaded with a capacitor.

Gyrators are available as discrete integrated circuits, e.g., TCA 580 and TESLA SN150 10, but they have not had a commercial success due to their relative limited market compared with operational amplifiers. However, a gyrator can be implemented as a circuit with only transistors and capacitors with about the same number of transistors as an operational amplifier. Hence, the gyrator-based approach is well suited and commonly used in integrated active filters.

Example 8.2 Realize the same filter as in Example 8.1 but with gyrators.

According to Equation (8.4), the inductors can be replaced with gyrators, which are loaded with capacitors according to Fig. 8.4.

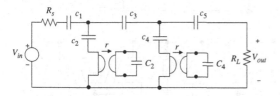

Fig. 8.4 Gyrator-C HP filter

We get the following realization where $C_2 = r^2 l_2$ and $C_4 = r^2 l_4$. An advantage of this approach is that it is easier to implement good capacitors compared to resistors in a digital CMOS technology.

8.3.1 Transconductor-Based Gyrator-C Filters

In the immitance simulation approach, some components are replaced with an equivalent network, e.g., an inductor is replaced with a gyrator that is loaded with a capacitor. A grounded inductor can be realized using a capacitor and a gyrator that is realized by two transconductors as shown in Fig. 5.55. However, many structures like the Cauer filter shown in Fig. 8.5 have floating inductors.

Figure 8.6 shows how floating inductors in the ladder network shown in Fig. 8.5 can be simulated with gyrators and capacitors. The floating inductors

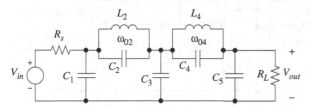

Fig. 8.5 Fifth-order Cauer filter

are realized by the circuit shown in Fig. 5.58. Note that the resulting circuit has two floating capacitors with top and bottom plate parasitic capacitances, which are added to the capacitances C_1, C_3, and C_5.

Floating immitances that are connected to high-impedance nodes suffer from parasitic capacitances that affect the frequency response. As was shown in Example 5.3, a floating impedance can be replaced with a shunt admittance between two gyrators as shown in Fig. 8.7 where $Z = r^2 Y$.

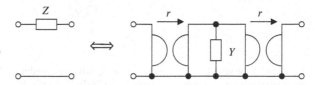

Fig. 8.7 Equivalent networks

A better and more general method is to use the equivalent network shown in Fig. 8.7 to convert the series admittances to grounded impedances as shown in Fig. 8.8.

An LP filter with finite zeros and series arms of parallel resonance type is converted to grounded series resonance circuits according to Figs. 5.59 and 5.61, where $Z_i' = Y_i/(g_{m2}g_{m3})$ and $g_{m1} = g_{m3}$.

8.3.2 CCII-Based Gyrator-C Filters

In this section, we show how an LC filter can be simulated using a gyrator-C circuit that is realized using current conveyors.

Example 8.3 The Cauer LC filter shown in Fig. 8.9 meets the requirements $\omega_c = 10$ Mrad/s, $\omega_s = 52.4$ Mrad/s, $A_{max} = 0.1$ dB, and $A_{min} = 50.5$ dB with the element values $R_s = R_L = 50$ Ω, $C_1 = C_3 = 2.020486$ pF, $C_2 = 49.048$ pF, $L_2 = 5.59284$ μH.

Figure 8.10 shows the corresponding realization with current conveyors. Conveyor 1 realizes a VCCS and conveyor 2 the source resistor R_s. A pair of symmetric gyrators is

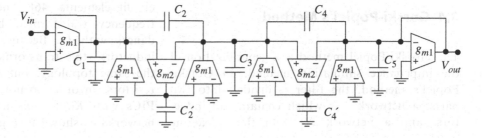

Fig. 8.6 Gyrator-C realization

Fig. 8.8 Improved gyrator-C realization

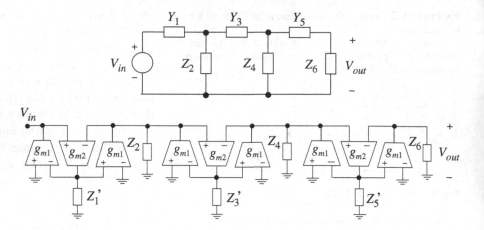

obtained if we select $r = r_3 = r_4 = r_5 = r_6$, which according to Equation (5.65) corresponds to a series inductor with the inductance

$$L = r^2 C$$

with $r = 2158.87\ \Omega$ and $C = 1.2$ pF yielding $L_2 = 5.59284$ μH. We may select $r_1 = \alpha r_L$ in order to increase the gain with a factor α. For example, $\alpha = 2$ compensates for the attenuation $R_L/(R_s + R_L) = 1/2$ when $R_s = R_L$.

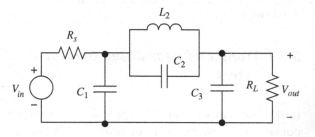

Fig. 8.9 Reference filter

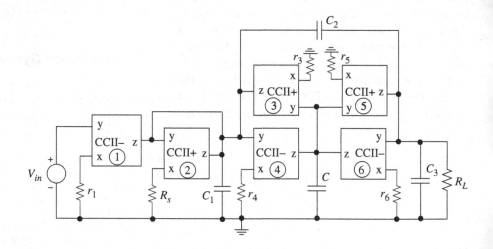

Fig. 8.10 Gyrator-C filter with current conveyors

8.4 Gorski-Popiel's Method

The Gorski-Popiel method is a generalization of the impedance simulation method. In Gorski-Popiel's method, the filter is divided into two parts, a network N_L, which contains all inductors, and a network N_1 with the remaining circuit elements [46]. The network N_L is then frequency transformed by dividing all impedances with k_s. Because N_L only consists of inductors, the transformed network will have the same topology, but it will only consist of resistors. In order to not affect the network N_1, PICs with $K(s) = ks$ are placed between the networks as shown in Fig. 8.11.

Fig. 8.11 Principle of the Gorski-Popiel method

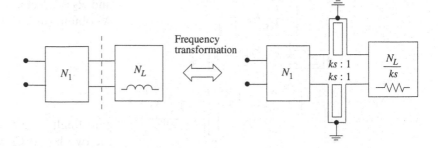

The Gorski-Popiel method is suitable for networks with floating inductors, e.g., LP and BP filters. Note that the number of PICs is in many cases fewer than the number of inductors.

It is useful to perform the above method in steps. Consider therefore the network shown in Fig. 8.12 where only N_L contains inductors.

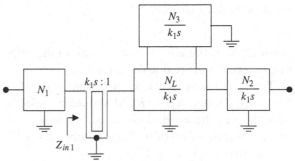

Fig. 8.13 Network after step 1

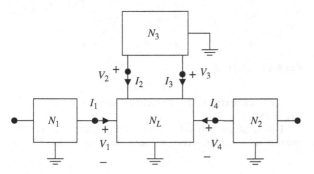

Fig. 8.12 Extraction of the inductor network, N_L

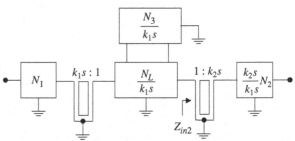

Fig. 8.14 Network after step 2

In the first step, we insert a PIC between N_1 and N_L. The input impedance Z_{in1} is according to Equation (5.20)

$$Z_{in1} = \frac{A(s)}{D(s)} Z_L = k_1 s Z_L$$

where Z_L is the input impedance of the left port of the N_L network. In order to not change Z_{in1}, we divide all impedances in the networks N_L, N_2, and N_3 with $k_1 s$. The resulting network is shown in Fig. 8.13. In the modified network, N_L, the inductors have now become resistors while in N_2 and N_3 the resistors and capacitors have become capacitors and supercapacitors, respectively.

In the second step, we insert a PIC between N_L and N_2 with the ports as shown in Fig. 8.14. The

input impedance Z_{in2} is according to Equation (5.21)

$$Z_{in2} = \frac{D(s)}{A(s)} Z_{load} = \frac{1}{k_2 s} Z_{load}$$

where Z_{load} is the input impedance of the left port of the N_2 network. In order to not change Z_{in2}, we multiply all impedances in the network N_2 with $k_2 s$. Hence, the impedances in N_2 are the same as in the original network, except for the real scale factor k_2/k_1. However, original resistors and capacitors in N_3 have become capacitors and supercapacitors, respectively.

In the last step, we therefore insert a layer of PIC between N_L and N_3 as shown in Fig. 8.15. The

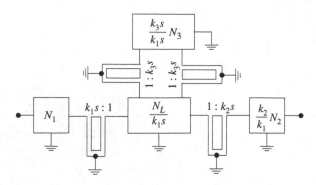

Fig. 8.15 Network after step 3

network N_3 is thereby converted back to the original, except for the real scale factor k_3/k_1.

The conversion factor, k, of Antoniou's GIC is $k = R_1 C_4$ when $R_2 = R_3$, it should be selected according to the method discussed below in order to minimize the effect of the finite bandwidths of the operational amplifiers [112].

The starting point is an impedance analysis of the network NL in Fig. 8.16. We first select [18, 112]

$$R_1 = \left|\frac{V_1}{I_1}\right|_{\omega = \omega_c}.$$

and R_2 is selected so the impedances in the network N_2 obtain suitable values. Next we select

$$C_3 = \frac{R_1 C_1}{|V_2/I_2|_{\omega = \omega_c}}, \quad C_4 = \frac{R_1 C_1}{|V_3/I_3|_{\omega = \omega_c}}$$

and, finally, we must select $R_3 C_3 = R_4 C_4$ because the two last PICs must have the same conversion factor, k_3.

Example 8.4 Consider the ladder in Fig. 3.68 where we had eliminated the right-most shunt inductor. The resulting ladder is shown in Fig. 8.17 where $C_4 = n(n-1)C_{2b} + n^2 C_3$.

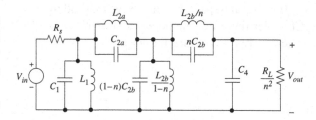

Fig 8.17 BP ladder

First we extract the inductive network as shown in Fig. 8.18.

The inductive network has only three ports connected to the rest of the ladder. Hence, we need three positive impedance

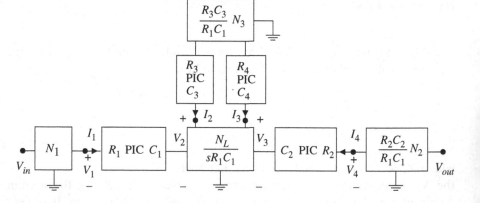

Fig. 8.16 Scaling of impedances according to the Gorski-Popiel method

Then C_1 is selected so the resistors in the network N_L obtain suitable values. Next, we select

$$C_2 = \frac{R_1 C_1}{|V_4/I_4|_{\omega = \omega_c}}.$$

converters. In this case, they must have the same conversion factor, $K(s) = ks$. The resulting realization is shown in Fig. 8.19 where

$$R_1 = \frac{L_1}{k} \quad R_2 = \frac{L_{2a}}{k} \quad R_3 = \frac{L_{2b}}{k(1-n)} \quad R_4 = \frac{L_{2b}}{kn}.$$

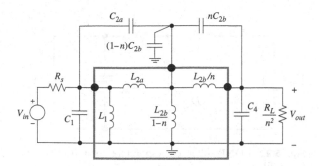

Fig 8.18 BP ladder with extracted inductive network

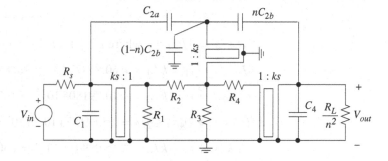

Fig 8.19 Gorski-Popiel's realization of a BP ladder

8.5 Bruton's Method

Bruton's method is a generalization of Gorski-Popiel's method [18, 48]. Instead of only performing a frequency transformation of the element in the inductive network, all circuit elements are transformed by multiplying the impedances with $1/s$ or s. The transfer function is not affected by the transformation of the impedances because the (voltage) transfer function is a ratio between the different impedances.

Theorem 8.1 *If all impedances in a network are multiplied with the same function, $g(s)$, all transfer functions of the type V_{out}/V_{in} and I_{out}/I_{in} as well as all input and output impedances are invariant.*

If all impedances are multiplied with $1/s$, the circuit elements are converted in the originating network in the following way.

$R \Rightarrow 1/s^2 D$ (Capacitor with the value[1] $C \leftarrow \frac{1}{R}$)
$Ls \Rightarrow R$ (Resistor with the value $R \leftarrow L$)
$1/sC \Rightarrow 1/s2D$ (FDNR; supercapacitor with the value $D \leftarrow C$ [Fs])

---

[1]The same numerical value, i.e., if $R = 10\,\text{k}\Omega$, we get $C = 10^{-4}\,\text{F}$.

As an alternative, all impedances can be multiplied with s at which the circuit elements in the originating network are converted in the following way:

$R \Rightarrow Ls$ (Inductor with the value $L \leftarrow R$)
$Ls \Rightarrow Es2$ (FDNR; superinductor with the value $E \leftarrow L$ [Hs])
$1/sC \Rightarrow R$ (Resistor with the value $R \leftarrow \frac{1}{C}$).

The original filter is often called *reference filter*. The new circuit elements with the impedances $1/s^2 D$ and Es^2 are called *supercapacitor* and *superinductor*, respectively. These elements can only be realized with the help of active circuit elements. Figure 8.20 shows corresponding symbols.

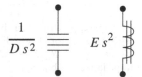

Fig. 8.20 Supercapacitor and superinductor

Superinductors are, however, not usable in practice because they represent a second-order derivative. The magnitude function for a differentiator increases linearly with frequency and even more rapidly for a superinductor, i.e., high-frequency noise will be strongly amplified and any circuit containing such elements will have a poor signal-to-noise ratio.

Bruton's method is suitable for networks with non-grounded inductors. We demonstrate Bruton's method using an example.

Example 8.5 Realize the ladder network shown in Fig. 8.21 using Bruton's method. The filter shall meet the following requirements: $A_{max} = 1$ dB, $A_{min} = 40$ dB, $\omega_c = 22$ krad/s, and $\omega_s = 38$ krad/s.

The denormalized element values are $R_s = R_L = 1$ kΩ, $L_1 = 60.86877$ mH, $L_2 = 10.791805$ mH, $C_2 = 55.841824$ nF, $\omega_{02} = 40.7355$ krad/s, $L_3 = 65.099016$ mH, and $C_4 = 70.125962$ nF.

Because the T network contains three inductors and only two capacitors, it is suitable to convert the network

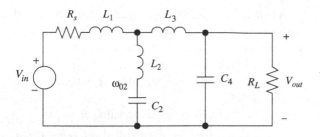

Fig. 8.21 Fourth-order Cauer filter

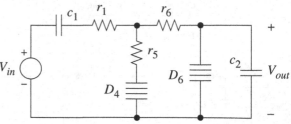

Fig. 8.22 FDNR filter corresponding to a fourth-order Cauer filter

by multiplying all impedances with k/s, which yields a filter structure with supercapacitors. The constant k is selected so the elements obtain suitable values; select for

example $k = 105$ [1/s]. The element values shown in Fig. 8.22 become

$$R_s k/s \Rightarrow 1/(c_1 s) \quad \text{where } c_1 = 1/(R_s k) = 1/(1000\,k) = 10 \text{ nF}$$

$$s L_1 k/s \Rightarrow r_1 = k L_1 = k\,60.86877\,10^{-3} = 6086.88\,\Omega$$

$$s L_2 k/s \Rightarrow r_5 = k L_2 = k\,10.791805\,10^{-3} = 1079.18\,\Omega$$

$$k/(C_2 s^2) \Rightarrow D_4 = C_2/k = 55.841824\,10^{-9}/10^5 = 55.84182410^{-14} \text{ Fs}$$

$$s L_3 k/s \Rightarrow r_6 = k L_3 = k\,65.099016\,10^{-3} = 6509.9\,\Omega$$

$$k/(C_4 s^2) \Rightarrow D_6 = C_4/k = 70.125962\,10^{-9}/10^5 = 70.12596210^{-14} \text{ Fs}$$

$$R_L k/s \Rightarrow 1/(c_2 s) \quad \text{where } c_2 = 1/(R_L k) = 1/(1000\,k) = 10 \text{ nF}.$$

Supercapacitors can be realized with a GIC loaded with a capacitor. From Equation (5.63) with $Z_1 = 1/sC$, $Z_2 = Z_3 = R$, $Z_4 = R_4$, and $Z_5 = 1/sC$, the input impedance is obtained

$$Z_{in}(s) = 1/s^2 C^2 R_4 = 1/s^2 D \quad \text{where} \quad D = C^2 R_4.$$

Suitable design for the supercapacitor D_4 is $\omega_c C_4 R_4 = 1$, which yields $C_4 = D_4 \omega_c$. Select $R_2 = R_3 = R = 10$ kΩ, which yields $C_4 = D_4 \omega_c = 12.2852$ nF and $R_4 = 3.6999$ kΩ, and in the same way we obtain $C_6 = D_6 \omega_c = 15.4277$ nF and $R_6 = 2.9463$ kΩ for the supercapacitor D_6.

Both inputs to the operational amplifiers require a DC path to ground or to an output of an operational amplifier for the bias current. Such a DC path to ground does not exist in Fig. 8.22 due to the capacitors c_1 and c_2.

One way of solving this problem is to parallel connect c_1 and c_2 with two resistors R_{1p} and R_{2p}. These can be selected according to the following principle. The gain for the reference filter at $\omega = 0$ is $|H| = 0.5$, which also shall be valid for the FDNR filter. For $\omega = 0$ we shall have

$$\frac{R_{2p}}{R_{1p} + R_{2p} + r_1 + r_6} = 0.5.$$

Furthermore, both R_{1p} and R_{2p} should be much larger than r_1 and r_6. Select, i.e., $R_{1p} = 56$ kΩ and $R_{2p} = 81$ kΩ. These resistors have a small effect on the filter frequency response at very low frequencies. A buffer amplifier is also required for the output, as the output may not be loaded.

It is, if possible, suitable to select a reference filter such that the FNDR elements become grounded. Filters that contain floating inductors can be realized with this method, which is one of the best for realization of lowpass filters even if certain problems can occur with bias currents.

Bruton's method can also be used when transconductors or current conveyors are used to realize the supercapacitors.

8.6 Problems

8.1 Design a second-order notch-filter (very narrow bandstop filter) for 50 Hz. The reference filter is a grounded series resonance circuit.

Fig. 8.23 FDNR filter

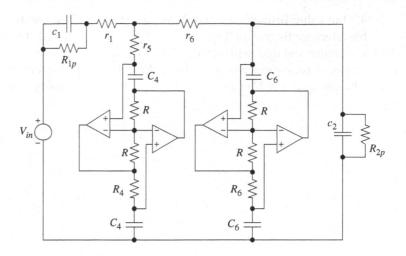

8.2 Design first an LC filter that meets the following requirements: $\omega_s = 2$ krad/s, $A_{max} = 2$ dB, $\omega_c = 4$ krad/s, and $A_{min} = 20$ dB. Use a Chebyshev I approximation and a T ladder with $R_s = 0\,\Omega$ and $R_L = 600\,\Omega$ as reference filter. Is this a good choice? Realize the reference filter using the immitance simulation method where inductors are realized by using PICs. Select, if possible, capacitances to 10 nF. Assume that $\tau_{gmax} \approx \tau_g(\omega_c)$.

8.3 Realize the same transfer function as in Problem 8.2, but use instead a doubly resistively terminated reference filter.

8.4 Design first an LC filter of Chebyshev I type with a minimum number of inductors and $R_s = R_L = 600\,\Omega$. Replace the inductors with PIC circuits. Assume for sake of simplicity the largest group delay occurs at the passband edge. The requirements are $f_c = 200$ kHz, $f_s = 50$ kHz, $A_{max} = 1$ dB, and $A_{min} = 35$ dB.

8.5 Realize the transfer function to a Chebyshev I HP filter that meets the requirement in Example 5.3 using the immitance simulation method (PIC).

8.6 Realize an active filter based on a third-order π ladder using gyrators. When is this approach appropriate?

8.7 Realize the ladder shown in Fig. 8.9 using gyrators when $R_s = R_L = 1$ kΩ, $C_1 = 209.4$ nF, $C_2 = 33.06$ nF, $C_3 = 209.4$ nF, and $L_2 = 83.47$ mH.

8.8 Realize an active filter based on a third-order π ladder structure using Gorski-Popiel's method.

8.9 Realize the ladder structure shown in Fig. 8.24 using Gorski-Popiel's method.

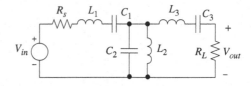

Fig. 8.24 Bandpass ladder in problem 8.9

8.10 Realize the transfer function to a Cauer filter that meets the requirement in Example 4.7 using Gorski-Popiel's method. Select an appropriate reference filter.

8.11 Realize the filter in Problem 8.8 using Gorski-Popiel's method.

8.12 Realize the filter in Problem 8.8 using Bruton's method, but use instead the corresponding T ladder.

8.13 Realize the transfer function of a Cauer filter that meets the same requirement as in Example

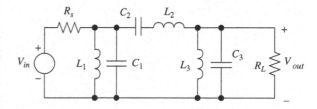

Fig. 8.25 Bandpass filter in problem 8.16

4.7 but using Bruton's method. Select a suitable reference filter structure.

8.14 Determine suitable element values in a GIC to realize a supercapacitor with $D = 10^{-6}$ Fs when $\omega_{critical} = 1$ Mrad/s.

8.15 Show that a single supercapacitor can be used to realize a third-order Butterworth filter.

8.16 Realize the ladder structure shown in Fig. 8.25 using as few PIC as possible.

Chapter 9
Wave Active Filters

9.1 Introduction

Wave active filters represent another approach, proposed by H. Wupper and K. Meerkötter [142, 143], to simulate LC filters where the components are described using waves. That is, instead of using the port current and voltage to describe a port, incident and reflected waves are used. The incident and reflected waves are a linear combination of current and voltage. Hence, we have performed a linear coordinate transformation from the variables current and voltage to incident and reflected waves.

Wave active filters that simulate doubly resistively terminated LC filters inherit the low sensitivity properties of the reference filter [13]. The same approach is also used for digital wave filters, but in this case the doubly resistively terminated network consists only of lossless commensurate transmission lines [35, 135].

We begin with a description of generalized wave filters and later describe wave active filters with voltage waves in more detail. Note that this description includes, as special case, several other structures, i.e., leapfrog filters [32].

9.2 Generalized Wave Variables

Figure 9.1 shows a general two-port with lumped elements. The case of commensurate transmission lines is analogous, but the transfer functions will not be rational functions in s.

Usually, we use currents and voltage to describe two-ports. Here, however, we will use incident and reflected waves.

The *generalized wave variables* x and y are a linear combination of input current and voltage of a port.

We define for each port

$$\begin{bmatrix} x_1 \\ y_1 \end{bmatrix} = P \begin{bmatrix} V_1 \\ I_1 \end{bmatrix} \qquad (9.1)$$

where

$$P = \begin{bmatrix} p_{11} & p_{12} \\ p_{21} & p_{22} \end{bmatrix} \quad \det(P) = p_{11}p_{22} - p_{12}p_{21} \neq 0 \qquad (9.2)$$

and for port 2

$$\begin{bmatrix} x_2 \\ y_2 \end{bmatrix} = Q \begin{bmatrix} V_2 \\ I_2 \end{bmatrix} \qquad (9.3)$$

where

$$Q = \begin{bmatrix} q_{11} & q_{12} \\ q_{21} & q_{22} \end{bmatrix} \quad \det(Q) = q_{11}q_{22} - q_{12}q_{21} \neq 0. \qquad (9.4)$$

We can interpret the generalized wave variables x and y, shown in Fig. 9.2, as *incident* and *reflected waves*, respectively. A wave description is equivalent

Fig. 9.1 Two-port

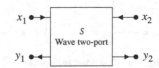

Fig. 9.2 Wave two-port

L. Wanhammar, *Analog Filters using MATLAB*, DOI 10.1007/978-0-387-92767-1_9,
© Springer Science+Business Media, LLC 2009

with describing a network with currents and voltages.

It is essential that the $\mathbb{P}$ and $\mathbb{Q}$ matrices are selected so the power relations are retained, as low-sensitive LC filters obtain their low element sensitivity through maximal power transfer from the source to the load.

Several very useful matrix representations for two-ports have earlier been discussed. It is also useful to define similar matrices, but in terms of wave variables.

9.2.1 Wave Transmission Matrix

The *transmission matrix* is defined as

$$\begin{bmatrix} V_1 \\ I_1 \end{bmatrix} = \begin{bmatrix} A - B \\ C - D \end{bmatrix} \begin{bmatrix} V_2 \\ I_2 \end{bmatrix} = T \begin{bmatrix} V_2 \\ I_2 \end{bmatrix} \quad (9.5)$$

where A, B, C, and D are the elements in the K matrix. By eliminating currents and voltage, we obtain

$$\begin{bmatrix} x_1 \\ y_1 \end{bmatrix} = P T Q^{-1} \begin{bmatrix} x_2 \\ y_2 \end{bmatrix} = \mathbb{F} \begin{bmatrix} x_2 \\ y_2 \end{bmatrix} \quad (9.6)$$

where $\mathbb{F}$ is the wave transmission matrix of the two-port.

9.2.2 Chain Scattering Matrix

Recall that the K matrix (*chain matrix*) is

$$\begin{bmatrix} V_1 \\ I_1 \end{bmatrix} = \begin{bmatrix} A & B \\ C & D \end{bmatrix} \begin{bmatrix} V_2 \\ -I_2 \end{bmatrix} = K \begin{bmatrix} V_2 \\ -I_2 \end{bmatrix}. \quad (9.7)$$

The chain matrix is useful to compute the matrix for two cascaded two-ports. If two two-ports are cascaded, according to Fig. 9.3, the resulting chain

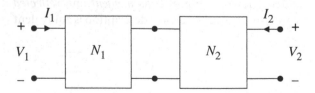

Fig. 9.3 Cascaded two-ports

matrix becomes $\mathbb{K} = \mathbb{K}_1 \mathbb{K}_2$. It is therefore useful to define a similar matrix for wave two-ports.

The *chain scattering matrix* is defined as

$$\begin{bmatrix} y_1 \\ x_1 \end{bmatrix} = C \begin{bmatrix} x_2 \\ y_2 \end{bmatrix} \quad (9.8)$$

where

$$C = \begin{bmatrix} 0 & 1 \\ 1 & 0 \end{bmatrix} F. \quad (9.9)$$

The chain scattering matrix is used in the same way as the chain matrix, i.e., the chain scattering matrix for two, cascaded wave two-ports equals the product of the corresponding chain scattering matrices.

9.2.3 Generalized Scattering Matrix

We define the *generalized scattering matrix* as

$$\begin{bmatrix} y_1 \\ y_2 \end{bmatrix} = S \begin{bmatrix} x_1 \\ x_2 \end{bmatrix} = \begin{bmatrix} s_{11} & s_{12} \\ s_{21} & s_{22} \end{bmatrix} \begin{bmatrix} x_1 \\ x_2 \end{bmatrix}. \quad (9.10)$$

The output y_2, when the second input x_2 is zero, is $y_2 = s_{21}x_1$, and s_{21} behaves as the normal transfer function. A two-port with $s_{11} = s_{22}$ is symmetric whereas a two-port with $s_{11} = -s_{22}$ is antimetric.

An advantage with the scattering matrix is that it exists for all circuit elements contrary to impedance and admittance matrices, which become singular for certain circuit elements.

The relation between the C and S matrices is

$$C = \frac{1}{s_{21}} \begin{bmatrix} s_{21}s_{12} - s_{11}s_{22} & s_{11} \\ -s_{22} & 1 \end{bmatrix} \quad (9.11)$$

and

$$S = \frac{1}{c_{22}} \begin{bmatrix} c_{12} & c_{22}c_{11} - c_{12}c_{21} \\ 1 & -c_{21} \end{bmatrix}. \quad (9.12)$$

9.2.4 Voltage Scattering Matrix

Henceforth we will only discuss the voltage scattering matrix unless otherwise explicitly stated. The voltage scattering matrix is a special case of the generalized scattering matrix. The voltage scattering matrix of voltage type[1] is obtained with the following selection of the parameters

$$P = \begin{bmatrix} 1 & R_1 \\ 1 & -R_1 \end{bmatrix} \quad (9.13)$$

and

$$Q = \begin{bmatrix} 1 & R_2 \\ 1 & -R_2 \end{bmatrix}. \quad (9.14)$$

This can also be written as

$$\begin{cases} x_1 = V_1 + R_1 I_1 \\ y_1 = V_1 - R_1 I_1 \end{cases} \quad (9.15)$$

and

$$\begin{cases} x_2 = V_2 + R_2 I_2 \\ y_2 = V_2 - R_2 I_2 \end{cases} \quad (9.16)$$

where R_1 and R_2 are two positive constants, so-called *port resistances*.

The variables x_1 and x_2 can be interpreted as *incident voltage waves* and y_1 and y_2 as *reflected voltage waves*. With this choice, the power relationships are retained, which is a prerequisite for inheriting the low-sensitivity of the reference filter.

For a two-port, the chain scattering matrix C is obtained from Equations (9.6) and (9.8)

$$\begin{bmatrix} y_1 \\ x_1 \end{bmatrix} = \begin{bmatrix} 0 & 1 \\ 1 & 0 \end{bmatrix} \begin{bmatrix} 1 & R_1 \\ 1 & -R_1 \end{bmatrix} \begin{bmatrix} A & -B \\ C & -D \end{bmatrix} \begin{bmatrix} 1 & R_2 \\ 1 & -R_2 \end{bmatrix}^{-1} \begin{bmatrix} x_2 \\ y_2 \end{bmatrix} \quad (9.17)$$

and with $R_1 = R_2 = R = 1/G$, which usually is the case, we obtain

$$C = \begin{bmatrix} c_{11} & c_{12} \\ c_{21} & c_{22} \end{bmatrix} = \frac{1}{2} \begin{bmatrix} A - BG - RC + D & A + BG - RC - D \\ A - BG + RC - D & A + BG + RC + D \end{bmatrix} \quad (9.18)$$

and

$$S = \frac{1}{\Delta} \begin{bmatrix} A + BG - RC - D & 2(AD - BC) \\ 2 & -A + BG - RC + D \end{bmatrix} \quad (9.19)$$

where $\Delta = A + BG + RC + D$. For a reciprocal two-port, we have $AD - BC = 1$.

Henceforth we will denote the incident and reflected voltage waves with $A(s)$ and $B(s)$, respectively, in accordance with what is customary in the literature for wave digital filters, i.e., $x = A$ and $y = $ B. Note the conflict in notation with the scattering matrix parameters and A, B, C, and D.

Example 9.1 Determine the scattering matrix parameters of voltage type for a two-port consisting of a series impedance according to Fig. 9.4.

We first determine the K matrix. We have

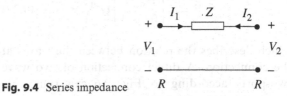

Fig. 9.4 Series impedance

---

[1]Power waves, which normally are used in the literature for microwave filters, are defined as $x_1 = V_1/\sqrt{R_1} + \sqrt{R_1}I_1$ and $y_1 = V_1/\sqrt{R_1} - \sqrt{R_1}I_1$ [70]. The name comes from the fact that x^2 and y^2 have the dimension power.

$$V_1 = ZI_1 + V_2$$

$$I_1 = -I_2$$

i.e., $K = \begin{bmatrix} 1 & Z \\ 0 & 1 \end{bmatrix}$. Insertion of $A = 1$, $B = Z$, $C = 0$, and $D = 1$ in Equation (9.19) yields the voltage scattering matrix for a series impedance

$$S = \begin{bmatrix} s_{11} & s_{12} \\ s_{21} & s_{22} \end{bmatrix} = \frac{1}{Z + 2R} \begin{bmatrix} Z & 2R \\ 2R & Z \end{bmatrix}. \quad (9.20)$$

The voltage chain scattering matrix becomes according to Equation (9.18)

$$C = \begin{bmatrix} c_{11} & c_{12} \\ c_{21} & c_{22} \end{bmatrix} = \frac{1}{2R} \begin{bmatrix} 2R - Z & Z \\ -Z & 2R + Z \end{bmatrix}. \quad (9.21)$$

9.3 Interconnection of Wave Two-Ports

Kirchhoff's laws are valid for the two connected ports shown in Fig. 9.5, i.e.,

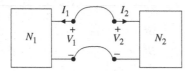

Fig. 9.5 Interconnection of two networks

$$\begin{cases} V_2 = V_1 \\ I_2 = -I_1. \end{cases}$$

These equations correspond to the matrices

$$K = \begin{bmatrix} 1 & 0 \\ 0 & 1 \end{bmatrix} \text{ and } T = \begin{bmatrix} 1 & 0 \\ 0 & -1 \end{bmatrix}.$$

Insertion in Equation (9.6) yields

$$F = P \begin{bmatrix} 1 & 0 \\ 0 & -1 \end{bmatrix} Q^{-1} \quad (9.22)$$

which describes the relation between the waves at the connection. A direct connection of two wave two-ports according to Fig. 9.6 can thus only occur if $x_1 = y_2$ and $x_2 = y_1$, i.e., if F for the connection is

$$F = \begin{bmatrix} 1 & 0 \\ 0 & 1 \end{bmatrix}.$$

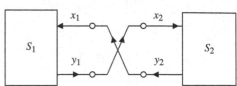

Fig. 9.6 Direct interconnection of two wave two-ports

Several selections of the parameters in the P and Q matrices are possible, but only some of these retain the power concept and yield low sensitive filters and at some time result in low-complexity realizations. Here, we therefore only discuss the selection that corresponds to the voltage scattering parameters.

9.4 Elementary Wave Two-Ports

Consider a two-port that consists of a series inductor and that both port resistances are equal, i.e., $R_1 = R_2 = R$. The scattering matrix becomes according to Equation (9.20)

$$S_{SeriesL} = \begin{bmatrix} s_{11} & s_{12} \\ s_{21} & s_{22} \end{bmatrix} = \frac{1}{2R + sL} \begin{bmatrix} sL & 2R \\ 2R & sL \end{bmatrix}$$

which can be written as

$$S_{SeriesL} = \frac{1}{1 + s\tau} \begin{bmatrix} s\tau & 1 \\ 1 & s\tau \end{bmatrix} \quad (9.23)$$

where $\tau = L/2R$.

Note that the parameters in the voltage scattering matrix for a series inductor are first-order rational functions, i.e., $s_{11} = s_{22} = s\tau/(1 + s\tau)$ and $s_{12} = s_{21} = 1/(1 + s\tau)$ have highpass and lowpass characteristics, respectively.

For simplicity, we will henceforth use the symbol shown in Fig. 9.7 for a wave two-port with a voltage scattering matrix according to Equation (9.23), i.e., for a series inductor.

Fig. 9.7 Wave two-port for a series inductor

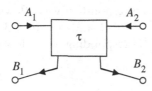

The corresponding chain scattering matrix for a series inductor becomes

$$C_{SeriesL} = \begin{bmatrix} c_{11} & c_{12} \\ C_{21} & c_{22} \end{bmatrix} = \begin{bmatrix} 1 - \tau_1 s & \tau s \\ -\tau_1 s & \tau_1 s - 1 \end{bmatrix}. \quad (9.24)$$

The scattering matrix for a two-port consisting of a series capacitor is in the same way obtained from Equation (9.20)

$$S_{SeriesC} = \frac{1}{2R + \frac{1}{sC}} \begin{bmatrix} \frac{1}{sC} & 2R \\ 2R & \frac{1}{sC} \end{bmatrix}$$

$$= \frac{1}{1 + s\tau_1} \begin{bmatrix} 1 & s\tau_1 \\ s\tau_1 & 1 \end{bmatrix} \quad (9.25)$$

where $\tau_1 = 2RC$.

Also in this case, the scattering matrix parameters are first-order rational functions with either lowpass or highpass characteristic. The corresponding chain scattering matrix for a series capacitor is

$$C_{SeriesC} = \begin{bmatrix} \frac{\tau_1 s - 1}{\tau_1 s} & \frac{1}{\tau_1 s} \\ \frac{-1}{\tau_1 s} & \frac{\tau_1 s + 1}{\tau_1 s} \end{bmatrix}. \quad (9.26)$$

The voltage scattering matrices for a series inductor and series capacitor are

$$\begin{bmatrix} B_{1\,SeriesL} \\ B_{2\,SeriesL} \end{bmatrix} = \begin{bmatrix} \frac{s\tau}{1+s\tau} & \frac{1}{1+s\tau} \\ \frac{1}{1+s\tau} & \frac{s\tau}{1+s\tau} \end{bmatrix} \begin{bmatrix} A_{1\,SeriesL} \\ A_{2\,SeriesL} \end{bmatrix} \quad (9.27)$$

and

$$\begin{bmatrix} B_{1\,SeriesC} \\ B_{2\,SeriesC} \end{bmatrix} = \begin{bmatrix} \frac{1}{1+s\tau} & \frac{s\tau}{1+s\tau} \\ \frac{s\tau}{1+s\tau} & \frac{1}{1+s\tau} \end{bmatrix} \begin{bmatrix} A_{1\,SeriesC} \\ A_{2\,SeriesC} \end{bmatrix}. \quad (9.28)$$

Note that the series capacitor matrix can be obtained from the series inductor matrix by changing the rows and vice versa. That is, $B_{1\,SeriesC}$ has the same form as $B_{2\,SeriesL}$ and $B_{2\,SeriesC}$ has the same form as $B_{1\,SeriesL}$.

This means that if we have a realization of a wave two-port for a series inductor, a realization of a wave two-port for a series capacitor is obtained by changing the reflected waves B_1 and B_2 according to Fig. 9.8 where $\tau = L/2R = 2RC$. This corresponds to multiplying the $\mathbb{C}$ matrix with the matrix

Fig. 9.8 Wave two-port for a sereis capacitor

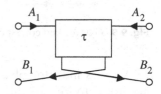

$$\begin{bmatrix} 0 & 1 \\ 1 & 0 \end{bmatrix}.$$

Theorem 9.1 *If the reflected waves of a wave two-port to a series admittance, Y, are changed according to Fig. 9.8, a wave two-port corresponding to a series impedance $Z = 4R^2 Y$ is realized.*

Example 9.2 Show that a series resonance circuit with the impedance $Z_1 = sL_1 + 1/sC_1$ in a series arm corresponds to a series admittance with a parallel resonance circuit if the waves on the wave two-ports are changed according to Fig. 9.8.

According to Theorem 9.1, we have $Y_2 = Z_1/4R^2 = sL_1/4R^2 + 1/sC_1 4R^2 = sC_2 + 1/sL_2$, which corresponds to a series arm with a parallel resonance circuit with the elements $L_2 = C_1 4R^2$ and $C_2 = L_1/4R^2$.

In the same way can the wave two-ports for shunt arms be derived. Here instead we shall use an equivalent technique that is based on gyrators.

A gyrator is described by Equation (5.17) by the chain matrix

$$K \begin{bmatrix} 0 & r_1 \\ \frac{1}{r_2} & 0 \end{bmatrix} \quad (9.29)$$

where $r_1 = r_2$. Insertion in Equation (9.19) yields

$$S_{Gyrator} = \begin{bmatrix} 0 & -1 \\ 1 & 0 \end{bmatrix} \quad (9.30)$$

and

$$C_{Gyrator} = \begin{bmatrix} -1 & 0 \\ 0 & 1 \end{bmatrix}. \quad (9.31)$$

A gyrator, with $r_1 = R$ and $r_2 = 1/R$, and corresponding wave-flow graph are shown in Fig. 9.9.

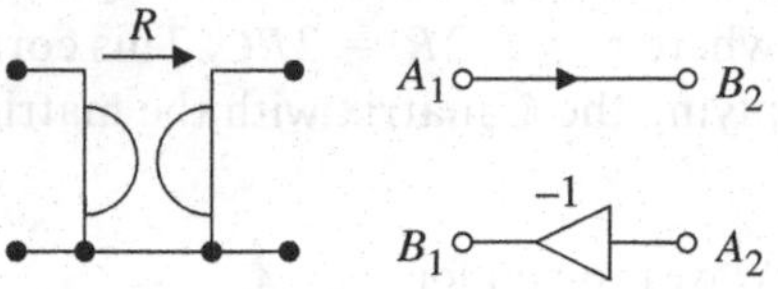

Fig. 9.9 Gyrator and corresponding wave two-port

A series inductor placed between two gyrators according to Fig. 8.7 thus corresponds to inverting B_1 and A_1 of the wave two-ports for a shunt inductor. Gyrators are, thus, cheap to implement as they only require one inversion.

Table 9.1 shows some of the most common two-ports with series elements and the corresponding voltage wave two-ports. Table 9.2 shows some wave-flow diagrams for shunt elements and the

corresponding wave-flow diagrams with $\tau = L/2R = RC/2$, and in Table 9.3 some useful one-ports and corresponding wave-flow diagrams are shown.

9.5 Higher-Order Wave One-Ports

Figure 9.10 shows the symbol for a three-port circulator and its corresponding wave-flow diagram. Note that an incident wave to port 1 is reflected to port 2, and an incident wave to port 2 is reflected to port 3, and so on. Hence, the circulator "circulates" the incident waves to the next port in order.

The scattering matrix is

$$
\begin{bmatrix} B_1 \\ B_2 \\ B_3 \end{bmatrix} = \begin{bmatrix} 0 & 0 & 1 \\ 1 & 0 & 0 \\ 0 & 1 & 0 \end{bmatrix} \begin{bmatrix} A_1 \\ A_2 \\ A_3 \end{bmatrix} = S \begin{bmatrix} A_1 \\ A_2 \\ A_3 \end{bmatrix}. \quad (9.32)
$$

Table 9.1 Series impedances

Two-port	Wave two-port	Element values
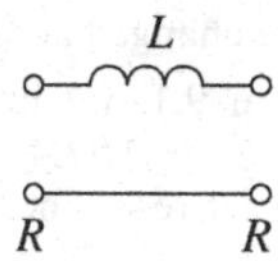	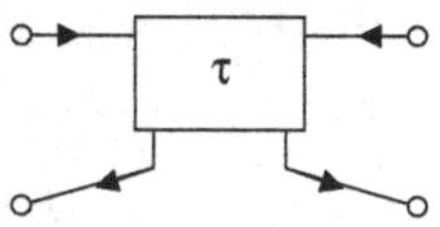	$\tau = L/2R$
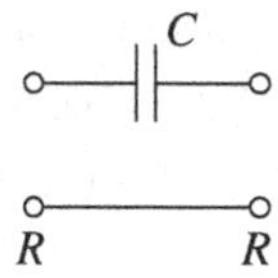	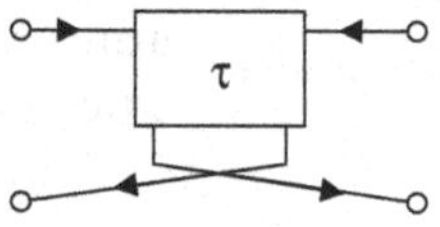	$\tau = 2RC$
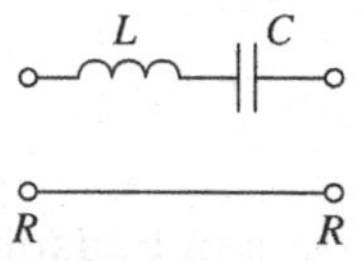	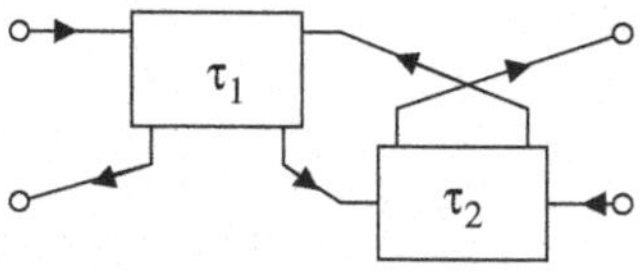	$\tau_1 = L/2R$ $\tau_2 = 2RC$
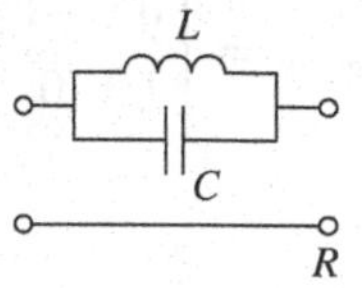	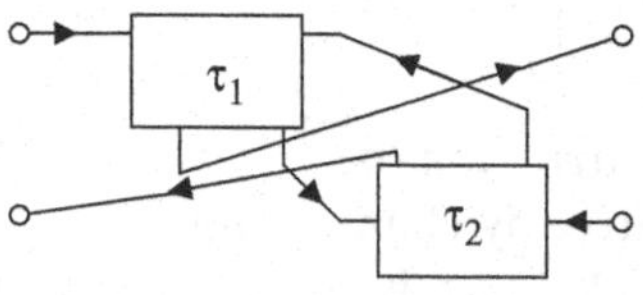	$\tau_1 = 2RC$ $\tau_2 = L/2R$

Table 9.2 Shunt impedances

Two-port	Wave two-port	Element values
		$\tau = RC/2$
		$\tau = 2L/R$
		$\tau_1 = RC/2$ $\tau_2 = 2L/R$
		$\tau_1 = 2L/R$ $\tau_2 = RC/2$

The input impedance to port 1 in Fig. 5.12 is

$$Z = \frac{R^2 + Z_2 Z_3}{Z_2 + Z_3} \qquad (9.33)$$

where Z_2 and Z_3 are the impedances connected to port 2 and 3, respectively. The corresponding reflectance at port 1 is

$$S = \frac{Z - R}{Z + R} = \frac{\dfrac{R^2 + Z_2 Z_3}{Z_2 + Z_3} - R}{\dfrac{R^2 + Z_2 Z_3}{Z_2 + Z_3} + R} = \left(-\frac{Z_2 - R}{Z_2 + R} \right)\left(-\frac{Z_3 - R}{Z_3 + R} \right) = (-S_2)(-S_3). \qquad (9.34)$$

Hence, the resulting reflectance is the product of the two reflectances. Higher-order reflectances can therefore be realized by connecting several circulators. For example, replacing Z_3 with a second circulator that is terminated with the impedances Z_3 and Z_4 yields a reflectance consisting of three reflectances

$$S = (-S_2)(-S_3)(-S_4)$$

and so on. The order of the scattering matrices is arbitrary. The resulting structure is shown in Fig. 9.11 for the case of four cascaded circulators. Usually, only first- and second-order reactances are used for sensitivity reasons.

Table 9.3 One-ports

One-port	Wave one-port
$I = 0$ $+$ V $-$	A B
$V = 0$ I R	A -1 B
$+$ I V R $-$ R	A $B = 0$
R I V_{in} V R	$B = V_{in}$ A

9.6 Circulator-Tree Wave Active Filters

A method for synthesis of lossless two-ports can be based on factorization of the scattering matrix into a product of lower-order scattering matrices. The factored scattering matrices correspond to a two circulator trees as illustrated in Fig. 9.12 for the special case with four branches.

The scattering matrix S for the complete filter is

$$S = S_4 S_3 S_2 S_1 \qquad (9.35)$$

where S_i is the scattering matrix corresponding to the network N_i. This decomposition into lower-order scattering matrices corresponding to elementary reactance two-ports is only possible if S corresponds to a symmetrical or antisymmetrical two-port. Hence, only odd-order lowpass filters are feasible.

Circulator-tree structures have, similar to lattice structures, high sensitivity in the stopband, because the transmission zeros are formed by cancellation instead of by reflection.

Fig. 9.10 Three-port circulator and its corresponding wave-flow diagram

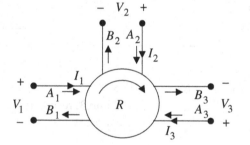
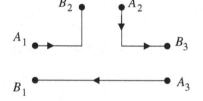

Fig. 9.11 Circulator structure

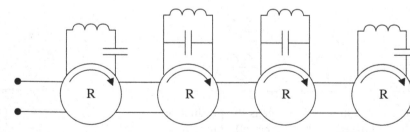

These structures are of less interest in the analog domain, but they are very useful in the digital domain. In fact, they are the basic building block used in lattice wave digital filters [35, 134, 135].

An odd-order lowpass Cauer filter typically consists of a two-port series inductor, and the remaining two-ports consist of series and parallel resonance circuits. The element values for odd-order

Fig. 9.12 Circulator-tree structure

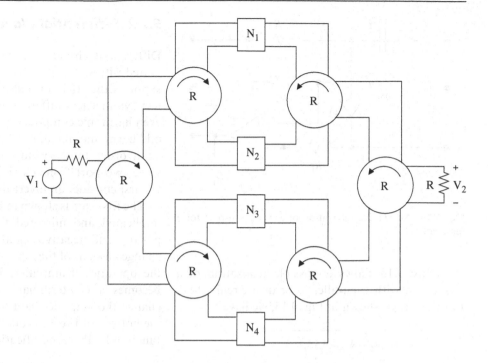

Cauer filters can be determined with the function CIRCULATOR_THREE_LP and CIRCULATOR_THREE_BP for geometrically symmetric bandpass filters.

These structures are of no interest in the analog domain, but they are very useful in the digital domain. In fact, they may be used as analog prototype filters for efficient wave digital filters [35, 135].

9.7 Realization of Wave Two-Ports

To realize a wave two-port, different physical signal carriers can be used, i.e., voltages and currents. In this section, we use voltages, i.e., a voltage will represent a wave and the scattering parameters will be represented by transfer functions of the type output voltage/input voltage.

There are methods for implementation of wave filters that are suitable for integrated circuit techniques, where the waves instead are represented by currents [126, 127]. Here, however, the two incident waves to the wave two-port are represented by input voltages to an active RC network and the reflected waves are represented by output voltages.

9.7.1 Realization of a Generic Wave Two-Port

In Fig. 9.13 and 9.14, two alternative realizations of a wave two-port for a series inductor are shown.

The time constant $\tau = 2RC$ can easily be determined by measuring the frequency for which the phase difference between the input and output is 45°. This can be done with a higher accuracy than when resistances and the capacitances are measured individually. The resistor and the capacitor values can be selected arbitrary, as only their product is important. Note that the $+1$ amplifiers have high input impedances and do not load the driving circuits [90].

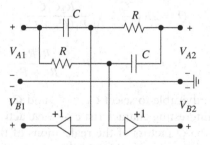

Fig. 9.13 Realization of a wave two-port for a series inductor

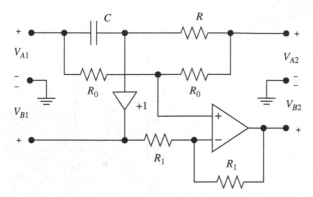

Fig. 9.14 Alternative realization of a wave two-port for a series inductor

Figure 9.15 shows a possible realization of a series arm with a parallel resonance circuit [143]. For the circuit shown in Fig. 9.15, we have

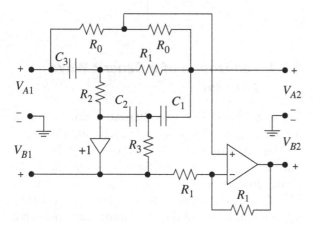

Fig. 9.15 Realization of a wave two-port for a parallel resonance circuit in the series arm

$$\omega_0^2 = \frac{1}{LC} = \frac{1}{R_1 R_2 C_1 C_2}$$

$$Q = 2R\sqrt{\frac{C}{L}} = \sqrt{\frac{R_2 C_1 C_2}{R_1(C_1 + C_2)}}$$

$$C_3 = C_1 + C_2 \qquad R_3 = \frac{R_1 R_2}{R_1 + R_2}.$$

It is advisable to select $C_1 = C_2$ and $C_3 = 2C_1$.

An interesting characteristic of wave active filters is that the Q factors of the realizations of the wave two-ports become considerably lower than the corresponding realization with sections in cascade form.

9.7.2 Differential Wave Two-Port

Differential circuits are commonly used in integrated filters. The main reason is that they tend to suppress the effect of nonlinearities. In fact, a perfect symmetrical differential circuit suppresses the even harmonic components of a sinusoidal, but the odd harmonics remain.

In order to show how to derive a differential wave two-port from its single-ended counterpart, we first consider an inverting amplifier.

The first step is shown in Fig. 9.16 where we have duplicated and mirrored the amplifier to obtain positive and negative signal paths. In addition, we change all signs of the input and output terminals of the operational amplifier. Hence, a (+)-terminal becomes a (−)-terminal and vice versa and we change the amplifier gain to −A. The amplifier at the bottom of Fig. 9.16 does not change its transfer function by these modifications.

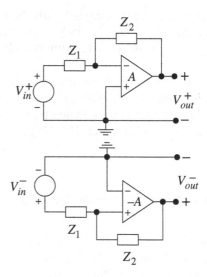

Fig. 9.16 First step in the derivation of a differential vave two-port

Notice that none of the grounded terminals are needed, and we can merge the two operational amplifiers in the two paths each into one fully differential operational amplifier as shown in Fig. 9.17.

Now, consider the inverting amplifier shown in Fig. 9.18. The output is

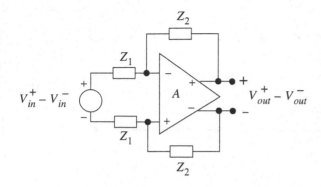

Fig. 9.17 Differential inverting amplifier

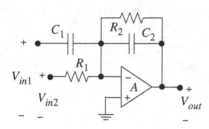

Fig. 9.18 Inverting amplifier

$$V_{out} = -\frac{Z_2}{1/sC_1} V_{in1} - \frac{Z_2}{R_1} V_{in2}$$

where $Z_2 = \frac{R_2}{R_2 C_2 s + 1}$. We get by selecting $R_1 C_1 = R_2 C_2 = \tau$

$$V_{out} = \frac{\tau s}{\tau s + 1} V_{in1} - \frac{1}{\tau s + 1} V_{in2}. \qquad (9.36)$$

Comparing Equation (9.36) with Equation (9.27), we find that the circuit in Fig. 9.18 may be used to realize a wave two-port corresponding to a series inductor.

Using the technique discussed above, we convert the single-ended amplifier with a fully differential

operational amplifier with a differential output port. We need two such circuits, one for B_1 and another for B_2. The resulting circuits are shown in Fig. 9.19 where we have reversed the output voltages to compensate for the minus signs in Equation (9.36). A drawback of this realization is that four time constants, τ, have to be matched.

9.8 Realization Of Wave Active Filters

We demonstrate the design of a wave active filter by the means of an example. Wave active filters have been implemented using current-mode circuits [126, 127].

Example 9.3 Realize a third-order LC ladder network of the type mid-shunt with an active wave filter. Figure 9.20 shows the reference filter where the elementary two-ports have been marked.

From Tables 9.1, 9.2 and 9.3 the corresponding wave two-ports can be identified. The two-ports can be connected directly because the port resistance has been selected equal for all wave two-ports. Figures 9.21 and 9.22 show the wave active filter and the corresponding realization.

The input signal is the voltage wave A_1, which corresponds to the input voltage V_{in}, and the normal output wave is VB_2, which corresponds to the output voltage V_B2. Note that for $\omega = 0$ we have $H(0) = V_{B2}/V_{in} = 1$.

9.9 Power Complementarity

For a lossless two-port, we have the scattering parameters

$$|s_{11}|^2 + |s_{21}|^2 = 1. \qquad (9.37)$$

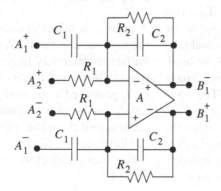
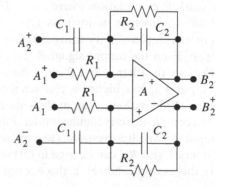

Fig. 9.19 Realization of a differential wave two-port

Fig. 9.20 Reference filter, third-order *LC* filter

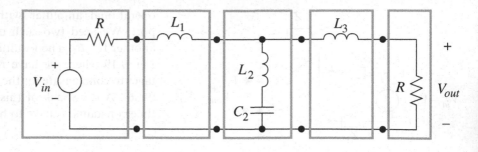

Fig. 9.21 Wave active filter corresponding to a third-order *LC* filter

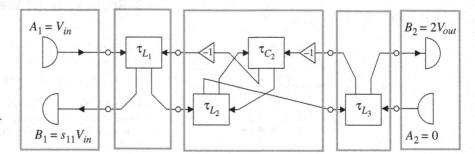

Fig. 9.22 Realization of a wave active filter corresponding to a third-order *LC* filter

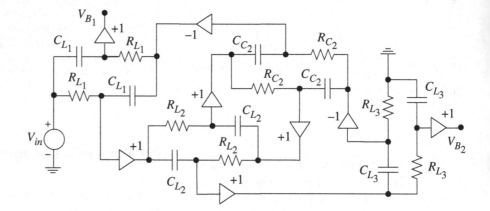

Equation (9.37) has a direct correspondence to Feldtkeller's equation where $s_{21} = H(s)$. If s_{21} is a transfer function of lowpass type, then s_{11} is the power complementary transfer function of highpass type. Even the output signal B_1 can, thus, be used. Because the filter is an LP filter, the complementary output V_{B1} has highpass characteristics. The wave active filter realizes, thus, at the same time a lowpass and complementary highpass filter. Furthermore, the input A_2, which corresponds to a second signal source in series with R_L, can be used in certain applications. In this example, however, this has not been done.

9.10 Alternative Approach

A drawback of the above-discussed method is that the number of amplifiers, i.e., positive and negative buffers, is relatively large; that is, one per reactive element in the reference filter. Haritantis has, however, proposed a method to reduce the number of amplifiers using a combination of Gorski-Popiel's method and the wave active approach [47].

Wave active filters can also be implemented with transconductors and current conveyors [126, 127].

Fig. 9.23 Magnitude and reflection function

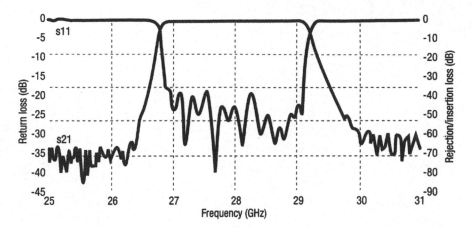

9.11 Problems

9.1 Determine the scattering parameters for a two-port consisting of a shunt admittance Y.

9.2 Show that the power into a port is $P = G(|a|^2 - |b|^2)$.

9.3 Use the LC ladder filter in Fig. 3.36 as reference filter and find the corresponding wave active filter.

9.4 Use an active wave filter structure shown in Fig. 3.41 to realize the crossover filter pair.

9.5 Propose an operational amplifier based circuit realization of a high-performance buffer with high input impedance and gain $= -1$.

9.6 Determine the passband ripple and stopband attenuation from the measurements shown in Fig. 9.23.

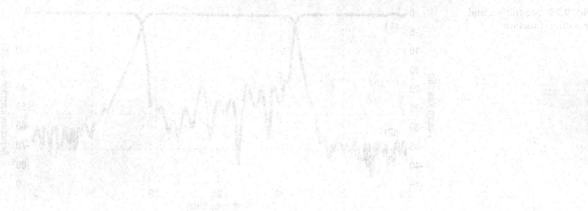

Chapter 10
Topological Simulation

10.1 Introduction

The best filter structures, from a sensitivity point of view, simulate doubly resistively terminated LC filters. Simulation can be made in several different ways. Previously, we have discussed several variants of immitance simulation and wave active filters. In this section, methods where the voltage-current relationships of the ladder network are preserved, i.e., the networks topology, is simulated [14, 15, 16, 18, 32]. The resulting filter structures have as expected low sensitivity.

10.2 LP Filters Without Finite Zeros

A simulating method proposed by F.E.I. Girling and E.F. Good (1969) [45] simulates node voltages and branch currents in the *reference filter*, i.e., the LC filter. The resulting filter structure is called a leapfrog filter, as the signal-flow graph can be drawn so that it can be associated with the well-known children's game "leapfrog."

A leapfrog filter realizes a signal-flow graph that is obtained from the current and voltage equations and that describes an LC filter. By modifying the equations, we can derive a signal-flow graph that only contain integrators, adders, and inverters. Good integrators can be implemented using only operational amplifiers, resistors, and capacitors [42] but

also with switched-capacitor (SC) techniques. Leapfrog filter is the structure that is most commonly used in SC filters, as good SC integrators can be implemented in digital CMOS technologies. The design method for leapfrog filters is easily described with the help of a few examples.

Example 10.1 It is instructive to derive the leapfrog structure as was done originally. Consider therefore the fifth-order T ladder shown in Fig. 10.1 and compute the currents in the series arms and voltages across shunt arms.

We have

$$\begin{cases} I_1 = \dfrac{V_{in} - V_2}{R_s + sL_1} & V_2 = \dfrac{I_1 - I_3}{sC_2} \\ I_3 = \dfrac{V_2 - V_4}{sL_3} & V_4 = \dfrac{I_3 - I_5}{sC_4} \\ I_5 = \dfrac{V_4 - V_{out}}{sL_5} & V_{out} = R_L I_5. \end{cases}$$

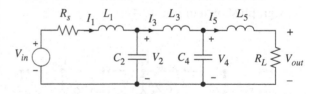

Fig. 10.1 Fifth-order T ladder

The equations above can be represented with the block diagram shown in Fig. 10.2. The name leapfrog is obvious from the diagram. Notice that there are four interconnected two-integrator loops. Hence, we may conclude that the leapfrog structure has similar sensitivities as the previously discussed two-integrator loops.

Fig. 10.2 Leapfrog structure for a fifth-order T ladder

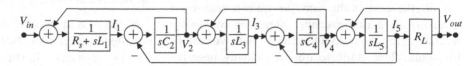

L. Wanhammar, *Analog Filters Using MATLAB*, DOI 10.1007/978-0-387-92767-1_10,
© Springer Science+Business Media, LLC 2009

10.2.1 Lowpass Leapfrog Filters

Consider the doubly resistively terminated lowpass ladder network shown in Fig. 10.3, which is designed to have optimally low sensitivity in the passband. We will later see that it is favorable to select a π ladder instead of a T ladder, as the former in some cases will require fewer amplifiers.

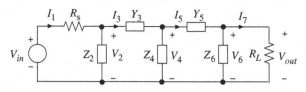

Fig. 10.3 LP ladder

The shunt and series arms are denoted as impedances and admittances, respectively.

We use as (state) variables the currents in the series arms and the voltages across the shunt arms. We get the following equations from left to right, where $G_s = 1/R_s$ and $G_L = 1/R_L$:

$$
\begin{cases}
I_1 = G_s(V_{in} - V_2) & V_2 = Z_2(I_1 - I_3) \\
I_3 = Y_3(V_2 - V_4) & V_4 = Z_4(I_3 - I_5) \\
I_5 = Y_5(V_4 - V_6) & V_6 = Z_6(I_5 - I_7). \\
I_7 = G_L V_6
\end{cases}
\tag{10.1}
$$

It is convenient to multiply the currents with an arbitrary positive constant, R_0, in order to obtain only voltage equations. This is not necessary but practical because it becomes easier to identify different circuit elements. We get

$$
\begin{cases}
V_1 = R_0 I_1 = R_0 G_s(V_{in} - V_2) & V_2 = G_0 Z_2(V_1 - V_3) \\
V_3 = R_0 I_3 = R_0 Y_3(V_2 - V_4) & V_4 = G_0 Z_4(V_3 - V_5) \\
V_5 = R_0 I_5 = R_0 Y_5(V_4 - V_6) & V_6 = G_0 Z_6(V_5 - V_7) \\
V_7 = R_0 I_7 = R_0 G_L V_6
\end{cases}
\tag{10.2}
$$

where $G_0 = 1/R_0$.

For a lowpass filter without finite zeros, the immitances Z_i and Y_i are capacitors and inductors, respectively. Hence, the factors $G_0 Z_i$ and $R_0 Y_i$ correspond to integrators, i.e., of the form k/s.

Equation (10.2) can be represented by the signal-flow graph shown in Fig. 10.4.

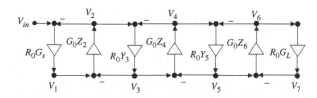

Fig. 10.4 Signal-flow graph

Inverting integrators are in many cases less expensive to implement. However, in a differential circuit, the cost is the same as we can realize an inverting or a noninverting integrator by simply interchanging the output terminals. Here we modify the signal-flow graph to obtain as many negative

integrators as possible, assuming that they are less expensive to implement.

First we recognize that every loop contains a minus sign. In the first step, we change the sign of every other transmittance (integrator), except for the terminating branches. That is, we move the market minus signs across the nodes V_2, V_4, and V_6, which are split into two minus signs, as shown in Fig. 10.5. Hence the signs of V_2, V_4, and V_6 change as well. Check that all loops still have an odd number of minus signs.

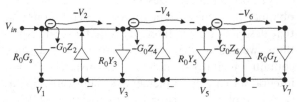

Fig. 10.5 Signal-flow graph after the first design step

In the second step, we move the minus signs, as indicated in Fig. 10.6. Note that two minus signs cancel in each case. Starting from the right, we find

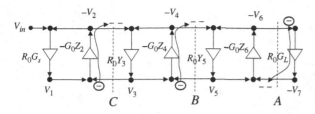

Fig. 10.6 Signal-flow graph after the second design step

that all nodes to the right of the cut A change sign. Hence, the sign of V_7 as changed. Next, all nodes to

the right of cut B also change sign, i.e., we get the new nodes V_6, V_7, and $-V_5$. Finally, all nodes to the right of cut C change sign, i.e., we get the new nodes $-V_3$, V_4, V_5, $-V_6$, and $-V_7$. The final signal-flow graph is shown in Fig. 10.7.

The resulting structure has two positive and three negative integrators that have as inputs the sum of two node voltages. The signal-flow graph shown in Fig. 10.7 corresponds to the Equations (10.3), which of course are equivalent with Equations (10.2).

$$
\begin{aligned}
V_1 &= R_0 G_s(V_{in} + (-V_2)) & -V_2 &= -G_0 Z_2(V_1 + (-V_3)) \\
-V_3 &= R_0 Y_3((-V_2) + V_4) & V_4 &= -G_0 Z_4((-V_3) + V_5) \\
V_5 &= R_0 Y_5(V_4 + (-V_6)) & -V_6 &= -G_0 Z_6(V_5 + (-V_7)). \\
-V_7 &= R_0 G_L(-V_6)
\end{aligned}
\tag{10.3}
$$

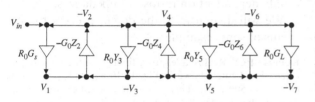

Fig. 10.7 Signal-flow graph after the third design step

Note that the output voltage often, as in this case, obtains an extra phase shift of π rad, i.e., we get $V_{out} = -V_6$ instead of $V_{out} = V_6$. In most applications this is of little concern.

The nodes in the lower part of the figure correspond to the voltages across the shunt branches whereas the nodes in the upper part correspond to the currents in the series branches.

Finally, we modify the signal-flow graph in Fig. 10.7 by moving the input signal to the input of the left-most integrator, as shown in Fig. 10.8. Moreover, we rename the input transmittance $R_0 G_s$ to $R_0 G_{in}$ for reasons to be explained later.

The vertical branches in a lowpass filter contain either constant transmittances or functions of the type k/s where k is a real constant, i.e., integrators. Negative integrators correspond to capacitors in the reference filter whereas positive integrators correspond to inductors. Hence, we prefer reference filters with few inductors.

Note that every loop, except for the two outer loops, consist of interconnected two-integrator loops that have the positive integrator in common. Hence, it may be advisable to realize the negative integrators using Miller integrators and the positive integrators with the phase-lead integrator, as the phase errors in the integrators tend to cancel [42]. The inner loops have lossless integrators whereas the two outer loops have lossy integrators.

A similar signal-flow graph can be derived for highpass filters without finite zeros.

10.2.2 Realization of the Signal-Flow Graph

The element values in the active realization can directly be obtained through identification with the modified signal-flow graph. The signal-flow graph can be realized by the circuit shown in Fig. 10.9. Note that each integrator has two inputs. For example, the integrators corresponding to the series admittances Y_3

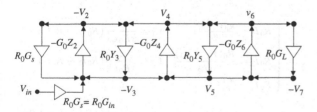

Fig. 10.8 Modified signal-flow graph

Fig. 10.9 Fifth-order
leapfrog filter

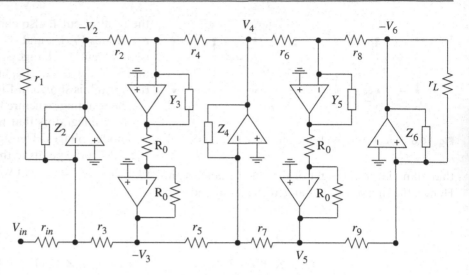

and Y_5 have inputs from V_4 and $(-V_2)$ and V_4 and $(-V_6)$, respectively. The left-most integrator has inputs from V_{in}, $(-V_2)$, and $(-V_3)$. Note that the terminating resistors yields lossy integrators.

We use lowercase letters for the elements in the leapfrog realization in order to differentiate from the elements in the reference filter and the transmittances, which use capital fonts. The element denoted R_0 can be chosen arbitrarily. The relationships between the reference filter and the leapfrog filter can be derived by comparing the voltage equations for the nodes. See Example 10.2.

The reader is recommended to derive the relationships. See Problem 10.1.

If the reference filter is denormalized in frequency and has the desired frequency response, the corresponding leapfrog filter will automatically have the same frequency response. On the other hand, if the reference filter is normalized in frequency, the leapfrog filter has to be denormalized. This can be done by letting $S = s/k$, where S is the frequency variable of the reference filter. Before and after frequency scaling, we have for the integrators, which are the only frequency-dependent components in the leapfrog filter,

$$\frac{\pm 1}{SRC} = \frac{\pm k}{sRC}.$$

Hence, we must make the RC products k times smaller in order to increase the frequency by a factor k. However, there is no need for scaling the

impedance level in the reference filter as the leapfrog filter is independent of the impedance level; it is only dependent on ratios of impedances.

We will demonstrate the design process by the means of some examples.

Example 10.2 Realize a leapfrog filter from the third-order Chebyshev I filter shown in Fig. 10.10 when $A_{max} = 1$ dB and $\omega_c = 500$ rad/s. This is the same specification as was used for the lowpass filter in Example 7.8.

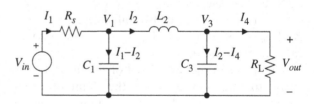

Fig. 10.10 Third-order LC ladder

The normalized element values are $C_1 = C_3 = 2.0236$, $L_2 = 0.9941$, and $R_s = R_L = 1$. The equations for the currents in the series arms and voltages across the shunt arms, when all currents have been multiplied with a positive constant R_0, are

$$\begin{cases} R_0 I_1 = \frac{R_0(V_{in}-V_1)}{R_s} & V_1 = \frac{1}{sR_0C_1}(R_0I_1 - R_0I_2) \\ R_0 I_2 = \frac{R_0(V_1-V_3)}{sL_2} & V_3 = \frac{1}{sR_0C_3}(R_0I_2 - R_0I_4). \\ R_0 I_4 = \frac{R_0 V_3}{R_L} \end{cases}$$

The equations above are represented by the signal-flow graph shown in Fig. 10.11.

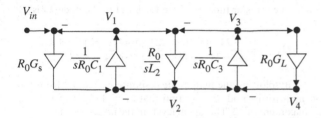

Fig. 10.11 Signal-flow graph for a third-order *LC* filter

By propagating the minus signs as was done above, we obtain the signal-flow graph shown in Fig. 10.12. The corresponding circuit realization is shown in Fig. 10.13

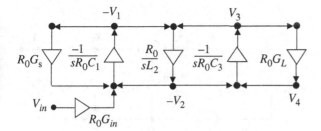

Fig. 10.12 Modified signal-flow graph for a third-order *LC* filter

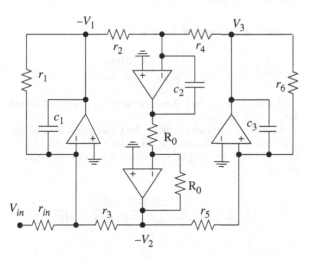

Fig. 10.13 Realization of a third-order leapfrog filter

The element values in the active realization can be determined by comparising the circuits shown in Fig. 10.12 and 10.13. We calculate for each integrator starting from left to right. For the left-most integrator we get

$$\begin{cases} -V_1 = \dfrac{-1}{sR_0C_1}[R_0G_s(-V_1) + R_0G_{in}V_{in} + (-V_2)] \\ -V_1 = \dfrac{-1}{sc_1}\left[\dfrac{(-V_1)}{r_1} + \dfrac{V_{in}}{r_{in}} + \dfrac{(-V_2)}{r_3}\right] \end{cases}$$

and by identifying corresponding terms we get

$$\frac{G_s}{C_1} = \frac{1}{r_1c_1} \quad \frac{G_{in}}{C_1} = \frac{1}{r_{in}c_1} \quad \frac{G_0}{C_1} = \frac{1}{r_3c_1}$$

and

$$R_sC_1 = r_1c_1 \quad R_{in}C_1 = r_{in}c_1 \quad R_0C_1 = r_3c_1. \quad (10.4)$$

For the second integrator we get

$$\begin{cases} -V_2 = \dfrac{R_0}{sL_2}[(-V_1) + V_3] \\ -V_2 = \dfrac{-1}{sc_2}\left[\dfrac{(-V_1)}{r_2} + \dfrac{V_3}{r_4}\right]\left(\dfrac{-R_0}{R_0}\right) \end{cases}$$

and

$$L_2/R_0 = r_2c_2 \quad L_2/R_0 = r_4c_2. \quad (10.5)$$

Finally, for the last integrator we get

$$\begin{cases} -V_3 = \dfrac{-1}{sR_0C_3}[R_0G_L(-V_3) + (-V_2)] \\ -V_4 = \dfrac{-1}{sc_3}\left[\dfrac{(-V_4)}{r_6} + \dfrac{(-V_3)}{r_5}\right] \end{cases}$$

and

$$R_L/C_3 = r_6c_3 \quad R_0C_3 = r_5c_3. \quad (10.6)$$

In order to check the validity of the result, it is advisable to compare the loop gains between the signal-flow graph and the circuit.

With these constraints, we find the following normalized resistors when all capacitors have been chosen to unity, $c_1 = c_2 = c_3 = 1$ and $r_1 = r_3 = r_5 = r_6 = 2.0236$, $r_2 = r_4 = 0.9941$, and $R_0 = 1$ while $r_{in} = 2.0236R_{in}$ is free to be chosen to a suitable positive value.

The realization that is shown in Fig. 10.13 has two coupled two-integrator loops where the positive integrator is shared between the two loops. The two-integrator loops are similar to a Tow-Thomas section. Hence, the realization is sensitive for the operational amplifiers' finite bandwidth, which, as discussed in Section 6.8, leads to increased Q factors. It is therefore better to use a positive phase-lead integrator that counteracts the phase errors that are introduced by the Miller integrators.

Note that leapfrog filters are sensitive for errors in the phase response of the inverters and integrators, which is due to the operational amplifiers' finite bandwidth. The phase errors can be compensated with a resistor in series with the integrators' capacitors and a capacitor in parallel with the input resistor to the inverters, as discussed in Section 5.9.2 [83, 116].

A drawback is that these filters often obtain relatively many operational amplifiers. Another drawback is that the filters in practice are limited to transfer functions that can be realized with ladder networks, i.e., transfer functions with minimum phase.

Example 10.3 Scale the element values in Example 10.2 so that the cutoff frequency becomes $\omega_c = 500$ rad/s.

The element values derived in Example 10.2 correspond to a normalized cutoff frequency, $\omega_{c\ norm} = 1$. We need to make all RC products ω_c times smaller.

First, we select three equal size capacitors, e.g., $c_1 = c_2 = c_3 = 10$ nF. The denormalized RC products are $RC = r_{norm} c_{norm}/\omega_c$ where we previously selected $c_{norm} = 1$. Hence, with $C = 10$ nF, the resistors should therefore be multiplied with a factor $1/C\omega_c = 2 \cdot 10^5$. The denormalized element values are $c_1 = c_2 = c_3 = 10$ nF, $r_1 = r_3 = r_5 = r_6 = 404.720$ kΩ, and $r_2 = r_4 = 198.820$ kΩ. The resistor r_{in} does not affect the frequency response, it only affects the gain.

10.2.3 Scaling of Signal Levels

An advantage with leapfrog filters is that the signal levels are easy to scale so that the signal dynamic becomes optimal, i.e., so that the available dynamic signal range of the operational amplifiers are efficiently utilized [112]. We continue the previous example and show how the signal levels can be optimally scaled.

Example 10.4 Consider the leapfrog filter that is shown in Fig. 10.12. The leapfrog filter has three critical nodes, i.e., amplifier outputs, that need to be scaled. We assume that the gain of the filter shall be equal to 12.

We successively introduce scaling coefficients, starting form the node that is closest to the input. Using scaling coefficient k_1, shown in Fig. 10.14, we can scale the node X_1. Note that we have split the transmittance $R_0 G_s$, shown in Fig. 10.8, into an input transmittance $R_0 G_{in}$ and a feedback factor $R_0 G_s$, as the former may be used for adjusting the gain whereas the latter must be unchanged because it affects the poles.

The transfer function from the input to the node X_1 is

$$H_{X_1}(s) = \frac{-247.0844(s^2 + 247.0844s + 124275.4)k_1}{(s + 247.0844)(s^2 + 247.0844s + 248550.9)}.$$

By plotting the magnitude response, measured from the input to the node X_1, we find that $k_1 = 18.727015$ yields a maximum of 12. The signal levels in the nodes X_2 and X_3 are, of course, affected by the scaling coefficient k_1.

Next, we introduce the scaling coefficient k_2 in order to scale node X_2. However, in order to not affect the feedback loop, we must introduce a coefficient $1/k_2$, as shown in Fig. 10.15. The transfer function from the input to the node X_2 must, of course, be recomputed because we have scaled all nodes with a factor k_1. In fact, X_2 and X_3 are k_1 times larger.

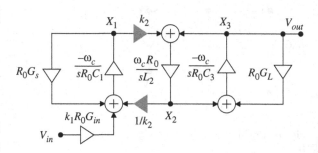

Fig. 10.15 Scaling of node X_2

The transfer function from the input to the node X_2 is

$$H_{X_2}(s) = \frac{-2327430k_2}{(s + 247.0844)(s^2 + 247.0844s + 248550.9)}.$$

A maximum of the magnitude function of 12 is obtained with $k_2 = 0.615346$.

Finally, we introduce the scaling coefficient k_3 and the compensating coefficient $1/k_3$ in order to scale node X_3, as shown in Fig. 10.16. The re-computed transfer function to the node X_3 is

$$H_{X_3}(s) = \frac{3.5384946 \cdot 10^8 k_3}{(s^2 + 247.0844)(s^2 + 247.0844s + 248550.9)}.$$

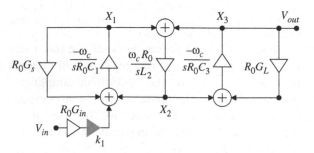

Fig. 10.14 Scaling of node X_1

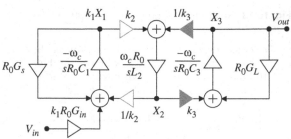

Fig. 10.16 Scaling of node X_3

Fig. 10.17 Scaled magnitude functions

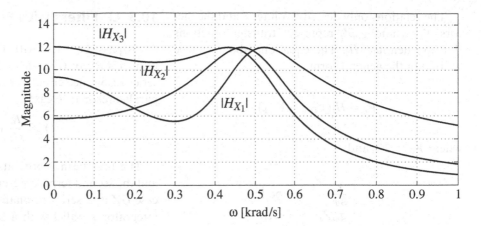

A maximum of the magnitude function of 12 is obtained with $k_3 = 2.0824$. The three scaled magnitude functions are shown in Fig. 10.17.

To scale the circuit that is shown in Fig. 10.13, we first scale the node X_1 (corresponding to the node marked $-V_2$ in Fig. 10.13) by increasing the gain by a factor k_1. Hence, we divide input resistance r_{in} = by k_1, i.e., we get $r_{in} = 404.720/k_1 = 12.161$ kΩ. The resistor $r_1 = 404.72$ kΩ is unchanged.

Next, we scale the node X_2 (corresponding to the node marked $-V_3$ in Fig. 10.13) by increasing the gain by a factor k_2, i.e., we get $r_2 = 198.820/k_2 = 323.10$ kΩ and $r_3 = 404.720k_2 = 249.043$ kΩ.

Finally, we scale the node X_3 (corresponding to the node marked V_4) by increasing the gain by a factor $k_3 = 2.082684$, i.e., we get $r_4 = 198820k_3 = 414.079$ kΩ and $r_5 = 404.720/k_3 = 194.326$ kΩ while $r_6 = 404.720$ kΩ is unchanged.

10.3 Geometrically Symmetric BP Leapfrog Filters

A geometrically symmetric bandpass leapfrog filter can easily be realized by first synthesizing the signal-flow graph to the corresponding LP filter and then performing the LP-BP transformation, see Example 7.8. The integrators in the LP filter are transformed to second-order transfer functions, which can be realized per some of the previously discussed sections.

10.4 Lowpass Filters Realized with Transconductors

Leapfrog structures are well suited for implementation using transconductors. The basic principle is the same as described before, but we have some

additional constraints to consider. It is desired that all transconductors have the same transconductance, which simplifies the tuning. Inaccuracies due to the bottom plate parasitics are reduced as all capacitors are grounded.

Consider therefore the ladder network shown in Fig. 10.18 where $Y_1 = 1/R_s$ and $Z_6 = R_L$.

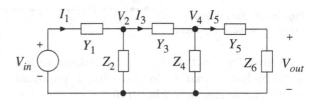

Fig. 10.18 Reference filter

For the reference filter, we have the following relations

$$\begin{cases} I_1 = Y_1(V_{in} - V_2) & V_2 = Z_2(I_1 - I_3) \\ I_3 = Y_3(V_2 - V_4) & V_4 = Z_4(I_3 - I_5) \\ I_5 = Y_5(V_4 - V_{out}) & V_{out} = Z_6 I_5 \end{cases} \quad (10.7)$$

In a realization with active RC technique, the current relations are multiplied with an arbitrary constant R_0 so all relations describe voltages. In the same way, we here divide with a common constant g_m. We get

$$\begin{cases} V_1 = \frac{Y_1}{g_m}(V_{in} - V_2) & V_2 = g_m Z_2(V_1 - V_3) \\ V_3 = \frac{Y_3}{g_m}(V_2 - V_4) & V_4 = g_m Z_4(V_3 - V_5) \\ V_5 = \frac{Y_5}{g_m}(V_4 - V_{out}) & V_{out} = g_m Z_6 V_5. \end{cases}$$

The relations only describe voltages, and the factors Y_i/g_m and $g_m Z_i$ represent transfer functions that are denoted H_i. The equations above can be written in the general form

$$V_i = H_i(V_{i-1} - V_{i+1})$$

where $V_0 = V_{in}$ and

$$H_i = \begin{cases} \dfrac{Y_i}{g_m} & i = \text{odd} \\ g_m Z_i & i = \text{even.} \end{cases}$$

These equations are represented by the signal-flow graph shown in Fig. 10.19.[1] Note that in this case, all feedback coefficients are equal to −1, which is simple to implement.

10.5 LP Filters with Finite Zeros

A lowpass LC filter with finite zeros will contain branches RY or GZ, where Y and Z are parallel and series resonance circuits, respectively. Consider the parallel resonance case,

$$RY = \frac{R}{sL} + sRC.$$

The first term represents an integrator parallel and the second term represents a differentiator. The case GZ and series resonance circuit also yields an integrator parallel with a differentiator. Hence, a lowpass LC filter with finite zeros cannot directly be realized with the above method because the corresponding leapfrog will contain differentiators, which are unsuitable to use as their gain increases with frequency [18, 112].

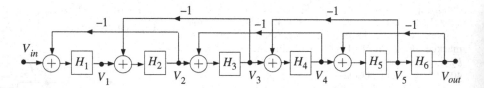

Fig. 10.19 Signal-flow graph for a leapfrog filter

A transfer function H_i can be realized with a transconductor that is loaded with a grounded impedance as shown in Fig. 10.20.

Geometric symmetric BP filters without finite zeros can, however, be realized by replacing the integrators to second-order sections, i.e., perform-

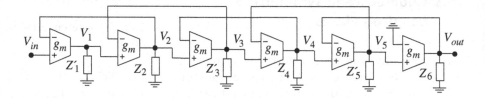

Fig. 10.20 Leapfrog filter with grounded impedances

Note that the grounded impedances Z_i' for $i = \text{odd}$ are not the same as the elements in the original reference filter and that it is possible to use the same transconductance for all transconductors. Hence, we may use a single control circuit to control all transconductors. Moreover, several of the impedances will be grounded capacitors, which is favorable. The remaining impedances must, however, be realized by active circuits.

ing an LP-to-BP transformation as was done in Example 7.8. By modifying the above method, which will be discussed later, filters with finite zeros can be realized.

Leapfrog filters with finite zeros can, however, be realized by, instead of using the currents through the series arms as state variables, only using the currents in corresponding series inductors[2].

---

[1] From the signal-flow graph that is drawn in this form, it is clear why the graph is associated with the children's game leapfrog.

[2] Proposed by Sven Eriksson, Linköping University.

10.5.1 Odd-Order Lowpass Filters with Finite Zeros

We explain the method by the means of examples. Consider the third-order π ladder shown in Fig. 10.21.

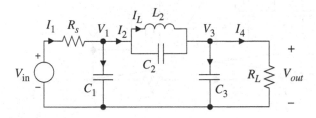

Fig. 10.21 *LC* filter with finite zeros

The currents in the series arms and the voltages over the shunt arms are

$$
\begin{cases}
RI_1 = \dfrac{R(V_{in} - V_1)}{R_S} & V_1 = \dfrac{1}{sRC_1}(RI_1 - RI_2) \\[2mm]
RI_2 = RY_2(V_1 - V_3) & V_3 = \dfrac{1}{sRC_3}(RI_2 - RI_4) \\[2mm]
RI_4 = \dfrac{RV_{out}}{R_L}
\end{cases}
\tag{10.8}
$$

where

$$
Y_2 = sC_2 + \frac{1}{sL_2}. \tag{10.9}
$$

Let I_L represent the current through the series inductor. We have for the series arm

$$
RI_2 = RY_2(V_1 - V_3) = \left(sC_2 + \frac{1}{sL_2}\right)R(V_1 - V_3)
$$
$$
= sRC_2(V_1 - V_3) + RI_L
$$

where $I_L = (V_1 - V_3)/sL_2$. We get from Equation (10.8) and by eliminating the current I_3

$$
V_1 = \frac{1}{sRC_1}\left(RI_1 - (sRC_2(V_1 - V_3) + RI_L)\right)
$$
$$
= \frac{1}{sRC_1}(RI_1 - RI_L) - \frac{C_2}{C_1}V_1 + \frac{C_2}{C_1}V_3
$$

and

$$
V_1 = \frac{1}{sR(C_1 + C_2)}(RI_1 - RI_L) + \frac{C_2}{(C_1 + C_2)}V_3. \tag{10.10}
$$

In the same way we get

$$
V_3 = \frac{1}{sR(C_2 + C_3)}(RI_L - RI_4) + \frac{C_2}{(C_2 + C_3)}V_1. \tag{10.11}
$$

Equations (10.8) through (10.11) correspond to the circuit shown in Fig. 10.22. Each of the shunt arms contains a VCVS and a capacitor with modified capacitance value. The series arm has been replaced with an inductor L_2 with the current I_L.

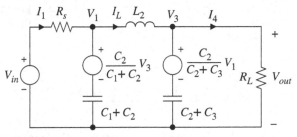

Fig. 10.22 Ladder with finite zeros

The network shown in Fig. 10.22 is represented by the signal-flow graph that is shown in Fig. 10.23. The two VCVS are represented by the branches with the transmittances $k_1 = -C_2/\alpha_1$ and $k_2 = -C_2/\alpha_2$, where $\alpha_1 = C_1 + C_2$ and $\alpha_2 = C_2 + C_3$, respectively.

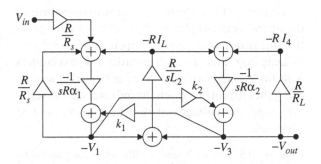

Fig. 10.23 Leapfrog filter with finite zeros

The circuit shown in Fig. 10.24 can be used to realize Equation (10.10) and (10.11). We have for the circuit

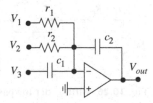

Fig. 10.24 Miller integrator with additive inputs

Fig. 10.25 Realization of a third-order leapfrog filter with finite zeros

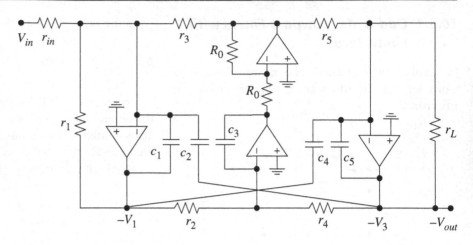

$$V_{out} = \frac{-V_1}{sr_1c_2} + \frac{-V_2}{sr_2c_2} + \frac{-c_1V_3}{c_2}.$$

Figure 10.25 shows the realization of the leapfrog filter corresponding to a lowpass filter with finite zeros.

The capacitances needed to realize the zeros are

$$c_1 = C_1 + C_2, c_2 = C_2/(C_1 + C_2),$$
$$c_4 = C_2 + C_3 \text{ and } c_5 = C_2/(C_2 + C_3).$$

To reduce the requirement on the bandwidth of the operational amplifiers, it is better to use a positive integrator with active compensation.

Leapfrog filters for higher-order lowpass ladders with finite zeros can be derived in the same way by using the currents through the inductors in the series arms and the voltages of the shunt arms as state variables.

Example 10.5 Fig. 10.26 shows a fifth-order lowpass ladder with finite zeros. Use this ladder to design a leapfrog filter that meets the specification in Example 3.7, i.e., $N = 5$, $\rho = 50\%$, $A_{min} = 40.3$ dB, $\omega_c = 10$ krad/s.

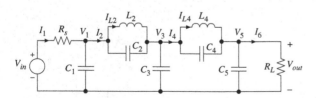

Fig. 10.26 Fifth-order lowpass filter with finite zeros

The following element values were earlier determined

$R_s = R_L = 1000$
$C_1 = 2.00056 \text{ e-}07$

$C_2 = 0.3992140 \text{ e-}07$ $\qquad \omega_{02} = 17367.63$
$L_2 = 0.830449 \text{ e-}01$

$C_3 = 2.247705 \text{ e-}07$

$C_4 = 1.188883 \text{ e-}07$ $\qquad \omega_{04} = 12401.59$
$L_4 = 0.546898 \text{ e-}01$

$C_5 = 1.537829 \text{ e-}07$

We proceed as before by writing the equations for the currents in the series arms and voltages over the shunt arms. We get for the LC network

$$\begin{cases} RI_1 = \frac{R(V_{in}-V_1)}{R_s} & V_1 = \frac{1}{sRC_1}(RI_1 - RI_2) \\ RI_2 = RI_{L2} + sRC_2(V_1 - V_3) & V_3 = \frac{1}{sRC_3}(RI_2 - RI_4) \\ RI_4 = RI_{L4} + sRC_4(V_3 - V_5) & V_5 = \frac{1}{sRC_5}(RI_4 - RI_6) \\ RI_6 = \frac{RV_{out}}{R_L} z \end{cases}$$

Eliminating I_3 and I_5 yields

$$V_1 = \frac{1}{sR(C_1 + C_2)}(RI_1 - RI_{L2}) + \frac{C_2}{C_1 + C_2}V_3$$

$$V_3 = \frac{1}{sR(C_2 + C_3 + C_4)}(RI_{L2} - RI_{L4})$$
$$+ \frac{C_2}{C_2 + C_3 + C_4}V_1 + \frac{C_4}{C_2 + C_3 + C_4}V_5$$

$$V_5 = \frac{1}{sR(C_4 + C_5)}(RI_{L4} - RI_6)\frac{C_4}{C_4 + C_5}V_3$$

where $RI_{L2} = \frac{R(V_1 - V_3)}{sL_2}$ and $RI_{L4} = \frac{R(V_3 - V_5)}{sL_4}$.

Fig. 10.27 Fifth-order network with finite zeros

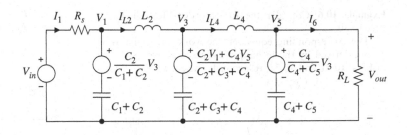

Figure 10.27 shows the corresponding network where the currents in the series inductors have been used as state variables.

Finally, Fig. 10.28 shows the corresponding signal-flow graph for the leapfrog filter where $\alpha_1 = C_1 + C_2$, $\alpha_2 = C_2 + C_3 + C_4$, $\alpha_3 = C_4 + C_5$, $k_1 = -C_1/\alpha_1$, $k_2 = -C_2/\alpha_2$, $k_3 = -C_4/\alpha_2$, and $k_4 = -C_4/\alpha_3$.

The cross-coupling coefficients k_1 through k_4 can be realized by using the circuit shown in Fig. 10.24 without any extra cost in terms of operational amplifiers compared with an allpole filter. The element values in the circuit shown in Fig. 10.29 are obtained by comparing with the signal-flow graph in Fig. 10.28.

Leapfrog filters with finite zeros are, thus, slightly more complex to design, but they do not require any additional operational amplifiers. In [18, 19, 24, 112],

a general method for designing leapfrog filters that use more complicated structures as reference filter is presented.

10.5.2 Even-Order Lowpass Filters with Finite Zeros

The technique discussed above can directly be used for higher-order odd filters, but for even-order filters some modification is required. Once again, we demonstrate the design method by the means of an example.

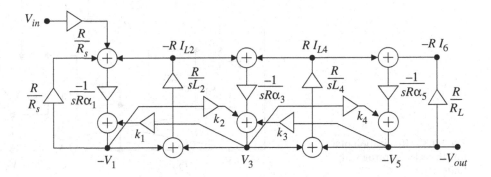

Fig. 10.28 Fifth-order leapfrog filter with finite zeros

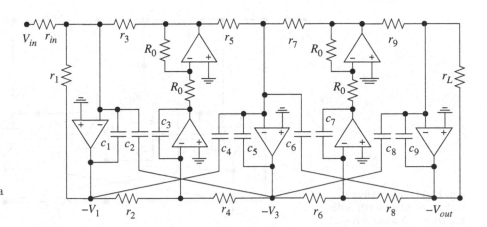

Fig. 10.29 Realization of a fifth-order leapfrog filter with finite zeros

Example 10.6 Consider the fourth-order filter shown in Fig. 10.30.

The corresponding equivalent network and signal-flow graph, derived by the technique discussed above, is shown in Figs. 10.31 and 10.32, respectively, where $\alpha_1 = C_1 + C_2$ and $\alpha_2 = C_2 + C_3 + C_4$.

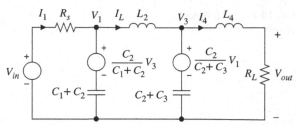

Fig. 10.31 Fourth-order network with finite zeros

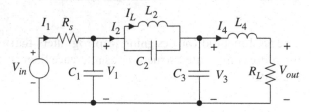

Fig. 10.30 Fourth-order LC ladder

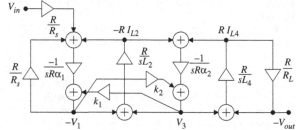

Fig. 10.32 Fourth-order leapfrog with finite zeros

Figure 10.33 show the corresponding leapfrog realization of the signal-flow graph in Fig. 10.32. An alternative realization with phase-lead integrators is shown in Fig. 10.34. Notice how the resistor r_8 is connected.

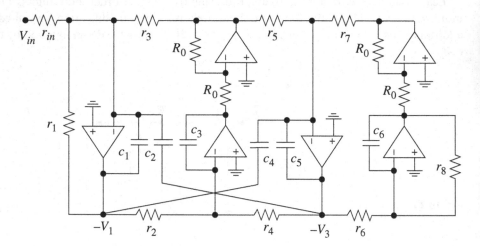

Fig. 10.33 Realization of a fourth-order leapfrog with finite zeros

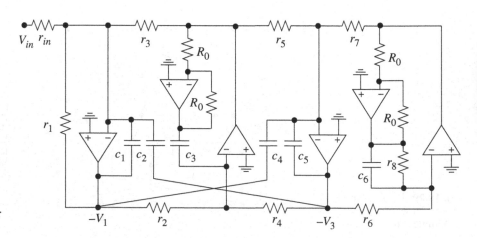

Fig. 10.34 Alternative realization of a fourth-order leapfrog with finite zeros

10.6 Problems

10.1 Derive the relationships between the elements in the reference filter and the leapfrog filer in Fig. 10.9.

10.2 Compute the Q factors in the two-integrator loops and compare these with the Q factors of the Chebyshev I filter in Example 10.2.

10.3 Use an LC ladder of π type as reference filter and synthesize an active leapfrog filter of Chebyshev I type that meets the following specification. Passband: $0 \leq f \leq 4$ kHz, $A_{max} = 0.1$ dB. Stopband: $f > 12$ kHz, $A_{min} = 20$ dB. Use the following element values: $R_s = R_L = 1$ kΩ, $C_1 = C_3 = 41$ nF, and $L_2 = 45.7$ mH.

10.4 Use the filter in Fig. 10.35 as reference filter and synthesize the corresponding active leapfrog filter. $R_s = R_L = 1$ kΩ, $L_1 = 647$ mH, and $C_2 = 57$ nF. Use if possible capacitors with 10 nF and resistors with 10 kΩ.

10.5 Realize the Butterworth filter in Example 3.4 as a leapfrog filter.

10.6 Show how the signal-flow graph in Example 10.2 can be transformed into a symmetric geometric BP filter with center frequency 10 kHz and a bandwidth of 2 kHz.

10.7 Determine the element values in the leapfrog filter shown in Fig. 10.33 when the filter shall

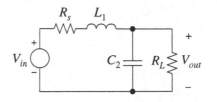

Fig. 10.35 Resonance circuits in problem 10.4

realize the transfer function $C040522b$ and the cutoff edge is $\omega_c = 100$ krad/s.

10.8 Derive a leapfrog realization of third-order T ladder and compare the result with a leapfrog filter that is derived from a corresponding π ladder.

10.9 Determine suitable element values for the realization derived in Example 10.5.

10.10 Determine suitable element values for the realization derived in Example 10.6.

10.11 Use the circuit shown in Fig. 10.25 to realize the Cauer filter $C031516$ with $\omega_c = 150$ krad/s. Determine first suitable element values and then perform scaling of the signal levels in order to maximize the dynamic signal range.

10.12 Derive a leapfrog filter from a third-order HP ladder with all zeros at $s = 0$.

10.13 Determine suitable output nodes in the circuits in Figs. 10.33 and 10.34.

Chapter 11
Tuning Techniques

11.1 Introduction

Analog filters are typically implemented using matched components in order to reduce the influence of component errors as well as temperature and power supply voltage variations, etc. For example, special symmetric layouts of the amplifiers are used to reduce the spread of the performance parameters and achieve tracking over large temperature and power supply voltage ranges. Moreover, the temperature coefficient for the resistors and capacitances are chosen equal, but with opposite sign, so that the RC products become invariant with temperature.

Tuning of the parameter values is still necessary in most filters and in particular when they are implemented as integrated circuits, because the component element values vary strongly between different chips and also with temperature and power supply voltage. It requires one or more control circuits and a scheme where the filter properties are measured and adjusted so that the performance requirements are met. Even for filters with moderate performance requirements is therefore some form of on-chip tuning required [24, 56, 117, 119]. The circuits for tuning of an integrated filter may occupy 10–20% of the chip area and may have a considerable power consumption. Hence, the design and implementation of such control circuitry is a non-trivial problem. Here we will only discuss the most commonly used tuning methods. It is beyond the scope of this book to discuss more advanced methods that are based on digital signal processing techniques [69]. However, as technology improves, these method may become more viable alternatives.

In practice, we have two different situations: on-line and off-line tuning. In some applications, the tuning must be done continuously, i.e., while the filter is performing its filtering. Hence, it is not possible to apply special test signals. This case is referred to as on-line tuning.

In other applications, e.g., read channel in a hard disk, the tuning can be performed off-line as the filter only has valid input signals during finite periods of time. In between these active time slots, it is possible to apply test signals and tune the filter because the tuning procedure may only take a small amount of time to perform.

11.2 Component Errors

The values of both passive and active components affect the frequency response of the filter. The size of these errors depends on the technology used to implement the components.

11.2.1 Absolute Component Errors

Twenty percent or larger errors in the absolute values of integrated resistors and capacitors can be expected. Tolerances of the gain-bandwidth product of 50% in an integrated amplifier are common. Hence, an integrated filter will require postfabrication tuning. Discrete resistors can have errors as low as 0.1%, but are expensive, whereas capacitors typically have larger errors. Moreover, the resistors may in some technologies be trimmed.

L. Wanhammar, *Analog Filters Using MATLAB*, DOI 10.1007/978-0-387-92767-1_11,
© Springer Science+Business Media, LLC 2009

11.2.2 Ratio Errors

Fortunately, carefully designed components can be integrated with very accurate component ratios. For example, for capacitors using a polysilicon-oxide-polysilicon layout, accuracies in the capacitance ratios as low as 0.1% or better can be achieved. Special layout techniques based on arrays of unit size capacitors are used, see Fig. 5.54, to achieve high accuracy ratios. Similar accuracies can be achieved for resistors, see Fig. 5.44.

The success of switched-capacitor filters and other switched-capacitor circuits, e.g., ADCs, is due to the fact that integrated capacitors can be manufactured with very accurate component ratios. Hence no trimming is needed.

The ratio of the transconductances of two identical transconductors may be better than 1% for moderate transconductances. To achieve high accuracy in the transconductance ratios and simplify tuning, the transconductances are quantized into small integers times a common transconductance, e.g., ng_{m0}. A transconductor with $g_m = 3g_{m0}$ can be implemented as illustrated in Fig. 11.1.

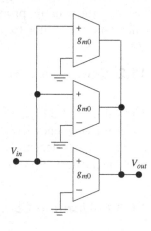

Fig. 11.1 Transconductor implemented with unit-size transconductors

11.2.3 Dummy Components

The input and output capacitances of the transconductor have large variations depending on process variations and operating conditions. These capacitances affect the frequency response of the filter. For high-frequency applications requiring large g_m/C ratios, the capacitors become small in order to realize a pole pair with a large radius. In addition,

the transconductors should have high output current in order to charge and discharge the capacitive load faster. This requires larger transistors, which yield larger parasitic capacitances. Hence, for high-frequency applications we have to accept that the parasitic capacitances are no longer negligible. In such cases are dummy transconductors used to make the effective capacitive load independent of these capacitances. We demonstrate the idea by the means of an example.

Example 11.1 Consider the circuit in Fig. 11.2, which is derived from a parallel resonance RLC circuit. Assume $Q = 0.5$ and $H_{LP}(0) = 1$.

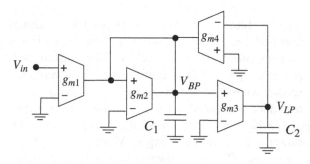

Fig. 11.2 Second-order section

$$H_{BP}(s) = -\frac{g_{m1}}{C_1}\frac{s}{s^2 + \frac{g_{m2}}{C_1}s + \frac{g_{m3}g_{m4}}{C_1C_2}} \quad (11.1)$$

$$H_{LP}(s) = -\frac{g_{m1}g_{m3}}{C_1C_2}\frac{1}{s^2 + \frac{g_{m2}}{C_1}s + \frac{g_{m3}g_{m4}}{C_1C_2}}. \quad (11.2)$$

We select $C_1 = C_2 = C$, $g_{m1} = g_{m3} = g_{m4} = g_{m0}$, and $g_{m2} = 2g_{m0}$. Hence, $r_p = g_{m0}/C$ and, hence, we may tune the circuit by g_{m0} as C is assumed to be fixed [58].

Consider the capacitances at the LP and BP nodes. We have

$$C_{LP} = C_2 + C_{out3} + C_{in4}$$

$$C_{BP} = C_1 + C_{out1} + C_{out2} + C_{out4} + C_{in2} + C_{in3}.$$

Because $g_{m2} = 2g_{m0}$, i.e., two identical transconductors with g_{m0} in parallel, we have $C_{out2} = 2C_{out1}$ and $C_{in2} = 2C_{in1}$, which simplifies to

$$C_{LP} = C + C_{out} + C_{in}$$

$$C_{BP} = C + 4C_{out} + 3C_{in}.$$

Obviously, the effective capacitance in the two nodes is not the same.

In order to make the effective capacitances equal, we add dummy transconductors so that the same fraction, f, of the effective capacitance is due to the parasitic capacitances in all nodes. Hence

$$f = \frac{C_{p1}}{C_{BP}} = \frac{C_{p2}}{C_{LP}} \qquad (11.3)$$

where C_{p1} and C_{p2} are the parasitic capacitances for nodes 1 and 2, respectively.

If Equation (11.3) holds, we have $C_{BP} = \frac{C}{C-f}$, $C_{LP} = \frac{C}{C-f}$, and $C_{BP} = C_{LP}$.

Hence, the effective capacitances in the two nodes are independent of the parasitic capacitances and their variation. Hence, using this approach the relative positions of the poles and zeros are independent of the parasitics. Their absolute positions are set by tuning g_{m0}.

Because C_{LP} is smaller than C_{BP}, we select to add the dummy transconductors to the LP node. The resulting circuit with dummy transconductors is shown in Fig. 11.3. Of course, the dummy transconductors can be significantly simplified. They need only to have either the correct input or output capacitance. Hence, the required chip area, power consumption, and noise contribution may be reduced.

The BP node has the same effective capacitance as before, but the LP node now has the effective capacitance $C_{LP} = C + 4C_{out} + 3C_{in}$.

Hence, making all g_{m0}/C ratios equal simplifies the tuning because only one tuning signal is required. In addition, any tuning error will cause a shift in the frequency response, i.e.,

$$H(s) \rightarrow H((1 + \varepsilon)s) \qquad (11.4)$$

where ε is the tuning error in the g_{m0}/C ratio.

11.3 Trimming

Here we use the term "trimming" for prefabrication adjustment of the circuit and the term "tuning" for postfabrication adjustments. The poles and zeros are in operational amplifier-based structures set by RC products and in transconductor-based structures by g_m/C ratios. The accuracy of the RC products relies on the absolute values of resistors and capacitors.

In discrete circuits, we may perform trimming by adjusting a few components, usually resistors, until the circuit functions as specified. Often, the

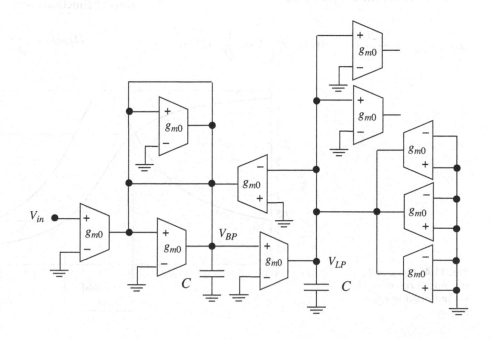

Fig. 11.3 Second-order section with dummy transconductors

trimming procedure for discrete component implementations becomes complicated because the resistance can only be increased by cutting away material from a resistor using a laser. However, the method permits us to use, for example, low-tolerance capacitors, but the trimming procedure is of course expensive.

11.3.1 Trimming of Second-Order Sections

Trimming of second-order sections is usually done by successively applying a sinusoidal input signal and measuring either ratio of the magnitude of the output/input signal or their phase difference. It is here useful to once again discuss the suitable selections of these reference frequencies for second-order sections.

11.3.1.1 Lowpass Section

The transfer function corresponding to Fig. 11.4 is

$$H_{LP}(s) = \frac{Gr_p^2}{s^2 + \frac{r_p}{Q}s + r_p^2} \qquad (11.5)$$

where $r_p = 1$, $Q = 5$, and $G = 1$.

We have for the lowpass case

$$|H(j\omega)|_{max} = \frac{2GQ^2}{\sqrt{4Q^2 - 1}} \text{ at } \omega = r_p\sqrt{1 - \frac{1}{2Q^2}} \quad (11.6)$$

which for high Q factors tends to

$$|H(j\omega)|_{max} \approx GQ \text{ at } = r_p. \qquad (11.7)$$

Of course, there is no peak in the magnitude response if $Q \le 1/\sqrt{2}$.

Several trimming and tuning schemes are based on this fact, i.e., using a sinusoidal input signal with the frequency given in Equation (11.6), the lowpass section is trimmed until the desired maximum in the magnitude function is obtained. A similar scheme is to use a sinusoidal input signal with frequency $\omega_{in} = r_p$ and trim the section until the phase is $\Phi = -\pi/2$.

Other frequencies of interest for trimming are

$$\omega_{-45} = r_p\left[\sqrt{1 + \frac{1}{4Q^2}} - \frac{1}{2Q}\right] \qquad (11.8)$$

$$\omega_{-135} = r_p\left[\sqrt{1 + \frac{1}{4Q^2}} + \frac{1}{2Q}\right]. \qquad (11.9)$$

Hence, the Q factor may be trimmed so that the phase is $-45°$ and $-135°$ at ω_{-45} and ω_{-135}, respectively. Note that $\omega_{-45} - \omega_{-135} = 2\sigma_p$.

If the Q factor and r_p cannot be trimmed independently, the procedure may have to be repeated several times.

11.3.1.2 Highpass Section

The transfer function corresponding to Fig. 11.5 is

$$H_{BP}(s) = \frac{Gr_p^2}{s^2 + \frac{r_p}{Q}s + r_p^2} \qquad (11.10)$$

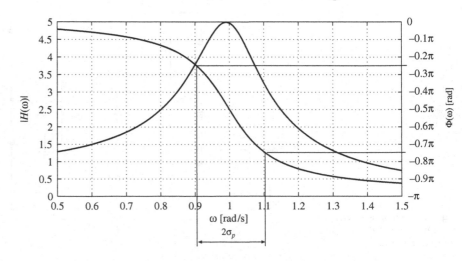

Fig. 11.4 Phase and magnitude response of a second-order lowpass section

Fig. 11.5 Phase and magnitude response of a second-order highpass section

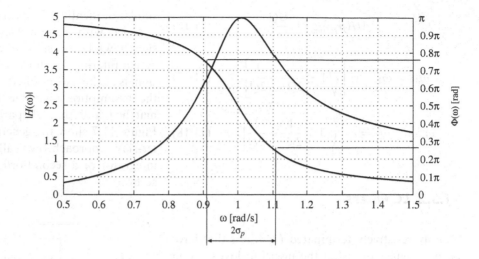

Figure 11.5 show the magnitude and phase response for a highpass section where $r_p = 1$, $Q = 5$, and $G = 1$. In the highpass case we have

$$|H(j\omega)|_{max} = \frac{2GQ^2}{\sqrt{4Q^2 - 1}}$$

$$\text{at } \omega = \frac{\sqrt{2}}{\sqrt{2Q^2 - 1}} Q r_p. \quad (11.11)$$

Other frequencies of interest for trimming are

$$\omega_{135} = r_p \left[\sqrt{1 + \frac{1}{4Q^2}} - \frac{1}{2Q} \right] \quad (11.12)$$

$$\omega_{90} = \frac{r_p}{\sqrt{1 - \frac{1}{2Q^2}}} \quad (11.13)$$

$$\omega_{45} = r_p \left[\sqrt{1 + \frac{1}{4Q^2}} + \frac{1}{2Q} \right]. \quad (11.14)$$

The trimming of a highpass section can be performed similarly to a lowpass section.

11.3.1.3 Bandpass Section

The transfer function corresponding to Fig. 11.6 is

$$H_{BP}(s) = \frac{Gs}{s^2 + \frac{r_p}{Q}s + r_p^2} \quad (11.15)$$

Figure 11.6 show the magnitude and phase response for a bandpass section where $r_p = 1$, $Q = 5$, and $G = 1$. In the bandpass case we have

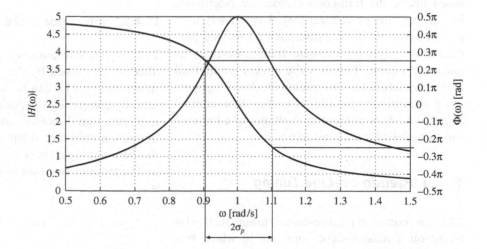

Fig. 11.6 Phase and magnitude response of a second-order bandpass section

$$|H(jr_p)|_{max} = \frac{GQ}{r_p} \qquad (11.16)$$

$$\omega_{45} = \omega_1 = r_p \left[\sqrt{1 + \frac{1}{2Q^2}} - \frac{1}{2Q} \right] \qquad (11.17)$$

$$\omega_{-45} = \omega_2 = r_p \left[\sqrt{1 + \frac{1}{2Q^2}} + \frac{1}{2Q} \right]. \qquad (11.18)$$

11.3.2 LC Filters

Doubly resistively terminated *LC* filters that have been designed by using the insertion loss method have optimally low element sensitivity in the passband. Moreover, if we use an approximation with small group delay, e.g., Cauer filter and small ripple in the passband, according to Equation (3.25), the specification can be met with high tolerance component. Hence, in practice it is often sufficient in, for example, a lowpass *T* ladder to use inductors with relatively high tolerances for the series branches and only tune the resonance frequencies of the shunt branches. Because of the low passband sensitivity, this scheme is often sufficient.

11.4 On-Line Tuning

In on-line tuning, an on-chip control loop is used to tune the integrated filter by electronically varying some component values. In the case of transconductor-based filters, the transconductances are determined by bias currents or voltages [58]. A commonly used practice is to realize all transconductors as a parallel connection of unit-size transconductors as discussed above. This has the advantage that all transconductors tend to track with changing operating conditions. In addition, the same control signal, i.e., bias current or voltage, can be used for all unit-size transconductors.

11.4.1 Pseudo-on-Line Tuning

The most common pseudo-on-line tuning method is based on a master-slave approach in which two similar or identical filters are used. The master filter is tuned with a control circuit using special test signals while the slave filter, which performs the actual filtering, is controlled with the same control signals as the master filter. This method is based on the assumption that the two filters are closely matched and that all parameters vary similarly. Figure 11.7 show the generic structure of the master-slave approach. Typically, the master filter is an integrator or a second-order lowpass or bandpass section.

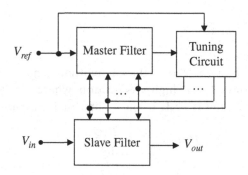

Fig. 11.7 The principle of master-slave tuning

11.4.2 Master-Slave Frequency Tuning

There exist several similar schemes to tune the *RC* products or C/g_m ratios of active filters. Note that, as discussed in Section 6.4.1, it is more important to reduce the error in the pole radius than errors in the *Q* factor.

11.4.2.1 Integrator-Based Tuning

The frequency response of a transconductor-based filter is determined by the ratio g_m/C. The circuit shown in Fig. 11.8 can be used to control the g_m/C ratio by comparing the peak level of the sinusoidal reference signal before and after it has passed through a reference integrator. The frequency of the reference signal is ω.

The transfer function of the left-most integrator is

$$H(s) = \frac{g_m}{sC}. \qquad (11.19)$$

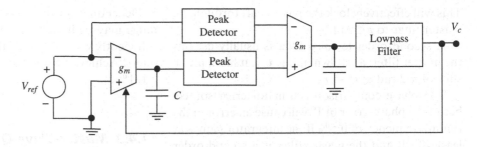

Fig. 11.8 Tuning using unity-gain frequency of the integrator

Solving for $|H(j\omega)| = 1$, we get

$$\omega = \frac{g_m}{C}. \tag{11.20}$$

If the g_m/C ratio is correct at ω_{ref}, the output from the integrator should have the same amplitude as the input. Any difference in amplitudes will be integrated over time by the second integrator, and the control signal V_c is changed to adjust the value of g_m until the correct g_m/C ratio is obtained. The signal V_c is then used to control the transconductances in the slave filter. Obviously, it is advantageous if all transconductors are realized by multiple identical transconductors. However, an input offset in the master integrator will appear at the output amplified by the DC gain of the integrator and the peak detector cannot distinguish between the DC offset and the AC signal. Hence, a significant DC offset will cause a tuning error.

An approach to alleviate this problem is shown in Fig. 11.9, where a lossy integrator is used. The transfer function is

$$H(s) = \frac{1}{1 + \frac{sC}{g_m}}. \tag{11.21}$$

Because $|H(0)| = 1$, the DC offset is not amplified. Here, it is convenient to, instead of using the unity gain frequency, use the -3 dB frequency.

Hence, we need to attenuate the input signal with a factor 0.5. A frequency error of less than 0.1% may be achieved using this approach.

11.4.2.2 Phase-Locked Filter

The main feature of this method is that good matching between master and slave filter is relatively easy to obtain, as both filters may be implemented using the same basic structures.

In Fig. 11.10, a sinusoidal reference signal, V_{ref}, is used as input to the master filter. The phase of the output signal from the filter is compared with that of the reference signal. The phase comparison is done by multiplying the signals, as the DC component of the product of two signals with the same frequency will depend on the phase difference. If there is a $\pi/2$ rad phase difference, the output will be zero. The output from the multiplier is integrated over time and used as the frequency control signal.

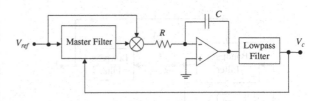

Fig 11.10 Tuning using a plase-locked filter

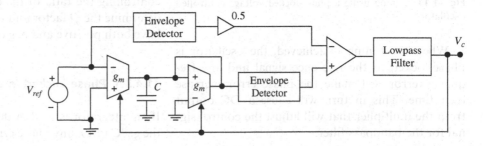

Fig 11.9 Tuning using a lossy integrator

This will effectively lock the phase-shift through the master filter to $\pi/2$ rad.

A second-order lowpass filter is usually used as the master filter; as shown in Fig. 6.7, it has a phase shift of $\pi/2$ rad at $\omega = r_p$.

The phase comparison is a major error source, because a phase-error of $1°$ will cause an error in the tuning frequency of 0.5% if the integrator gain is at least 40 dB and the master filter is a second-order lowpass with a Q factor of 2. Using a section with a higher Q factor will reduce this error. Hence, the phase-locked loop technique is accurate for frequency tuning.

11.4.2.3 Phase-Locked Oscillators

We may use phase-locking of an oscillator implemented with a structure similar to that of the slave filter to eliminate the requirement of a low-distortion sine wave reference signal and reduce the requirement on the accuracy of the phase-detector.

Figure 11.11 shows an oscillator that is formed by inserting a limiter in the feedback loop from the output to the input of a bandpass filter, which must have a passband gain larger than unity. The limiter will limit the peaks of the signal to ensure that the amplitude of the input signal is low enough for the filter to operate as a linear filter. For a too large input signal, the nonlinearities in the bandpass filter will be significant, which will affect the oscillation frequency.

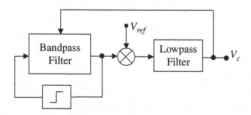

Fig 11.11 Tuning using a phase-locked voltage controlled oscillator

When the tuning is achieved, the oscillator is phase-locked to the reference signal and any frequency error will make the phase error increase over time. This in turn will cause a DC output from the multiplier that will adjust the control signal for the bandpass filter.

Depending on the phase-detector used, locking range may be limited to only one octave, which is sufficiently wide to handle the tuning range of most filters. Achievable frequency errors are in the range 0.1–1%.

11.4.3 Master-Slave Q Factor Tuning

As discussed in Section 6.4.1, the importance of an error in the Q factor for a second-order section is much less than in the pole radius. In addition, the quality factor for a pole pair is a dimensionless quantity that is determined by a ratio of components of the same type, e.g., g_{m1}/g_{m2} or C_1/C_2. Because the ratio of resistors, or capacitors, etc., can be tightly controlled in an integrated circuit, we can expect that the Q factor in many cases is sufficiently accurate, but various parasitic components may cause the Q factor to deviate unacceptably from its desired value. Depending on the size of these deviations, further tuning may be needed.

Many applications of integrated filters require only low Q factors as, for example, Bessel filters in hard disk drives. Only in a few such cases is Q tuning used, as the tuning circuitry tends to consume too much silicon area and power. In addition, signals from the tuning circuitry may leak into the filter and cause interference.

The situation is different in, for example, integrated high-frequency communications where highly selective filters with large Q factors are needed. In such cases, the errors in the Q factors need to be corrected using Q tuning. In addition, we showed earlier that phase errors in a two-integrator loop will have a significant effect on the Q factor. This is also the case in the structures discussed in Chapters 7 through 10.

Common methods to adjust the Q factors are controlling the ratio of the component values that determine the Q factor and using a resistor that can have both positive and negative resistance.

11.4.3.1 Phase-Locked Integrator

If an integrator is used in the master for control of the g_m/C ratio, any phase error will cause the phase difference over the integrator (after frequency tuning) to differ from $\pi/2$ rad.

In the Q tuning scheme shown in Fig. 11.12, the reference signal and the output signals are converted to logic levels and used as inputs to an XOR gate. If the phase difference is not $\pi/2$ rad, the output from the XOR gate will not have a 50% duty cycle. This output will be integrated by the second integrator and yield the control voltage V_c. The accuracy of this method will depend on the achievable phase accuracy of the phase-detector.

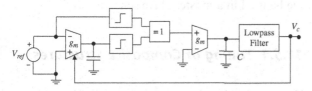

Fig. 11.12 Q factor tuning using the phase difference

11.4.3.2 Passband Gain–Based Q Tuning

Figure 11.13 shows one of the most common ways of implementing Q tuning. It uses the fact that the passband gain of a second-order bandpass section is GQ. A too low Q factor will produce an output signal that is smaller than that of the amplified reference signal; this difference will be integrated over time, until the control signal V_c has changed enough to correct the Q factor. This signal is also used to control the slave filters.

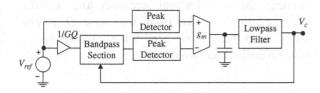

Fig 11.13 Passband gain–based Q tuning

If a second-order section is used as master in the phase-locked filter frequency-tuning loop, a bandpass filtered signal is usually already available in the circuit, otherwise, a separate Q factor tuning master is used.

The assumption that the Q factor is equal to GQ depends on the matching of the transconductors, capacitors, etc., and it ignores the effect of parasitics. Another drawback is that it usually requires a

very slow Q tuning loop to ensure stability. This requirement may not be satisfied at all times during the tuning operation because under certain circumstances the Q may become very sensitive to the changes in the control voltages.

11.4.3.3 LMS-Based Q Tuning

If a frequency-tuning error is present, the reference frequency will not be at the center of the passband. Hence, the measured gain will not be the passband gain of the filter. This will result in a Q factor tuning error, because the tuning circuit will make the measured gain equal to the desired passband gain. This error will be approximately proportional to the Q factor, as the passband width is inversely proportional to the Q factor. This error can be reduced by using the continuous-time adaptive LMS (least mean square) algorithm to generate the control signal. We have

$$V_c(t) = \alpha[V_{ref} - y(t)]\frac{\partial y(t)}{\partial V_c(t)} \qquad (11.22)$$

where $y(t)$ is the measured signal and α is the adaptation factor. Note that V_c becomes zero as the factor $[V_{ref} - y(t)]$ approaches zero. The derivative in Equation (11.22) provides the direction in which the tuning signal should be updated.

Consider the circuit in Fig. 11.14 where the output of the bandpass section should be V_{ref} at $\omega = r_p$. However, the tuning gradient is not available in the circuit, but the output of the bandpass section can be used as tuning gradient, see Problem 11.6. Thus, Equation (11.22) is approximated with

$$V_c(t) = \alpha[V_{ref} - V_{BP}(t)]V_{BP}(t). \qquad (11.23)$$

The filter will be tuned to the correct Q factor, even if the reference frequency is not in the exact center of the passband. This approach is much less

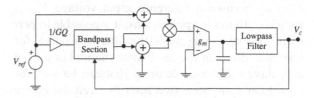

Fig 11.14 Improved passband gain–based Q tuning

sensitive to distortion of the sinusoidal signal V_{ref} than the scheme shown in Fig. 11.13.

A test of a similar circuit with a Q factor of 10 showed that a 3% frequency error would result in a 1.1% error in the Q factor. If the Q tuning circuit shown in Fig. 11.13 had been used, a 3% frequency-tuning error would have resulted in a 16% error in the Q factor. Hence, this later method is superior.

11.4.3.4 Combined Frequency and Q Factor Tuning

Figure 11.15 shows an improved version, which eliminates the requirement of a separate Q tuning master section using a phase-locked oscillator for frequency control. The circuit is less sensitive to offsets in the tuning circuit compared to the previous circuit.

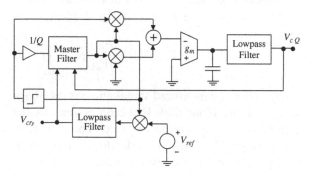

Fig 11.15 Combined frequency and Q tuning scheme

When the frequency and Q factor tuning are not entirely independent, the Q control loop is usually made an order of magnitude slower than the frequency control to make sure that the Q factor tuning is performed at the correct frequency.

11.5 Off-Line Tuning

During the tuning, the filter cannot be used for filtering because a reference input voltage V_{ref} is required. In many applications, it is possible to perform off-line tuning while the filter is inactive, for instance, filters in the read/write channel of a hard disk have sufficient idle times that can be used for tuning, and an integrated filter in a cellular phone can be tuned in a few milliseconds as soon as ringing

is detected or when the receiver is turned on. In other cases, it may only be possible when the filter is powered up. Depending on the complexity of the filter and the dynamics of the tuning loops, the time required for tuning the filter will typically be in the range of a few μs to a few ms.

An advantage of off-line tuning is that the actual filter is tuned, instead of a reference circuit. Hence, the accuracy of the tuning is not dependent on the matching of the filter and the reference circuit.

While the methods described in this section are mostly suited for off-line tuning, they can in principle be used in a master-slave scheme.

11.5.1 Tuning of Composite Structures

Tuning of composite structures, e.g., IFLF and leapfrog filters, is more complicated than tuning of isolated second-order sections. A common approach is therefore to isolate substructures in the filter and tune them individually. For example, leapfrog filters can be tuned by isolating integrators and two-integrator loops and tune them separately by any of the methods discussed before. However, this requires that switches are inserted into the filter structure that can degrade the performance. Moreover, a complication is that the input and output loads must be the same when a substructure is tuned and operated in the filter.

A tuning scheme that uses the phase information for both r_p and Q of high-order high-Q filters is described below. The scheme uses a digital control circuitry that enables high accuracy and stability while eliminating the need for slow Q tuning loops. Also, there is no need for using switches to tune individual sections.

11.5.1.1 Tuning of High-Order High-Q Filters

An accurate tuning scheme is proposed in [121] that can be used for digital tuning of high-order filter, e.g., cascade form and leapfrog filters.

For the sake of simplicity, we assume that the transfer function is realized within cascade form with only lowpass sections. The tuning of more complex circuits is similar.

The phase response of the outputs of the sections can easily be computed from the transfer function.

Without loss of generality, we assume that $r_{p1} < r_{p2} < r_{p3} < \ldots$ for the first, second, third section, and so on. The phase difference between input and output of section i is denoted $\Phi_i(\omega)$. The tuning circuit generates $2N$ reference frequencies (two per section), ω_{ai} and ω_{bi}, $1 < i < N$, to calibrate each output. The frequencies for a lowpass section are defined by

$$\Phi_{di}(\omega_{ai}) = \frac{\pi}{4} - \frac{\pi}{2} i$$

and

$$\Phi_{di}(\omega_{bi}) = \frac{\pi}{4} - \frac{\pi}{2} i$$

where Φ_{di} is the desired phase response of section i. Note that the outputs of the first section at ω_{ai} and ω_{bi} are 3 dB below the peak of the magnitude response. The reference frequencies, one for each tuning parameter, may be generated using a phase-locked VCO.

We define for the section 1 to $(i-1)$ the values of A_i and B_i

$$A_i = \begin{cases} 0, \ \Phi_i(\omega_{ai}) > \frac{\pi}{4} - \frac{\pi}{2} i \\ 1, \ \Phi_i(\omega_{ai}) < \frac{\pi}{4} - \frac{\pi}{2} i \end{cases} \tag{11.24}$$

$$B_i = \begin{cases} 0, \ \Phi_i(\omega_{bi}) > -\frac{\pi}{4} - \frac{\pi}{2} i \\ 1, \ \Phi_i(\omega_{bi}) < -\frac{\pi}{4} - \frac{\pi}{2} i \end{cases} \tag{11.25}$$

where $\Phi_i(\omega)$ is the current phase. The tuning can be performed for the ith section if the $(i-1)$th section has $B_{i-1} = 1$. Therefore, the tuning circuit should start with the first section and successively tune all sections using appropriate reference frequencies. The overall tuning procedure is

B_0 =1, i =1
repeat
  if $\Phi_i(\omega_{ai})$ > π /4 - π i/2 **then** A_i = 0; **else** A_i = 1; **end**
  if $\Phi_i(\omega_b)$ > -π /4 - π i/2 **then** B_i = 0; **else** B_i = 1; **end**
  if B_{i-1} = 1
    if A_i = 1 and B_i = 1 **then** Inc. r_{pi};
    else if A_i = 0 and B_i = 0 **then** Dec. r_{pi}; **end**
    else if A_i = 0 and B_i = 1 **then** Dec. Q_i; **end**
    else if A_i = 1 and B_i = 0 **then** Inc. Q_i; **end**
  end
  i = i + 1;
  if i > N **then** i = 1; **end**
until convergence

The tuning procedure is repeated until the desired responses have been obtained.

This approach is easily modified to support other types of second-order sections, or even more general

phase responses, by modifying the right-hand side of Equations (11.24) and (11.25) accordingly.

It the reference frequencies could be generated exactly at the desired values, theoretically the filter would be tuned with zero error. Digital frequency synthesizers can be used to generate very precise signals; however, their finite resolution will add some error to the proposed tuning scheme. An implementation of the corresponding tuning circuitry can be found in [121].

11.5.2 Parasitic Effects

In the absence of parasitic effects, but including the finite DC gain of integrators, it can be shown that filter structures previously discussed have sufficient degrees of freedom to tune the poles and zeros to their desired values. However, this is not the case when parasitic effects are present.

Several tuning techniques have been proposed, but few techniques directly address the problem of dealing with parasitic effects. The presence of parasitic elements, such as non-dominant poles in the integrators, increases the filter order by introducing parasitic poles and zeros, without adding any additional degrees of freedom for tuning. The problem becomes more severe for high-frequency filters, because the radii of the parasitic poles and zeros are not much greater than that of the poles and zeros of the filter.

Consider a filter with ideal integrators with the transfer function

$$H(s) = G \frac{\displaystyle\prod_{i=1}^{N}(s - s_{zi})}{\displaystyle\prod_{i=1}^{N}(s - s_{pi})} \tag{11.26}$$

where s_{pi} are s_{zi} are the tunable poles and zeros.

We model the non-ideal integrators by a single non-dominant pole, i.e., the integrators have the transfer function

$$H(s) = \frac{1}{s} \frac{1}{1 + \dfrac{s}{s_{p0}}} . \tag{11.27}$$

where s_{p0} is the non-dominant pole.

Each tunable pole (zero) in Equation (11.26) is denoted

$$s_{pt} = \sigma_t + j\omega_t. \tag{11.28}$$

The transfer function of the ideal integrator is $H(s) = \dfrac{1}{s}$. Hence, s in Equation (11.26) is with non-ideal integrators replaced by $s + \dfrac{s^2}{s_{p0}}$, i.e.,

$$\frac{s^2}{s_{p0}} + s - (\sigma_t + j\omega_t) = 0. \tag{11.29}$$

Solving for s yields

$$s = -\frac{s_{p0}}{2} \pm \frac{s_{p0}}{2}\sqrt{1 + \frac{4}{s_{p0}}(\sigma_t + j\omega_t)}. \tag{11.30}$$

Thus, the effect of non-ideal integrators is that each pole (zero) of the tunable filter is mapped to two poles (zeros). Moreover, the poles (zeros) are adjustable, but they are not independent, as both are tuned by varying σ_t and ω_t. Also, the tunability of the filter is further restricted because there is no control over the non-dominant poles of the integrators. Therefore, the number of tunable poles and zeros is doubled, but the number of free parameters remains the same. Consequently, the original Nth-order transfer function becomes a $2N$th-order transfer function

$$H(s) = G\frac{\displaystyle\prod_{i=1}^{N}(s - s_{zai})(s - s_{zbi})}{\displaystyle\prod_{i=1}^{N}(s - s_{pai})(s - s_{pbi})}. \tag{11.31}$$

Figure 11.16 shows how a complex pole s_{pi} is mapped to two dependent poles s_{pai} and s_{pbi}. The mapping of zeros is identical to that of the poles.

There are two possible mappings for a single pole on the real axis. If the initial pole is greater

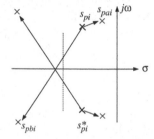

Fig 11.16 Mapping of a tunable pole pair due to a non-ideal integrator

than $-s_{p0}/4$, it will be mapped to two poles on the real axis as shown by the mapping of s_{p3} as shown at the left in Fig. 11.17. Otherwise, the pole will be mapped to a pair of complex conjugate poles on the dashed line at $-s_{p0}/2$. However, this case is unlikely because any real pole in the filter will have a significantly smaller radius than the parasitic pole of the integrator.

In each case, the pole s_{pai} has a higher Q factor than the original pole s_{pi}. A too high Q factor results

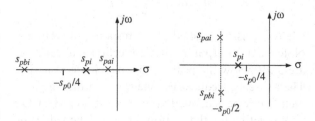

Fig 11.17 Mapping of a real pole due to a non-ideal integrator

in peaks in the passband and it becomes more severe at the passband edge and in higher-order filters.

If the non-dominant poles of the integrators, s_{p0}, is large in comparison to the frequencies of interest then the pole s_{pai} approaches the initial pole s_{pi}, and the other pole, s_{pbi}, moves far into the left-hand side of the s-plane so that their effects become insignificant. However, in high-frequency filters, the non-dominant pole is not much larger than the pole radius, s_{pi}, and the effects of the pole s_{pbi} cannot be ignored. Hence, the previously discussed tuning methods, which do not include the effects of parasitics, become less accurate.

11.6 Problems

11.1 Complete the second-order structures shown in Fig. 6.65 with dummy transconductors in order to reduce the effect of process variations.

11.2 Show that Equation (11.4) is valid.

11.3 Show that the signal after multiplier in Fig. 11.10 can be used as control signal, i.e., a small positive/negative deviation from $\pi/2$ in the phase difference yields a suitable control signal.

11.4 Estimate the error in the pole radius in the circuit shown in Fig. 11.10 if $Q = 10$ and the integrator gain is only 20 dB if the phase-error is $1°$.

11.5 Determine a suitable frequency for V_{ref} for the circuit shown in Figures 11.11 and 11.13.

11.6 Show that the output of the bandpass filter in Fig. 11.14 can be used as a tuning gradient.

References

1. Acar C., Anday F., and Kuntman H.: On the realization of OTA-*C* filters, *Intern. J. Circuit Theory Appl.*, Vol. 21, pp. 331–341, 1993.
2. Allan P.E. and Sanchez-Sinencio E.: *Switched Capacitor Circuits*, Van Nostrand Reinhold, London, 1984.
3. Andreani P. and Mattisson S.: On the use of Nauta's transconductor in low-frequency CMOS g_m–*C* bandpass filters, *IEEE J. Solid-State Circuits*, Vol. 37, No. 2, pp. 114–124, Feb. 2002.
4. Antoniou A.: Novel *RC*-active-network synthesis using generalized-immittance converters, *IEEE Trans. Circuit Theory*, Vol. 17, pp. 212–217, May 1970.
5. Antoniou A.: *Digital Filters: Analysis, Design and Applications*, 2nd ed., McGraw-Hill, New York, 1993.
6. Aparicio R. and Hajimiri A.: Capacity limits and matching properties of integrated capacitors, *IEEE J. Solid-State Circuits*, Vol. 37, No. 3, pp. 384–393, Mar. 2002.
7. Baher H.: *Synthesis of Electrical Networks*, John Wiley & Sons, New York, 1984.
8. Bahl I. and Bharitia P.: *Microwave Solid State Circuit Design*, 2nd Ed., John Wiley & Sons, New York, 2003.
9. Baker R.J., Li H.W., and Boyce D.E.: *CMOS Circuit Design, Layout, and Simulation*, IEEE Press Series on Microelectronics Systems, 1998.
10. Balkir S., Dündar G., and Ögrenci A.S.: *Analog VLSI Design Automation*, CRC Press, Boca Raton, Florida, 2003.
11. Blinchikoff H.J. and Zverev A.I.: *Filtering in the Time and Frequency Domains*, John Wiley & Sons, New York, 1976.
12. Belevitch V.: *Classical Network Theory*, Holden-Day, San Francisco, 1968.
13. Brackett P.O.: Circuit and limitations of wave filters, *Proc. IEEE Intern. Symp. Circuits Systems*, ISCAS-76, Munich, West Germany, pp. 69–72, 1976.
14. Brackett P.O. and Sedra A.S.: Applications of signal flow graphs to the synthesis of linear systems: Part I – a general theory of network simulation, *IEEE Intern. Symp. Circuits Syst.*, ISCAS-75, pp. 242–245, Newton, Mass., 1975.
15. Brackett P.O. and Sedra A.S.: Applications of signal flow graphs to the synthesis of linear systems: Part II – design of active ladder filters, *Proc. IEEE Intern. Symp. Circuits Syst.*, ISCAS-75, pp. 246–249, Newton, Mass., 1975.
16. Brackett P.O. and Sedra A.S.: Direct SFG simulation of *LC* ladder networks with application to active filter design, *IEEE Trans. Circuits Syst.*, Vol. 23, pp. 61–67, Feb. 1976.
17. Brackett P.O. and Sedra A.S.: Active compensation for high-frequency effects in op-amp circuits with applications to active *RC* filters, *IEEE Trans. Circuits Systems*, Vol. 23, No. 2; pp. 68–72, Feb. 1976.
18. Bruton L.T.: *RC-Active Circuits, Theory and Design*, Prentice Hall, Englewood Cliffs, New Jersey, 1980.
19. Budak A.: *Passive and Active Network Analysis and Synthesis*, Boston, Houghton Mifflin, 1974.
20. Budak A.: Active filters with zero transfer function sensitivity with respect to the time constants of operational amplifiers, *IEEE Trans. Circuits Syst.*, Vol. 27, No. 10, pp. 849–854, Oct. 1980.
21. Carlin H.J. and Civalleri P.P.: *Wideband Circuit Design*, CRC Press, Boston, 1998.
22. Chen W.K. and Chaisrakeo T.: Explicit formulas for the synthesis of optimum bandpass Butterworth and Chebyshev impedance-matching networks, *IEEE Trans. Circuits Systems*, Vol. 27, No. 10, pp. 928–942, Oct. 1980.
23. Chen W.K.: *Passive and Active Filters, Theory and Implementations*, John Wiley & Sons, New York, 1986.
24. Chen W.K. (Ed.): *The Circuit and Filters Handbook*, CRC Press, Boca Raton, Florida, 1995.
25. Chen W., Chen W., and Hsu K.: Three-dimensional fully symmetric inductors, transformer, and balun in CMOS technology, *IEEE Trans. Circuits Systems*, Part I, Vol. 54, No. 7, pp. 1413–1423, July 2007.
26. Christian E.: *LC Filters, Design, Testing, and Manufacturing*, John Wiley & Sons, New York, 1983.
27. Çiçekoglu O.: Current-mode biquad with a minimum number of passive elements, *IEEE Trans. on Circuits and Systems*, Part II, Vol. 48, No. 1, pp. 221–223, Jan. 2001.
28. Cutbert T.R. Jr.: *Circuit Design Using Personal Computers*, John Wiley & Sons, New York, 1983.
29. Daniels R.W.: *Approximation Methods for Electronic Filter Design*, McGraw-Hill, New York, 1974.
30. Darlington S.: A History of network synthesis and filter theory for circuits composed of resistors, inductors, and capacitors, *IEEE Trans. Circuits Syst.* Part I, Vol. 46, No. 1, pp. 4–13, Jan. 1999.

31. Daryanani G.: *Principles of Active Network Synthesis and Design*, John Wiley & Sons, New York, 1976.

32. Deliyannis T., Sun Y., and Fidler J.K.: *Continuous-Time Active Filter Design*, CRC Press, Boca Raton, Florida, 1999.

33. Ellis M.G.: *Electronic Filter Analysis and Synthesis*, Artech House, Boston, 1994.

34. Fabre A., Saaid O., Wiest F., Boucheron C.: High frequency applications based on a new current controlled conveyor, *IEEE Trans. Circuits Syst.*, Part I, Vol. 43, No. 2, pp. 82–91, Feb. 1996.

35. Fettweis A.: Wave digital filters: Theory and practice, *Proc. IEEE*, Vol. 74, pp. 270–327, Feb. 1986.

36. Fleischer P.E.: Sensitivity minimization in a single amplifier biquad circuit, *IEEE Trans. Circuits Syst.*, Vol. 23, No. 1, pp. 45–55, Jan. 1976.

37. Friend J.J.: A single operational amplifier biquadratic filter section, *Proc. Intern. Symp. Circuit Theory*, Atlanta, Georgia, pp. 179–180, 1970.

38. Ghausi M.S. and Laker K.R.: *Modern Filter Design, Active RC and Switched Capacitor*, Prentice Hall, Englewood Cliffs, New Jersey, 1981.

39. Geiger R.L.: Parasitic pole approximation techniques for active filter design, *IEEE Trans. Circuits Syst.*, Vol. 27, No. 9, pp. 793–799, Sept. 1980.

40. Geiger R.L. and Budak A.: Active filters with zero amplifier sensitivity, *IEEE Trans. Circuits Systems*, Vol. 24, No. 4, pp. 277–288, Apr. 1979.

41. Geiger R.L. and Budak A.: Design of active filters independent of first- and second order operational amplifier time constant effects, *IEEE Trans. Circuits Syst.*, Vol. 28, No. 8, pp. 749–757, Aug. 1981.

42. Geiger R.L. and Budak A.: Integrator design for high-frequency active filters applications, *IEEE Trans. Circuits Syst.*, Vol. 29, No. 9, pp. 595–603, Sept. 1982.

43. Gregorian R.: *Introduction to CMOS OP-AMPS and Comparators*, John Wiley & Sons, New York, 1999.

44. Groenewold G.: Noise and group delay in active filters, *IEEE Trans. Circuits Syst.*, Part I, Vol. 54, No. 7, pp. 1471–1480, July 2007.

45. Girling F.E.J. and Good E.F.: Active filters, *Wireless World*, Vol. 76, pp. 341–345 and 445–450, 1970.

46. Gorski-Popiel J.: *RC*-active synthesis using positive-immittance converters, *Electronics Letters*, Vol. 3, pp. 381–382, 1967.

47. Haritantis I., Constantinides G.A., and Deliyannis T.: Wave active filters, *Proc. IEE*, Vol. 123, No. 7, pp. 676–682, 1976.

48. Hassler M. and Neirynck J.: *Electric Filters*, Artech House, Boston, 1986.

49. Herrero J.L. and Willoner G.: *Synthesis of Filters*, Prentice Hall, Englewood Cliffs, New Jersey, 1966.

50. Hong S. and Lancaster M.J.: *Microstrip Filters for RF/Microwave Applications*, John Wiley & Sons, New York, 2001.

51. Huelsman L.P.: *Active and Passive Analog Filter Design*, McGraw-Hill, New York, 1993.

52. Huijsing J.H.: *Operational Amplifiers*, Kluwer, Boston, 2001.

53. Hunter I.: *Theory and Design of Microwave Filters*, The Institution of Electrical Engineers, London, 2001.

54. Ismail M. and Fiez T.: *Analog VLSI: Signal and Information Processing*, McGraw-Hill, New York, 1994.

55. Jewell M.L.: On high frequency narrow-band elliptic filters, *IEEE Trans. Circuits Syst.*, Vol. 37, No. 2, pp. 264–267, Feb. 1990.

56. Johns D.A. and Martin K.: *Analog Integrated Circuit Design*, John Wiley & Sons, New York, 1997.

57. Johansson R.A.: *Mechanical Filters in Electronics*, John Wiley & Sons, New York, 1983.

58. Kardontchik J.E.: *Introduction to the Design of Transconductance-Capacitor Filters*, Kluwer, Boston, 1992.

59. Kinsman R.G.: *Crystal Filters, Design, Manufacture, and Application*, John Wiley & Sons, 1987.

60. Kurth C.F.: Generation of single-sideband signals in multiplex communication systems, *IEEE Trans. Circuits Syst.*, Vol. 23, No. 1, pp. 1–17, Jan. 1976.

61. Laker K.R. and Ghausi M.S.: Synthesis of a low-sensitivity multiloop feedback active *RC* filter, *IEEE Trans. Circuits Syst.*, Vol. 21, No. 2, pp. 252–259, Mar. 1974.

62. Laker K.R., Schaumann R., and Ghausi M.S.: Multiple-loop feedback topologies for the design of low-sensitivity active filters, *IEEE Trans. Circuits Syst.*, Vol. 26, No. 1, pp. 1–21, Jan. 1979.

63. Laker K.R. and Sansen W.: *Design of Analog Integrated Circuits and Systems*, McGraw-Hill, New York, 1994.

64. Lam H.Y-F.: *Analog and Digital Filters, Design and Realization*, Prentice Hall, Englewood Cliffs, New Jersey, 1979.

65. Lampaert K., Gielen G., and Sansen W.: *Analog Layout Generation for Performance and Manufacturability*, Kluwer, Boston, 1999.

66. Leach W.M. Jr.: Fundamentals of low-noise analog circuit design, *Proc. IEEE*, Vol. 82, No. 10, pp. 1515–1538, Oct. 1994.

67. Li D. and Tsividis Y.: Active *LC* filters on silicon, *IEE Proc. Circuits Devices Syst.*, Vol. 147, No. 1, pp. 49–56, Feb. 2000.

68. Lindquist C.S.: *Active Network Design with Signal Filtering Applications*, Steward & Sons, Long Beach, California, 1977.

69. Ljung L.: *System Identification – Theory For the User*, 2nd ed., PTR Prentice Hall, Upper Saddle River, New Jersey, 1999.

70. Ludwig R. and Bretchko P.: *RF Circuit Design: Theory and Applications*, Prentice Hall, New Jersey, 2000.

71. Lutovac M.D., Tosic D.V., and Evans B.L.: *Filter Design for Signal Processing Using MATLAB and Mathematica*, Prentice Hall, Englewood Cliffs, New Jersey, 2001.

72. Löwenborg P., Johansson H., and Wanhammar L.: First-order sensitivity of complementary diplexers, *IEEE Trans. Circuits Syst.*, Part II, Vol. 51, No. 8, pp. 421–425, Aug. 2004.

73. Malherbe J.A.G.: *Microwave Transmission Line Filters*, Artech House, Boston, 1979.

74. Martin K. and Sedra A.S.: Optimum design of active filters using the generalized immittance converter, *IEEE Trans. Circuits Syst.*, Vol. 24, No. 9, pp. 495–503, Sept. 1977.

75. MATLAB *Signal Processing Toolbox*, Users's Guide, 1996.

76. Matthaei, G.L.: *Microwave Filters, Impedance-Matching Networks and Coupling Structures*, Artech House, Boston, 1980.

77. Minaei S., Çiçekoglu O., Kuntman H., and Türköz S.: New current-mode lowpass, bandpass and highpass filters employing CCCIIs, *Proc. Midwest Symp. on Circuits and Systems,* Dayton, Ohio, USA, pp. 106–109, Aug. 2001.

78. Minaei S., Sayin O.K., and Kuntman H.: A New CMOS electronically tunable current conveyor and its application to current-mode filters, *IEEE Trans. Circuits Syst.,* Part I, Vol. 53, No. 7, pp. 1448–1457, July 2006.

79. Mitra S.K.: *Analysis and Synthesis of Linear Active Neworks,* John Wiley & Sons, New York, 1968.

80. Mitra S.K. and Kurth C.F. (Eds.): *Miniaturized and Integrated Filters,* John Wiley & Sons, New York 1989.

81. Moschytz G.S. and Horn P.: *Active Filter Handbook,* John Wiley & Sons, New York, 1981.

82. Moreira J.P. and Silva M.M.: Limits to the dynamic range of low-power continuous-time integrators, *IEEE Trans. Circuits Syst.,* Part I, Vol. 48, No. 7, pp. 805–817, July 2001.

83. Mossberg K. and Signell S.: On the design of high frequency active filters, *Proc. European Conf. Circuits Theory Design,* ECCTD-83, Stuttgart, pp. 95–98, Sept. 6–8, 1983.

84. Nauta B.: *Analog CMOS filters for Very High Frequencies.* Kluwer, Boston, 1993.

85. Northrop R.B.: *Analog Electronic Circuits–Analysis and Applications,* Addison-Wesley, Reading, Massachosetts, 1990.

86. Orchard H.J.: Inductorless filters, *Electronics Lett.,* Vol. 2, p. 224, Sept. 1966.

87. Orchard H.J.: Filter design by iterated analysis, *IEEE Trans. Circuits Syst.,* Vol. 32, No. 11, pp. 1089–1096, Nov. 1985.

88. Paarmann L.D.: *Design and Analysis of Analog Filters – A Signal Processing Perspective,* Kluwer Academic Publishers, London, 2001.

89. Pallás-Areny R. and Webster J.G.: *Analog Signal Processing,* J. Wiley & Sons, New York, 1999.

90. Palmisano G., Palumbo G., and Pennisi S.: High-performance and simple CMOS unity-gain amplifier, *IEEE Trans. Circuits Syst.,* Part I, Vol. 47, No. 3, pp. 406–410, Mar. 2000.

91. Papananos Y.E.: *Radio-Frequency Mircoelectronic Circuits for Telecommunication Applications,* Kluwer, Boston, 2000.

92. Pavan S. and Tsividis Y.: *High Frequency Continuous Time Filters in Digital CMOS Processes,* Kluwer, Boston, 2000.

93. Pozar D.M.: *Microwave Engineering,* 3rd ed., J. Wiley & Sons, New York, 2005.

94. Premoli A.: A new class of equal-ripple filtering functions with low Q-factors: the MUCROER polynomials, *IEEE Trans. Circuits Syst.,* Vol. 21, No. 5, pp. 609–613, Sept. 1974.

95. Rhea R.W.: *HF Filter Design and Computer Simulation,* McGraw-Hill, New York, 1995.

96. Rhodes J.D.: *Theory of Electrical Filters,* J. Wiley & Sons, New York, 1976.

97. Rizzi P.A.: *Microwave Engineering,* Prentice Hall, Englewood Cliffs, New Jersey, 1988.

98. Roberts G.W and Leung V.W.: *Design and Analysis of Integrator-Based log-Domain Filter Circuits,* Kluwer, Boston, 2000.

99. Rosloniec S.: *Algorithms for Computer-Aided Design of Linear Microwave Circuits,* Artech House, Boston, 1990.

100. Saal R.: *Handbuch zum Filterentwurf, Handbook of Filter Design,* AEG-Telefunken, Berlin, 1979.

101. Sakurai S. and Ismail M.: *Low-Voltage CMOS Operational Amplifiers, Theory, Design and Implementation,* Kluwer, Boston, 1995.

102. Sallen R.P. and Key E.L.: A practical method of designing RC active filters, *IRE Trans. Circuit Theory,* Vol. 2, pp. 74–85, 1955.

103. Sansen W.M.C.: *Analog Design Essentials,* Springer, New York, 2006.

104. Schaumann R., Brand J.R., and Laker K.R.: Effects of excess phase in multiple feedback active filters, *IEEE Trans. Circuits Syst.,* Vol. 27, No. 19, pp. 967–970, Oct. 1980.

105. Schaumann R., Soderstrand M.A., and Laker K.R. (Eds.): *Modern Active Filter Design,* IEEE Press, New York, 1981.

106. Schaumann R., Ghausi M.S., and Laker K.R.: *Design of Analog Filters: Passive, Active RC and Switched Capacitor,* Prentice Hall, Englewood Cliffs, New Jersey, 1990.

107. Schaumann R. and van Valkenburg M.E.: *Design of Analog Filters,* Oxford University Press, New York, 2001.

108. Schmid H.: Circuit transposition using signal-flow graphs, *Proc. IEEE Intern. Symp. Circuits Syst.,* ISCAS 2002, Vol. 2, pp. 25–28, 26–29 May 2002.

109. Schmid H.: Approximating the universal active element, *IEEE Trans. Circuits Syst.,* Part II, Vol. 47, No. 11, pp. 1160–1169, Nov. 2000

110. Schmid H. and Moschyz G.S.: Active-MOSFET-C single-amplifier biquadratic filters for video frequencies, *IEE Proc. Circuits, Devices Syst.,* Vol. 147, No. 1, pp. 35–41, Feb. 2000.

111. Schmid H. and Moschyz G.S.: A tunable video-frequency, low-power single-amplifier biquadratic filter in CMOS, *Proc. IEEE Intern. Symp. Circuits Systems,* ISCAS-99, Vol. II, pp. 128–131, 1999.

112. Sedra A.S. and Brackett P.O.: *Filter Theory and Design: Active and Passive,* Pitman, London, 1978.

113. Sedra A.S.: Generation and classification of single amplifier filters, *Circuit Theory Appl.,* Vol. 2, pp. 51–67, 1974.

114. Sedra A. and Smith K.C.: A second-generation current conveyor and its applications, *IEEE Trans. Circuit Theory,* Vol. 17, No. 1, pp. 132–134, 1970.

115. Stephenson F.W.: *RC Active Filter Design Handbook,* John Wiley & Sons, New York, 1985.

116. Signell S. and Mossberg K.: Design of high-frequency active RC filters using passive compensation methods, *Proc. IEEE Intern. Symp. Circuits Systems,* ISCAS-88, Helsinki, Finland, 1988.

117. Silva-Martinez J., Steyaert M., and Sansen W.: *High-Performance CMOS Continuous-Time Filters,* Kluwer, Boston, 1993.

118. Soliman A.M.: Current conveyor filters: classification and review, *Microelectronics J.*, Vol. 29, pp. 133–149, 1998.

119. Sun Y. (Ed.): *Design of High Frequency Integrated Analogue Filters*, The Institution of Electrical Engineers, London, 2002.

120. Symbolic Math Toolbox, For Use with MATLAB, Users's Guide, 1994.

121. Sumesaglam T. amd Karsilay A.I.: Digital tuning of high-order high-Q continuous-time filters, *Electron. Lett.*, Vol. 38, No. 19, pp. 1076–1078, Sept. 2002.

122. Taylor J.T. and Huang Q.: *CRC Handbook of Electrical Filters*, CRC Press, Boca Raton, Florida, 1997.

123. Temes G.C. and Mitra S.K. (Eds.): *Modern Filter Theory and Design*, John Wiley & Sons, New York, 1973.

124. Temes G.C. and LaPatra J.W.: *Introduction to Circuit Synthesis and Design*, McGraw-Hill, New York, 1977.

125. Thomas L.C.: The biquad: Part I – Some practical design considerations, *IEEE Trans. Circuit Theory*, Vol. 18, pp. 350–357, May 1971.

126. Tingleff J. and Toumazou C.: Current mode continuous time wave active filters, *Electron. Lett.* Vol. 28, No. 5, pp. 463–465, 27 Feb. 1992.

127. Tingleff J. and Toumazou C.: Integrated current mode wave active filters based on lossy integrators, *IEEE Trans. Circuits Syst.*, Part II, Vol. 42, No. 5, pp. 237–244, May 1995.

128. Trofimenkoff F.N. and Onwuachi O.A.: Noise performance of operational amplifier circuits, *IEEE Trans. Education*, Vol. 32, No. 1, pp. 2–17, Feb. 1989.

129. Tsividis Y.P., Krishnapura N., Palaskas Y., and Toth L.: Internally varying analog circuits minimize power dissipation, *IEEE Circuits Devices Magazine*, pp. 63–72, Jan. 2003.

130. Tsividis Y.P. and Vooman J.O. (Eds.): *Integrated Continuous Time Filters*, IEEE Press, New York, 1993.

131. Van Valkenburg M.E.: *Analog Filter Design*, Qxford University Press, New York, 1982.

132. Wang H.-Y. and Lee C.-T.: Versatile insensitive current-mode universal biquad implementation using current conveyors, *IEEE Trans. Circuits Syst.*, Part II, Vol. 48, No. 4, pp. 409–413, Apr. 2001.

133. Wanhammar L.: A bound on the passband deviation for symmetric and antimetric commensurate transmission line filters, http://www.es.isy.liu.se/publications/, 1991.

134. Wanhammar L.: *DSP Integrated Circuits*, Academic Press, New York, 1999.

135. Wanhammar L. and Johansson H.: *Digital Filters*, Linköping University, Sweden, 2007.

136. Wambacq P. and Sansen W.M.C.: *Distortion Analysis of Analog Integrated Circuits*, Kluwer, Boston, 1998.

137. Williams A.B.: *Electronic Filter Design Handbook*, McGraw-Hill, New York, 1981.

138. Williams A.B. and Taylor F.J.: *Electronic Filter Design Handbook*, 3rd ed., McGraw-Hill, New York, 1995.

139. Wilson B.: Generalized analysis of gain-bandwidth independence in feedback amplifiers, *Intern. J. Elect. Eng. Educ.*, Vol. 39, pp. 20–30, Jan. 2002.

140. Weinberg L.: *Network Analysis and Synthesis*, McGraw-Hill, New York, 1962.

141. Wolfram S.: *The Mathematica Books*, Cambridge University Press, 1996..

142. Wupper H. and Meerkötter K.: New active filter synthesis based on scattering parameters, *IEEE Trans. Circuits Syst.*, Vol. 22, pp. 594–602, 1975.

143. Wupper H.: Scattering parameter active filters with reduced number of active elements, *IEEE Trans. Circuits Syst.*, Vol. 23, pp. 318–322, May 1976.

144. Yuce E., Minaei S., and Çiçekoglu O.: Limitations of the simulated inductors based on a single current conveyor, *IEEE Trans. Circuits Syst.*, Part I, Vol. 53, No. 12, pp. 2860–2087, Dec. 2006.

145. Zhu Y.-S. and Chen W.-K.: *Computer-Aided Design of Communication Network*s, World Scientific, Singapore, 2000.

146. Zverev A.I.: *Handbook of Filter Synthesis*, John Wiley & Sons, New York, 1967.

147. Åkerberg D. and Mossberg K.: A versatile active *RC* building block with inherent compensation for the finite bandwidth of the amplifier, *IEEE Trans. Circuits Systems*, Vol. 21, No. 1, pp. 75–78, Jan. 1974.

148. Özoguz S. and Günes E.O.: Universal filter with three inputs using CCII +. *Electron. Lett.*, Vol. 32, No. 23, pp. 2134–2135, 17 Nov. 1996.

149. Özoguz S., Toker A., and Çiçekoglu O.: High output impedance current-mode multifunction filter with minimum number of active and reduced number of passive elements, *Electron. Lett.*, Vol. 34 No. 79, pp. 1807–1809, Sept. 1998.

Toolbox for Analog Filters

The software discussed in this book has been developed and tested using MATLAB Version 5.2. The author believes that all routines should be compatible with newer versions of MATLAB. The software is maintained and regularly updated and can be downloaded from our Web site at http://www.es.isy.liu.se/publications/books/Analog_Filters/. The list below gives names of routines and pages where they are mentioned in the book.

Index